40 Years of Evolution

40 Years of Evolution

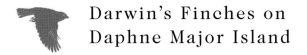

Darwin's Finches on
Daphne Major Island

Peter R. Grant and B. Rosemary Grant

Princeton University Press

Princeton and Oxford

Published by Princeton University Press, 41 William Street, Princeton, New Jersey
08540
In the United Kingdom: Princeton University Press, 6 Oxford Street, Woodstock, Ox-
fordshire OX20 1TW

press.princeton.edu
Library of Congress Cataloging-in-Publication Data

Grant, Peter R., 1936–
Forty years of evolution : Darwin's Finches on Daphne Major Island /
Peter R. Grant and B. Rosemary Grant.
pages cm
Includes bibliographical references and index.
ISBN 978-0-691-16046-7 (hardback : alk. paper)   1. Ground
finches—Evolution—Galapagos Islands.   2. Bird populations—Galapagos
Islands.   3. Birds—Evolution—Galapagos Islands.   I. Grant, B.
Rosemary. II. Title.
QL696.P246G7324 2014
598.072′32098665—dc23
2013018007

British Library Cataloging-in-Publication Data is available

This book has been composed in ITC Caslon 224

Printed on acid-free paper. ∞

Printed in the United States of America

2 4 6 8 10 9 7 5 3 1

To the next generation:
Nicola and Thalia

And the next one:
Rajul, Olivia, Anjali, and Devon

And succeeding ones

Daphne Major
She looms from the hyaline like some mutant
barnacle—frond once a pillar of smoke,
operculum blown off—that has assumed
a couchant pose, waiting for the searchers
with mist-nets and calipers to return each year
who pit the caltrop against magnirostris
and scratch protean generations from her flank,
until she blows again or sounds and sinks
back to Gondwanaland's deep ocean drift

(Weston 2005, p. 49)

The sight of Daphne Major conveys something like this
[passage of time] to us, even in the first glance over the
water, or in the last, as it revolves like a wood chip in
the wake of the boat. We know we are looking at a place
that was here before we came and will remain when we
are gone. The very island will sink someday, and
another will rise when it is drowned.

(Weiner 1994, p. 303)

Contents

Illustrations

Tables

Boxes

Preface

DARWIN'S *ORIGIN OF SPECIES* (1859, p. 1) begins: "When on board H.M.S. 'Beagle,' as naturalist, I was much struck with certain facts in the distribution of the inhabitants of South America, and the geological relations of the present to the past inhabitants of that continent. These facts seemed to me to throw some light on the origin of species—that mystery of mysteries, as it has been called by one of our greatest philosophers." We followed in his footsteps 138 years after his visit to the Galápagos, with the same quest in mind. Our book describes what we learned by studying Darwin's finches (fig. P.1) for 40 years on Daphne Major (fig. P.2), a small island in the center of the Galápagos archipelago (fig. P.3).

Genetic variation is the raw material for evolution. Although much has been learned about genetic variation from theoretical and laboratory research, long-term field studies in natural environment have been relatively neglected. Knowledge of how genetic and phenotypic variation interacts with environmental variation is fundamentally important for understanding evolution in nature. It can be achieved by long-term field studies of evolution with well-chosen organisms in well-chosen environments when coupled with the benefits of laboratory science. This is what we have attempted to do. By taking a multidimensional approach to questions of evolution and speciation, carefully documenting genetic, ecological, and behavioral factors responsible for evolution of finches across 40 years, we have made discoveries about speciation far beyond our initial expectations. And although finches are the focal organisms of the study, the principles of their evolution apply broadly to all organisms.

These are exciting times to be an evolutionary biologist. Changes in technology are proceeding at an ever-increasing pace, giving us the tools for expanding our knowledge and understanding of how evolution occurs in the natural world. In view of this it is worth reflecting on the state of knowledge in 1973 (Mayr 1970, Dobzhansky 1970), on the methods and tools available for evolutionary investigations at that time, and the transformation that studies of speciation have undergone since then. Techniques we take for granted now did not exist. Electrophoresis (Hubby and Lewontin 1966) was only just becoming widely available as a method of detecting allozyme variation if the appropriate buffers could be worked out. It required a power source, which ruled it out for most studies on uninhabited islands. Cladistics (Hennig 1966)

Fig. P.1 The four species of finches. **Upper left**: Small Ground Finch, *G. fuliginosa*. **Upper right**: Medium Ground Finch, *G. fortis*. **Lower left**: Large Ground Finch, *G. magnirostris*. **Lower right**: Cactus Finch, *G. scandens*. From Grant and Grant 2008a).

Fig. P.2 The two Daphnes. **Left**: Daphne Major, with Daphne Minor (Chica) in the background (D. Parer and E. Parer-Cook). **Right**: Daphne Minor, 1976. Daphne Minor has been climbed once, with ropes. Two immature *fortis* banded on Daphne Major were seen at the top (Grant et al. 1980).

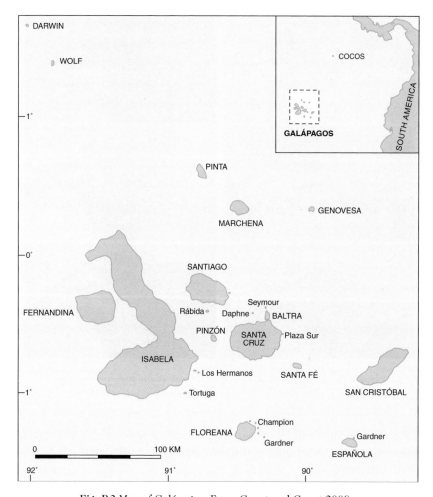

Fig. P.3 Map of Galápagos. From Grant and Grant 2008a.

had yet to shake the foundations of phylogenetic reconstruction and interpretation. Personal computers had yet to be invented; we were the handmaidens of university mainframes. Statistical programs for the analysis of complex data had not been invented either, and the revolution in molecular biology, including evodevo, lay far off in the future. Whole genomes were conceivable, but their sequences were not. Now we do things we could not anticipate doing at the outset; for example, we study genes, their expression patterns and regulation, and we infer from molecular data the time when events took place in

Fig. P.4 Phenotypic variation in the *G. fortis* population on Daphne.

the past. Our account of evolution on Daphne reflects the expanding knowledge brought about by the development of new methods as the study progressed.

In the absence of fossils, answers to questions about speciation in the past have to be sought with living organisms by looking backward in time. Typically this is done by using information on contemporary populations to test the assumptions of historical hypotheses: a retrospective analysis. We followed this path with a combination of field studies of short duration on many islands and a decade-long study on Genovesa. We concentrated almost entirely on the six ground finch species in the genus *Geospiza* because they are distributed widely in the archipelago, generally abundant, and easy to observe (Lack 1947). We wrote three books on our findings: two on the total finch radiation (Grant 1986 [1999], Grant and Grant 2008a) and one on the finch populations on Genovesa (Grant and Grant 1989a).

Here we describe what we have found by adopting a different approach to questions of speciation: studying populations through time,

Fig. P.5 Phenotypic variation in the *G. scandens* population on Daphne.

a prospective analysis. This provides insights into the process of speciation even if the progress we observe toward complete reproductive isolation is small. The present volume completes our program of converting research results into more accessible and synthetic book form, with new and unanticipated insights into Darwin's question on the origin of species.

We have designed this book for students, educators, and others to read for enjoyment and inspiration. Throughout we have tried to express our views in jargon-free language, with technical terms given only where necessary. We have adopted a few conventions. First, we have given numerical results of statistical tests only when not given before in the original papers we cite. Second, we have supplied details of methods of study and analysis in boxes and not in the main text. Third, we have used appendixes for some background information and discussion of topics that are relevant to the material in chapters but not central to it. In some cases these discussions have been supplemented with numbers in tables and figures.

Acknowledgements. Our research has been supported financially by the National Science and Engineering Research Council (Canada), National Science Foundation (USA), and McGill and Princeton Universities (Class of 1877 Fund), and logistically by the Charles Darwin Foundation and the Galápagos National Parks Service. More than twenty assistants helped us in fieldwork. Foremost among them were Ian and Lynette Abbott as postdoctoral fellows, and Peter Boag, Laurene Ratcliffe, Trevor Price, and Lisle Gibbs as PhD graduate students. They and we were helped by Phillip de Maynadier, James Gibbs, Lukas Keller, Greg Keys, Steve Latta, Irby Lovette, David McCullough, Stephen Millington, David Moore, Ken Petit, Ken Petren, Bob Podolsky, Wally Rendell, Gerry Retzlaf, Uli Reyer, Dan Rosenberg, Keith Tarvin, Ayse Unal, Carlos Valle, Jonathan Weiland, David Wiggins, and our two daughters, Nicola and Thalia. We thank Lisle Gibbs for reading chapter 7, and Trevor Price, Dolph Schluter, and an anonymous reviewer for reading previous drafts and offering numerous suggestions for improvement. They helped us write a significantly better book. For transforming the potential into the realized book we thank the editorial staff at Princeton University Press: Alison Kalett and Quinn Fusting for guidance in the early stages, Mark Bellis for shepherding it through production, and Dimitri Karetnikov for giving us the benefit of his artistic skills.

Data used in constructing the figures have been deposited in Dryad (doi:10.5061/dryad.g6g3h). Two educational films illustrate features of the book. One, made by the Howard Hughes Medical Institute for high schools, illustrates natural selection on Daphne. It is entitled *The Origin of Species: The Beak of the Finch* (www.biointeractive.org). We are preparing another that is designed to illustrate all the main topics of the book.

February 2013

PART 1

Early Problems, Early Solutions

Speciation, Adaptive Radiation, and Evolution

I should like to take some one family to study thoroughly,
principally with a view to the theory of the origin of
species. By that means I am strongly of opinion that
some definite results might be arrived at.

(Wallace 1847, letter to H. W. Bates)

Those forms which possess in some considerable degree
the character of species, but which are so closely similar
to some other forms, or are so closely linked to them by
intermediate gradations, that naturalists do not like to
rank them as distinct species, are in several respects the
most important to us.

(Darwin 1859, p. 47)

Introduction

MANY OF US ARE FASCINATED by the biological world around
us. We marvel at the diversity of color, pattern, form, shape,
size, ferocity, tameness, speed, and ingenious things that animals and plants do to find food and mates and avoid being eaten. Some
of us have peered into microscopes that have opened up a new and
wonderfully diverse world. Others have had the same thrilling experience in diving off coral reefs and being dazzled by the variety of fish.

Yet others have been simultaneously bewildered and stimulated by the overwhelming diversity of a tropical rain forest. All this is so enthralling that some of us not only want to know why the world is the way it is; we also want to explore, examine, and test ideas about it in order to make our own discoveries. We are evolutionary biologists.

As evolutionary biologists we ask, how do species form? If we can answer that question we have taken a large stride toward understanding the biological richness of the world. The question is old but remains unresolved because rarely is it possible to witness even a part of the process. It must generally be inferred from indirect evidence, and yet we have had the good fortune to be witnesses. In this book we describe what we have learned about speciation by tracking populations and measuring evolutionary changes across 40 years in contemporary time.

Our starting point is Darwin's *Origin of Species by Means of Natural Selection*. This is a manifesto of cardinal evolutionary principles. It laid out a theory of common descent of all organisms, represented evolutionary diversification as a branching pattern, and invoked the principle of natural selection as the driving agency that caused the divergence. Darwin argued that species formed by diverging in separate locations and then, when they came together, competed with each other for food and space, and diverged yet further. By this means, repeated, complex communities built up from simpler ones. Darwin had little hope of seeing evolution occur, but he did write that young radiations of species might provide windows through which we could view the steps involved in speciation. By an indirect pathway this led us to the Galápagos Islands, to Daphne Major in particular, to the finches named after him, to a fascination with them that lasted 40 years, and even to the origin of a new species.

Adaptive Radiation of Darwin's Finches

Finches on the Galápagos are a young radiation of ecologically diverse species that have evolved from a common ancestor (Lack 1947, Grant 1986). Other radiations of plants and animals are more spectacular in terms of both numbers of species and their diversity (e.g., Schluter 2000, Grant 2013), yet Darwin's finches have several advantages for the study of biological diversity (box 1.1). Many populations live in pristine environments, no species has become extinct as a result of human activities, and evolution can be studied as a contemporary process.

Box 1.1. The Choice of Darwin's Finches

When we began our Galápagos research, the best-known radiations of animals were the numerous species of cichlid fish in several of the African Great Lakes (Fryer and Iles 1972), *Anolis* lizards of the Caribbean (Williams 1972), *Drosophila* (Carson et al. 1967) and honeycreepers (Amadon 1950, Warner 1968) of Hawaii, and Darwin's finches (Lack 1945, 1947). The major features of morphological diversity were understood as feeding adaptations caused by natural selection in spatially segregated populations, and color and pattern variation as a result of sexual selection. Deepening this understanding required two things: a better estimation of phylogenies, which only became possible much later with the development of molecular genetic markers (Wagner and Funk 1996, Givnish and Sytsma 1997), and an analysis of contemporary populations to investigate the genetic basis and ecological causes of evolutionary change. This is where Darwin's finches had some advantages over the others. The subject was ripe for ecological analysis, and finches seemed suitable subjects because they could provide the missing focus on population biology.

Detailed population studies appeared to be feasible because some populations are small, the finches can be marked for individual recognition, and they are conspicuous and approachable, so their fates can be determined accurately. Ecological influences on their fates can be identified because the climate fluctuates strongly and the extremes are markedly different. In some years there is little or no rain (La Niña); in others there is an abundance of rain (El Niño). The change from one extreme to the other is caused by reversals in the gradient of atmospheric pressure and sea-surface temperature across the Pacific basin. It is known as the El Niño–Southern Oscillation or ENSO phenomenon. The climatic extremes occur somewhat predictably at approximately three- to seven-year intervals on average (Philander 1990, Chen et al. 2004), with multidecadal oscillations in amplitude (Schlesinger and Ramankutty 1994). Superimposed upon a normal annual cycle of hot-wet (January to April) and cool-dry (May to December) seasons, the interannual fluctuations create profound changes in both marine and terrestrial productivity. As we discovered, the swings from plenty to scarcity reveal the ecological forces that finches are subjected to, and the evolutionary consequences. These in turn help us to interpret the radiation because it is still in a natural state: no species is known to have become extinct through human agency, and several of the islands have scarcely or never been affected by human occupation or exploitation.

Once thought to be members of the passerine family Emberizidae (buntings and finches), Darwin's finches are now classified as tanagers (Thraupinae) (Burns 1997). This Neotropical family comprises about 400 species (Isler and Isler 1999) that evolved in the last 12 million years (Cracraft and Barker 2009). The drab-colored Darwin's finches are thus a small part of a much larger radiation of varied and often colorful birds. According to current understanding at least 13 species of finches evolved on the Galápagos in the last two to three million years, and another evolved on Cocos Island (Grant and Grant 2008a). The species are distinctive in morphology (box 1.2), especially in the size and shape of their beaks, as well as in their diets. How is all the variation to be explained?

Box 1.2. What Makes a Darwin's Finch Species?

Lack (1945), following Swarth (1931) and earlier taxonomists, classified species by their size, proportions, and to a lesser extent plumage. For example, four species of ground finches can be recognized morphologically by their differences on any one island and the consistency of the differences across islands. The Small Ground Finch (*fuliginosa*), Medium Ground Finch (*fortis*), and Large Ground Finch (*magnirostris*) differ principally in average size (fig. B.1.2, appendix 1.2, fig. P.1). As size increases from one species to the next, beak size becomes both larger and more blunt. We refer to them as the granivore group because they all feed extensively on seeds. The fourth species, the Cactus Ground Finch (*scandens*), is about the size of the Medium Ground Finch but has a proportionately longer and narrower beak than any of the other three. As the name implies, it is a cactus (*Opuntia*) specialist. In other respects the species are identical. As young birds they are brown and streaked. With successive molts the males, but not females, acquire partly black then completely black plumage (Salvin 1876, Grant 1986). The remaining 10 species of Darwin's finches, similarly recognized by morphology, have minor relevance to this book (they are briefly described in appendix 1.3). Lack confirmed the biological reality of species identified by morphology in the breeding season of 1938–39 on San Cristóbal and Santa Cruz islands. Without having the benefit of measurements of individuals, he observed members of each morphological group (aka species) pairing up and breeding with each other and not with members of another group. Later in the book (chapter

Box 1.2. (continued)

9) we discuss the additional relevance of song to the question of what constitutes a species.

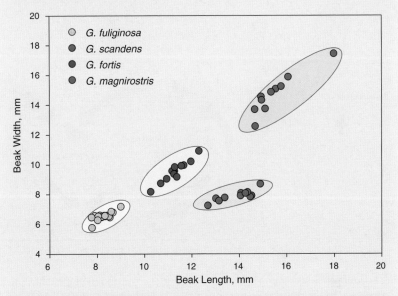

Fig. B.1.2 Morphological variation among four species of Darwin's ground finches (males) on several islands. Data are taken from Grant et al. 1985.

Lack (1945, 1947) made the first attempt to answer this question after studying Darwin's finches in the field. His explanation laid stress on three factors; natural selection, diversification on separate islands, and competition between species for food. Truly Darwinian! According to the Darwinian view, splitting of a species on Galápagos is initiated allopatrically when individuals disperse from one island to another and establish a new population. This is easy to visualize (fig. 1.1) because the archipelago has many islands. Colonists encounter new conditions, many die, and those surviving pass on to their offspring the heritable characteristics that contributed to their survival. In this way the population evolves by natural selection and becomes adapted to the new environment. There may be additional elements of randomness in how they evolve, if, for example, the founders are few in number or are not a representative sample of the original population and later diverge through genetic drift.

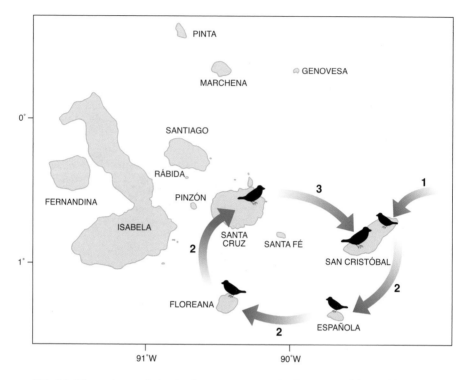

Fig. 1.1 Allopatric speciation in three stages: initial colonization (1), establishment of a second and additional populations (2), and secondary contact between two divergent populations (3). Choice of islands is arbitrary. Repetition of stages 2 and 3 in other parts of the archipelago gives rise to more species. From Grant 1981a, Grant and Grant 2008a.

The process of colonization and dispersal is repeated from one island to another until the two diverging lineages eventually come together on an island. As discussed in several chapters in this book, what happens at the point of secondary contact is crucially important in the speciation process. If members of the resident population and the immigrants do not interbreed, but the immigrants breed among themselves, two species will have been formed. Alternatively residents and immigrants might interbreed to some degree if prior divergence had not proceeded far. A tension might then exist between opposing tendencies: between fusion into a single population through interbreeding, and fission through divergent selection. The tendency to diverge would be expected if individuals produced by interbreeding had lower fitness than members of the parental populations, as would be the case if they suffered from an ecological disadvantage in competition for food or were physiologically weak. The end point of the divergence of the salient characters—character displacement—is reduced competition for food, a strengthened barrier to interbreeding, and enhanced prospects of long-term coexistence of these species (Grant 1981a, 1986).

Species and Speciation

Species differ (box 1.2). If they interbreed, they do so rarely yet remain distinct. They are said to be reproductively isolated from each other by behavioral barriers that prevent or inhibit interbreeding, or, if they do interbreed, by genetic barriers that prevent the formation of fertile offspring. To be more precise, it is individuals that interbreed, not species, and a species is a collection of individuals in one or more populations that are capable of interbreeding with little or no loss of offspring fitness. This is the biological species concept, with the essence of species being their complete or near-complete separation from each other.

Speciation is the evolutionary process that gives rise to the differences. It occurs when one species splits into two noninterbreeding populations or sets of populations (fig. 1.2). The challenge we face is to

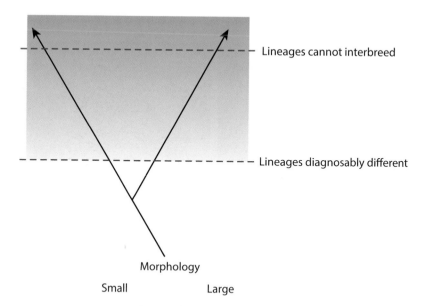

Fig. 1.2 Y diagram of speciation. The process is represented as a splitting and divergence of lineages. Opinions vary on when the lineages merit recognition as two species because divergence is gradual (discussed in Grant and Grant 2008a).

explain the product, species and their attributes, when the process by which they are produced is scarcely ever seen. To that end we need to answer the how, why, when, and where questions of speciation (Grant and Grant 2008a). What are the important factors that cause populations to diverge, how do they operate, and what are the circumstances? What prevents them from interbreeding and fusing into a single population? To answer these questions we chose to study finches on a small island: Daphne Major.

Daphne

Daphne (box 1.3, figs. P.2 and P.3, figs. 1.3 and 1.4) is centrally located in the main part of the archipelago about 8 km from the much larger islands of Santa Cruz to the south and Seymour to the east (fig. P.3). It is a pyroclastic or tuff cone that was formed explosively by underwater volcanic activity. Approximately three-quarters of a kilometer long and 120 m high, it has never had a human settlement. The center is a crater floor that is periodically occupied by breeding Blue-footed Boobies (*Sula nebouxii*). Topographically there are three vegetated habitats: an inner slope, an outer slope, and an area on the southern side with a more gentle slope of about 15 degrees that we refer to euphemistically as the

Fig. 1.3 Daphne Major landing. **Left**: The wave-cut, barnacle-covered platform used for landing and departing at low tide; arrows indicate "steps" (M. Wikelski). **Right**: Exit, leaving when the sea is calm.

plateau (figs. 1.4 and 1.5). All three habitats have shallow soils, seasonally deciduous annual and perennial plants, and clusters of *Opuntia* (prickly pear cactus) bushes where finches nest. The two main finch species are the Medium Ground Finch (*Geospiza fortis*) and the Cactus Finch (*G. scandens*).

Associated with its small size (0.34 km²), the most important feature of Daphne is its ecological simplicity. It has a community of fewer than 60 plant species, most of which are rare (appendix 1.1), and a similarly low diversity of insects and spiders. Breeding populations of finches rarely exceed 150 pairs. Small numbers make it relatively easy to determine how environmental factors affect morphological traits of finches, and how change in the environment brings about change in morphology. Their tameness makes them easy to observe. Of paramount importance for this study, they can be uniquely marked so that each individual can be identified by observation.

The Darwin's finch radiation is the macrocosm; Daphne finches are the microcosm. To throw light on the macrocosm, we studied the microcosm for 40 years, and witnessed evolution.

Fig. 1.4 Three habitats. **Upper:** Plateau, 1995. The fallen tree (near right) was present throughout the 40 years. **Middle:** Inner slope and plateau, 1983. **Lower:** Outer slope, 2012.

Fig. 1.5 Change in vegetation over 73 years. **Upper**: View across the crater floor in 1939 (L.S.V. Venables). **Middle**: 1973. Blue-footed boobies (*Sula nebouxii*) nesting on the crater floor. **Lower**: 2012. The pattern of vegetation on the crater floor, principally *Croton scouleri*, reflects drainage of rainwater to the lowest level. Notice in the left part of the figures that the extent of dark green cactus bushes increased over the years. Finches roost, nest, and in some years (e.g., 1992) die in large numbers in the bushes fringing the crater floor.

Evolution Observed

Figure 1.6 shows an intriguing pattern of change through time. The average beak size of the *fortis* population did not remain constant for 40 years but increased in 1978 and decreased in 2005. Beak size

Box 1.3. Recent History of Daphne

There is not one Daphne but two: Daphne Major and Daphne Minor (Chica) (figs. P.2 and 2.4). The Daphnes were named after the British naval vessel H.M.S. *Daphne* that visited Galápagos in 1846, possibly by Midshipman G.W.F. Edwardes, who was the first to show them on a map (K. T. Grant, pers. comm.). Beebe (1924) coined the terms Major and Minor (Woram 2013). The Daphnes differ in two respects. Whereas it is difficult to climb onto Daphne Major with camping supplies, it is impossible to do so on Daphne Minor without ropes. Daphne Minor has been climbed once. Also Daphne Minor has a crater lined with blocks of lava (Grant et al. 1980), indicating volcanic activity above the sea, whereas the Daphne Major crater lacks lava.

Daphne Major, hereafter Daphne, was put on the map ornithologically by Rollo Beck. He and companions collected specimens of finches in 1901 and 1905–6 for Walter Rothschild's museum in England and the California Academy of Sciences respectively (Gifford 1919). William Beebe (1924) reached a wide audience with an engaging description of a day on Daphne in his popular book *Galápagos: World's End*. David Lack never visited the island, but his assistant L.S.V. Venables did for one day in January 1939 (fig. 1.5), and reported that *fortis* (identified as *fuliginosa*!) were common all over the island whereas there were very few *scandens* in cactus clumps adjacent to the crater floor. He saw one *magnirostris*. Lack believed the Medium Ground Finch (*fortis*) was almost the sole occupant on the basis of these observations, those of the collectors (Gifford 1919), and 42 specimens in museums. In addition three Small (*fuliginosa*) and four Cactus (*scandens*) Ground Finches had been collected (Lack 1945), and Beebe (1924) had observed the breeding of a single pair of Large Ground Finches (*magnirostris*) in 1923. In the 1960s Daphne was a place for seabird research by David Snow and Michael Harris. Incidental observations on the finches (Harris 1973, 1974) provided the most up-to-date information on their status when we made our first visit to the island in April 1973. Providentially *The Flora of Galápagos* (Wiggins and Porter 1971), an invaluable field guide to the plants of Daphne, had been published just before.

evolved. A host of compelling questions arise when we confront a pattern of change like this. First, why did it occur? The strong and rapid transitions in average beak size at these two times implicate natural selection. How strong was selection, and what caused it? What is the source of beak-size variation, and what maintains it when selection

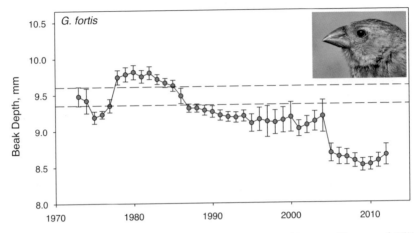

Fig. 1.6 Evolutionary trajectory of *fortis* beak size over 40 years. Means and 95% confidence limits are shown for all birds alive in each year. Parallel horizontal lines mark the upper and lower 95% confidence limits on the first estimate of a mean based on a large sample size ($n = 221$) in 1973.

occurs? To what extent is it genetically based? What is the relevance to speciation?

The most remarkable feature of the trajectory is the fact that *fortis* are no longer the same as they were 40 years ago. Change is not inevitable, however, as figure 1.7 shows. The trajectory of *scandens* is flat except for minor excursions; beak depth has remained the same for 40 years. Why did one species (*fortis*) change and the other (*scandens*) did not? Was *scandens* subject to natural selection but lacked the genetic variation to respond evolutionarily? Did *scandens* change in other traits? Did the two species interbreed, and if so with what result? Did they compete for food?

Among the several unexpected things that happened in the 40 years two events that are highly relevant to these questions stand out. One was the arrival of a hybrid from neighboring Santa Cruz Island. Years later the descendants were breeding among themselves: they were behaving as a new species! How could that happen in such a short time? Why did it happen? Why did they not breed with *fortis* or *scandens*?

The second event was the establishment of a breeding population of the Large Ground Finch (*G. magnirostris*) at the end of 1982. Thirty years later there were 50 pairs on the island. How could another species fit into the community? Did it compete with the residents for food, and if so what were the consequences, both evolutionary and ecological? How might the arrival of a new species throw light on speciation?

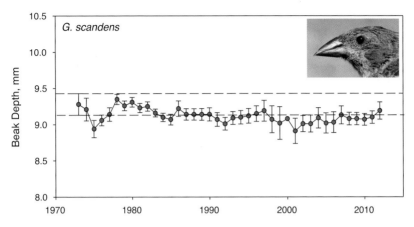

Fig. 1.7 Evolutionary trajectory of *scandens* beak size over 40 years. Means and 95% confidence limits are shown for all birds alive in each year. Parallel horizontal lines mark the upper and lower 95% confidence limits on the estimate of the mean in 1973 ($n = 71$).

All these questions are interdependent. By unraveling the dependencies we are able to reveal causes and complexities of evolution in contemporary time.

Chapters of the Book

We address these questions and describe the events that gave rise to them in the following chapters. The sequence is partly dictated by the nature of the study. In the first half we attempted to find every nest on the island and to mark every nestling uniquely. After we stopped doing this, our information on relatedness and biological success of individuals was reduced. On the other hand genetic data on relatedness became available in the second half. G. *magnirostris* became an important factor only in the second half. Thus the two halves of the study differ, and the organization of the chapters reflects this.

We start with an observation and a historical question in chapter 2. The Medium Ground Finch is exceptionally small on Daphne. What caused its evolution? This apparently simple problem was the magnet that drew us to Daphne in the first place. In contrast to the initial historical perspective all subsequent chapters are concerned with evolution as a contemporary process. Chapters 3 and 4 discuss the genetic basis of morphological variation and the evolutionary responses to natural selection in the first half of the study (fig. 1.6). Chapter 5 considers how the breeding component of fitness might affect the evolutionary trajectory. The arrival of *magnirostris* and its subsequent fate is described in chapter 6: how a population became established and why it prospered. It was a new factor, a competitive influence on *fortis*, and a cause of evolutionary change in the *fortis* trajectory (chapter 7). Chapters 8 to 10 discuss rare but persistent introgressive hybridization through backcrossing between *fortis* on the one hand and *scandens* and immigrant *fuliginosa* on the other, and the genetic and fitness effects of introgression on the trajectory depicted in figures 1.6 and 1.7. Chapter 11 surveys morphological evolution across the 40 years in both *fortis* and *scandens*, and in traits other than beak size, contrasting the relative influence of selection and introgressive hybridization on each of the two species. Chapter 12 quantifies the role of selection during the morphological transformation of one species into another, drawing upon knowledge of genes expressed during development as well as upon adult morphology. Chapter 13 discusses the events leading up to the formation of a new, reproductively isolated, lineage of finches. Chapter 14 speculates about two futures: the future of the finches if the anticipated environmental change occurs in Galápagos, and the future of phylogenetic understanding from genomic studies. Chapter 15 is a synthesis of the main evolutionary findings, and chapter 16 extends the discussion with some generalizations, and implications. An epilogue (chapter 17) completes the book. It stresses the value of continuous long-term study of ecology and evolution.

Summary

Darwin's finches on the Galápagos islands are a model system for the study of speciation and adaptive radiation, that is, the rapid evolution of morphologically and ecologically diverse species from an ancestor. Core ingredients of a theory to explain how and why the radiation occurred are natural selection, allopatric divergence, reproductive isolation, and interspecific competition. Our task is to determine how these ingredients occur, and how they are connected. In this book we describe what we learned about evolution by studying four species of ground finches on the single island of Daphne Major over a period of 40 years.

Daphne Finches: A Question of Size

If it is assumed that the various sizes and shapes of
bills amongst the Geospizae have been developed as
adaptation to differences in food habit, then it must
be shown that the different species of the genus
feed on different species of seeds.

(Snodgrass 1902, p. 381)

If these predictions and declarations are mistaken,
then the hypothesis must be discarded, or at least
modified. If, on the other hand, the predictions turn
out correct, then the hypothesis has stood up to trial,
and remains on probation as before.

(Medawar 1991, p. 232)

Introduction

ON FIRST ENTERING THE GALÁPAGOS ARCHIPELAGO, a naturalist is struck by the distribution of closely related species of
finches. Populations of the same species differ on average in
body and beak size from island to island, whereas on any one island
sympatric species differ discretely from each other in these characteristics. How did the morphological differences between species come
about? Two previous investigators confronted this question. Lack
(1947) laid stress on three factors—natural selection, diversification
on separate islands, and competition between species for food when

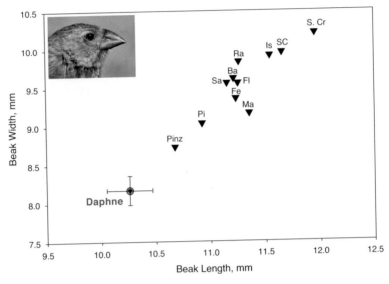

Fig. 2.1 Variation in average beak sizes among 12 populations of *fortis* (males), with 95% confidence intervals for the Daphne means. The Daphne samples are significantly smaller in both beak dimensions than all others including the most similar on Pinzón. Sample sizes vary from 11 (Rábida: Ra) to 278 (Floreana: Fl), with a median of 72. Other symbols refer to Balta (Ba), Fernandina (Fe), Isabela (Is), Marchena (Ma), Pinta (Pi), Pinzón (Pinz), Santiago (Sa), San Cristóbal (S. Cr), and Santa Cruz (SC). From data in Grant et al. 1985.

two closely related lineages came together on secondary contact (ch. 1)—whereas Bowman (1961) argued that available food on each island is sufficient to account for which particular species are present and their morphological characteristics, without having to invoke competition.

The focus of attention in this debate fell on Daphne, the reason being that on this island the Medium Ground Finch (*fortis*) is exceptionally small (fig. 2.1) and occurs there in the absence of the Small Ground Finch (*fuliginosa*). Elsewhere the Medium Ground Finch and Small Ground Finch coexist in the same habitat on 13 islands and differ substantially in size, but on the small island of Daphne the body size and beak size of *fortis* are intermediate between other members of its species elsewhere and *fuliginosa* (fig. 2.2). Lack explained the small size as character release, that is to say, expansion into the niche of an absent competitor, and hence the converse of character displacement described in chapter 1. Another situation strengthens the interpretation of character or competitive release. *G. fuliginosa* occurs alone on a group of four small islands called Los Hermanos (Crossmans) (figs. P.1

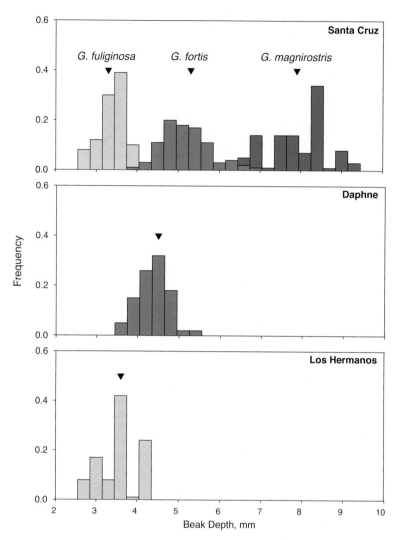

Fig. 2.2 Morphological intermediacy of solitary populations. Frequency distributions of beak depth (upper mandible) of adult males of three species of ground finches (*Geospiza*). Numbers of specimens: from Santa Cruz 134 *fuliginosa*, 156 *fortis*, and 26 *magnirostris*; from Daphne 89 *fortis*; and from Los Hermanos 12 *fuliginosa*. Solid triangles indicate average beak depths. From Grant 1986.

and 2.3), and here it is larger and intermediate in size between *fuliginosa* elsewhere and *fortis*. The two populations with intermediate body and beak sizes on Daphne and Los Hermanos are not identical; each is slightly more similar to the species to which it was assigned by Lack (1947).

Fig. 2.3 Seed sampling on other islands. **Upper**: Los Hermanos III. The study site is close to the top of the island. **Lower**: Ian and Lynette Abbott seed sampling at Bahía Borrero north shore of Santa Cruz.

For character release to be considered as the process resulting in the observed patterns, we need to know which is the original condition, on Santa Cruz or Daphne, and which is the derived one. In the case of *fortis* the allopatric state on Daphne is most likely to be the derived one. The island was once part of a large Santa Cruz–Baltra–Seymour Norte landmass in the last Ice Age (fig. 2.4), and only acquired its current island state about 15,000 years ago when the sea level rose (Grant and Grant 1996a, Geist et al. MS), probably abruptly (Blard et al. 2007). Thus *fortis*, in the absence of *fuliginosa*, could have been released from competition sometime after Daphne became an island. This has been described as the classical case of character release (Grant 1972, Boag and Grant 1984a).

David Lack made a convincing case for interspecific competition being an important factor in the evolution of Darwin's finches by concentrating exclusively on supporting evidence. However, observations elsewhere in the archipelago do not quite fit the pattern. For example, *fortis* is not intermediate in beak size on another small island (Champion) that lacks a population of *fuliginosa*. G. *fuliginosa* occurs in the

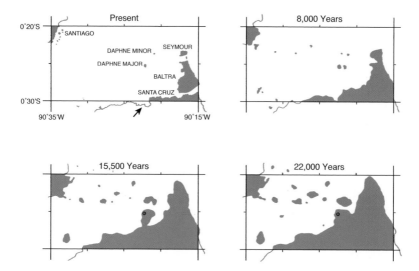

Fig. 2.4 Isolation of Daphne as a result of a rise in sea level associated with melting of glacial ice. From Grant and Grant 1996a, 2008a. The position of Borrero Bay is indicated by an arrow. The position of Daphne Major before 15,000 years ago is shown in red.

absence of *fortis* on eleven small islands and one large one (Española), yet has not become larger on any of them except on Los Hermanos. These facts raise doubts about the importance of competition.

Testing the Hypothesis

The character release hypothesis rests on five assumptions, which can be tested with *fortis* on Daphne. They are:

- Foods normally consumed by both *fortis* and *fuliginosa* are available on Daphne.
- Variation in diet is a regular function of variation in beak size. For seed-eating finches the obvious relationship is a mechanical or physical one: the average size of seeds in the diet increases as average beak-size increases among individuals.
- Natural selection is the process by which change takes place.
- There is heritable variation in the trait, or traits, that change.
- Finally *scandens*, the other finch species on Daphne, can be ignored because it contributes nothing to the morphological intermediacy of *fortis* by depleting resources.

Fig. 2.5 Map of Daphne, drawn from an aerial photograph. The study area used in 1973 for seed sampling, netting, and feeding observations is shown with a dotted-plus-solid line. This was extended in 1977 from the solid line to the outer slope. After 1978 the whole island was treated as the study area except for seed sampling. From Grant and Grant 1980a and Boag and Grant 1984a.

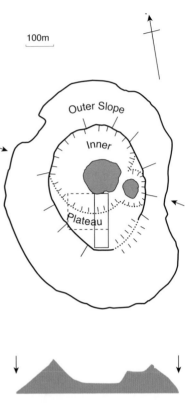

We test the first two and last assumptions in this chapter, the third assumption in chapter 4, and the fourth assumption in the next chapter.

AVAILABILITY OF FOOD

There was no published information on either food supply or diets on Daphne before we began. Testing the assumptions required a quantitative study of both. This was no simple task, because food supply depends strongly on rainfall, which varies markedly both within and between years (box 1.1). Quantifying available food required repeated sampling of seeds (box 2.1). We did this on Daphne (fig. 2.5) and, for comparison, at a coastal site (Borrero Bay, fig. 2.3) on neighboring Santa Cruz Island (fig. 2.4) at three times at the beginning of our study: April–May 1973, December 1973, and June–July 1975 (Abbott et al. 1977, Boag and Grant 1984a). The first and last samplings were in the early part of the dry season, rain and almost all seed production having ceased in the preceding three months, whereas the December sampling was at the end of the dry season when food is scarce.

DIETS OF *G. FORTIS* AND *G. FULIGINOSA*

Finches, like other birds, spend most of their lives searching for food and feeding. They eat seeds throughout the year and arthropods (caterpillars, aphids, etc.) predominantly in the wet season (fig. 2.6). To compare dry-season diets, we classified seeds by their size, measured as the second-longest dimension (depth, D), and hardness (H), mea-

Box 2.1. Estimating Food Supply

To quantify seed availability on Daphne, we established three sampling grids of 50 × 100 m on the inner slope, outer slope, and the plateau (slope 15°; figs. 1.4 and 2.5). On each sampling occasion we used a random-numbers table to select 50 1 m² quadrats distributed approximately equally among the three sampling grids (Abbott et al. 1977, Smith et al. 1978, Boag and Grant 1984a, 1984b). All buds, seeds, and fruits were counted on the vegetation. For counting seeds on the ground, we scooped out soil to a depth of ~2 cm in 25 × 25 cm areas in the northwest and northeast corners of each quadrat. We then multiplied the number of seeds on the ground and in the soil by 16 and added them to those on standing vegetation to give a total number per m². We sampled seeds twice per season each year and averaged them. We followed the same procedure on other islands (fig. 2.3) with only minor modifications and sampled on Santa Cruz only once per season each year.

To quantify diets we watched as many birds feeding as possible during mornings and, in the same areas, sampled food supply in afternoons of the same days. Initially finch diets were quantified by

Fig. B.2.1 Medium Ground Finches feeding. **Upper left**: Fruits of *Chamaesyce amplexicaulis*. **Upper right**: Capsules of *Sida salviifolia*. **Lower left**: Seeds and nectar of *Sesuvium edmonstonei*. **Lower right**: Spiderlings. From Grant and Grant 2008a.

Box 2.1. (continued)

recording identified foods eaten by a bird during a maximum of 300 seconds of observation (fig. 2.22). Then, in order to improve statistical analysis, we switched to recording just one food item per bird, the first item that could be identified, for as many birds as possible during morning hours. To compare quantitative features of the diets of *fortis* and *fuliginosa*, we classified seeds by their size, measured as the second-longest dimension (depth, D), and hardness (H), measured as the force needed to crack them (in kilogram force, kgf, then converted to newtons) in a specially constructed pliers device, the McGill seed cracker (fig. 2.7). A simple index $(D \times H)^{1/2} = (DH)^{1/2}$ was computed for each seed type (see appendix 1.1). Diets tell us what is meaningfully available to the finches. On Daphne finches consumed all seed types, from smallest to largest, except for two moderately large ones: *Merremia aegyptica* and *Ipomoea linearifolia* (appendix 1.1). They are both members of the Convolvulaceae and are possibly chemically protected, although doves eat both (Grant and Grant 1980a). They were deleted from our seed counts.

sured as the force needed to crack them (box 2.1, fig. 2.7), and then combined the two measures into the index $(DH)^{1/2}$. Our observations at the Santa Cruz site in the early dry season of 1973 showed that both species had broad diets (Abbott et al. 1977). *G. fuliginosa* fed on 13 different types of seeds and *fortis* fed on 11 types. At this time their diets were similar. On a scale of 0 to 1 the Whittaker similarity index was 0.87, as calculated by first determining the proportion (P) of each seed type (i) in the diet of each of the two species 1 and 2, and then summing the lowest values for all seed types: Σ min (P_{1i}, P_{2i}) (Whittaker 1960). Dietary differences were small but important. More *fuliginosa* fed on small and soft seeds than *fortis*, whereas only *fortis* fed on the largest and hardest seeds at this site, those of *Bursera graveolens* (fig. 2.8). Arthropods were absent from the diets at this time.

At the end of the dry season the dietary difference was more pronounced (Smith et al. 1978). A small number of temporary immigrants

Fig. 2.7 *(facing page)* Banding birds and measuring seeds. **Left**: Equipment for banding birds. **Right**: McGill seed cracker for determining the force necessary to crack a seed (Abbott et al. 1977). Finches can remove a split-ring band because the lower mandible is the same shape as the spoon used to apply the band to a leg. We changed from celluloid to plastic bands, applied by pliers. Colors are coded to correspond to numbers (box 2.2).

Fig. 2.6 Food for young finches. **Upper left**: Caterpillar on *Chloris virgata*. **Upper middle**: Caterpillar on *Sida salviifolia* (G. C. Keys). **Upper right**: Pair of sphingid caterpillar (*Hyles lineatus?*). **Upper middle left**: *Manduca* (?). **Lower middle left**: Caterpillar with warning coloration. **Middle center**: Green *sphingid* caterpillar. **Middle right**: Green *sphingid*. **Lower left**: Diptera larvae in *Opuntia* pad.

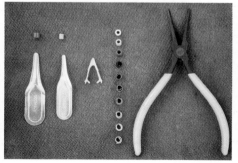

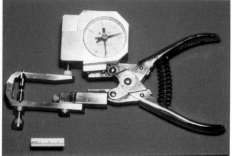

Fig. 2.8 Two important foods for finches. **Left:** *Bursera* berries. **Right:** *Croton* fruits. All finches are able to feed on the fleshy aril of *Bursera* berries, but only large-beaked birds can crack the woody stones and extract the kernels. *Croton* fruits are relatively easy to crack open (appendix 1.1).

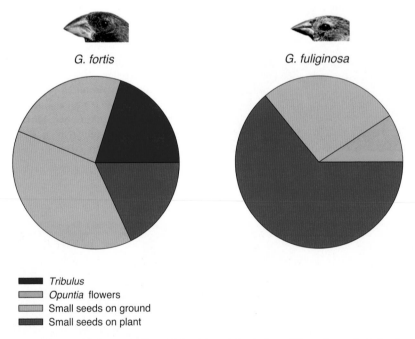

Fig. 2.9 Diets of *fuliginosa* (*n* = 11 birds) and *fortis* (*n* = 66) at the end of the dry season on Daphne in 1973. The small seeds on plants comprise *Chamaesyce*, *Tiquilia*, and *Heliotropium* (appendix 1.1). From Boag and Grant 1984a.

Fig. 2.10 *Tribulus cistoides*. **Upper**: Flowering plant. **Lower**: Fruits. When mature and on the ground, the fruits are hard for finches to crack open. From Grant and Grant 2008a.

(*fuliginosa*) to Daphne allowed us to compare their diets with the diets of *fortis* there. As on Santa Cruz both species fed on the smallest and softest seeds (fig. 2.9), but only *fortis* fed on the large and hard seeds of *Tribulus cistoides* (fig. 2.10) with a depth-hardness index of approximately 10 (appendix 1.1). The difference between species in consumption of small and large seeds was also observed on Santa Cruz, and has been sustained in all subsequent studies on Daphne, Santa Cruz, and other islands, regardless of the particular identity of the large seeds (Grant 1986).

DIFFERENCES IN SEED SUPPLY

To test the first assumption, that foods normally consumed by both *fortis* and *fuliginosa* on large islands such as Santa Cruz are present on Daphne, we compared profiles of seed abundances of different sizes on Daphne and on neighboring Santa Cruz Island (Boag and Grant 1984a). Small seeds of grasses and other herbs eaten by *fuliginosa* on Santa Cruz are certainly present on Daphne. When the data are expressed in terms of total volume per seed size-hardness category, they show a relative paucity of the smallest seed classes on Daphne compared with Santa Cruz in the early dry season (fig. 2.11), but not in the late dry season. At the latter time Daphne had a larger volume of small-soft seeds than Santa Cruz (fig. 2.11). Furthermore at all three sampling times the largest and hardest seeds (*Tribulus cistoides*) were on Daphne.

These observations show, first, *fuliginosa* is not absent because its normal foods are absent, and second, *fortis* has not become small on Daphne as a result of large seeds being absent. These findings are consistent with the first assumption.

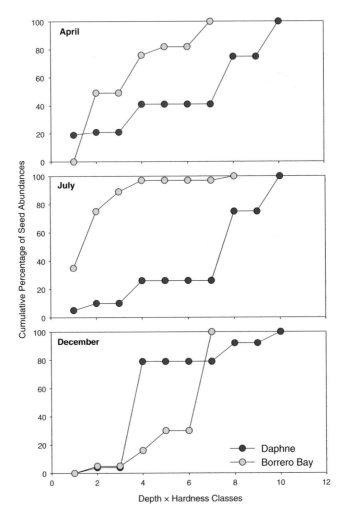

Fig. 2.11 Food distribution profiles in April and December 1973 and July 1975. Hardness classes are the square root of the product of seed depth and hardness (box 2.1). From Boag and Grant 1984a.

G. FULIGINOSA ON LOS HERMANOS

Feeding observations on Los Hermanos (fig. 2.3) give additional support to the hypothesis of character release. Late dry-season seed profiles are similar on Los Hermanos and Daphne in two respects: a scarcity of small seeds and prevalence of moderately hard seeds of *Cenchrus platyacanthus* (fig. 2.12) and even harder seeds of *Tribulus cistoides* (fig. 2.10). *G. fuliginosa* feed on both ripe and unripe seeds

Fig. 2.12 Plants that produce small seeds. **Upper left**: *Chloris virgata*. **Upper right**: *Eragrostis cilianensis*. **Middle left**: *Heliotropium angiospermum*. **Middle right**: Outer slope covered by *Portulaca howelli*. **Lower left**: Immature fruits of *Cenchrus platyacanthus*. **Lower right**: Mature fruits of *Cenchrus platyacanthus*. Small finches can open the soft immature fruits of *Cenchrus* but not the hard mature ones. Upper four figures are from Grant and Grant 2008a.

of *Cenchrus platyacanthus*, as *fortis* do elsewhere, whereas on other islands such as neighboring Isabela *fuliginosa* feed only on the soft, green, and immature seeds (Abbott et al. 1977, Boag and Grant 1984a, and fig. 2.12). Thus *fuliginosa* on Los Hermanos, in the absence of *fortis*, are convergently similar to *fortis* on Daphne in diet as well as in morphology.

Association between Beak Size and Diet

The second assumption is that diet varies as a function of beak size. It is supported by numerous observations of seeds eaten by banded birds of known beak size. Very small seeds are picked up and consumed whole. All others are husked or cracked before the kernels are extracted and eaten. The force applied to the seed is a function of beak size, especially in depth and width dimensions, and muscle mass (Bowman 1961, Herrel et al. 2005, 2009). Therefore average seed size in the diet is expected to increase as average beak-size increases among individuals. This is indeed so at the level of species: *fortis* crack larger and harder seeds than *fuliginosa*. It is also correct at the level of individuals within the *fortis* population on Daphne (Grant et al. 1976, Grant 1981b, Price 1987) and on Santa Cruz (De Léon et al. 2011). In 1973–78 we captured, banded, and measured beak dimensions (in mm) of many birds (box 2.2, figs. 2.7, 2.13, and 2.14), and later recorded what they ate (Boag and Grant 1984a, 1984b). Those that fed on the largest and hardest seeds had much deeper beaks on average (9.95 ± 0.08 standard error; $n = 82$) than those that fed only on the smallest-soft seeds (9.28 ± 0.06; $n = 199$) (fig. 2.15). Birds that fed on seeds of medium size-hardness values but never on the largest-hard seeds had an intermediate average beak depth (9.73 ± 0.12; $n = 39$). Since males are larger than females, on average, this could simply reflect a sex difference in diets, but this is not the case. The difference in average beak size among the three dietary categories was found separately in a sample of males and females. In a detailed follow-up study Price (1987) confirmed that large individuals tend to forage more on large seeds than small individuals do, they consume large seeds more rapidly than smaller individuals do, and the smallest individuals do not feed on large seeds at all.

Adaptive Landscapes

Given the relationship between beak size and seed size, average beak size of each species on an island can be predicted from the food supply under the contrasting conditions of competition and no competition.

Fig. 2.13 Capturing and measuring finches. **Upper left**: *scandens* in net, 1973. **Upper right**: Mist net line in sector 2 (fig. A.3.1). **Middle left**: Weighing a finch. **Middle right**: Weighing and taking a blood sample in a cave used for banding finches (K. T. Grant). **Lower left**: Cave used as a laboratory and kitchen.

Fig. 2.14 Measuring a finch. **Upper left**: Beak length. **Upper right**: Beak depth. **Lower left**: Beak width. **Lower right**: Tarsus length. From Grant and Grant (2008a). Illustrated with a large cactus finch on Genovesa. The same measurements are taken on Daphne finches.

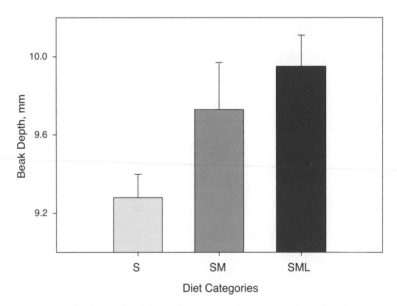

Fig. 2.15 Mean beak depths of *fortis* observed feeding on small seeds only (S), small and medium seeds (SM), or seeds of all sizes (SML). Lines represent 95% confidence intervals.

Box 2.2 Measuring and Marking Finches

For comparing diets with morphological features of individually marked birds, we captured finches in mist nets and measured and marked them with unique combinations of bands prior to release (figs. 2.7, 2.13, and 2.14). In the early part of the study we used mist nets mainly on the inner slope and plateau (Abbott et al. 1977, Smith et al. 1978, Boag and Grant 1984a). We measured six traits; four are illustrated in figure 2.14. We weighed birds to the nearest 0.1 g and measured wing length to the nearest mm with a wing ruler from the carpal joint to the wing tip, with feathers flattened and straightened. All other traits were measured to the nearest 0.1 mm (illustrated in Grant and Grant 2008a). Tarsus length was measured with dividers from the tarsometatarsal joint to the middle of the lowest undivided scute. Beak length was measured with dividers from the anterior edge of a nostril (nares) to the beak tip. Beak depth was measured with calipers in the plane of the anterior nares at right angles to the commissure, the line at which upper and lower mandibles meet. Beak width was also measured with calipers at the base of the lower mandible at the edge of the feathers. Later, when assistants measured birds, we applied correction factors to make their measurements comparable to ours.

Before releasing the finches, we marked them with unique combinations of leg bands for later identification. We used a numbered metal band (ring) and one colored band on one of the legs and two colored bands on the other leg. Colors were coded to correspond to the number (fig. 2.7). Starting alphabetically, we gave the number 0 to a black band, and so on with different colors for 1, 2, 3, etc, finishing with 9 for yellow. Bands were read in the sequence upper left leg, lower left, upper right, and lower right. Thus metal over yellow on the left leg and white over black on the right leg signified 980. To aid in identification, the metal band was placed in the upper position on either leg of *fortis*, and in the lower position on a leg of *scandens* or *fuliginosa*. We started by using celluloid bands (fig. 2.7). These were easily removed by the finches, and even when they weren't they became brittle and fell off. We switched to using split plastic (PVC) bands, first purchased commercially and then made by ourselves. These were much more successful, but some were not finch-proof. Gluing the split ends proved to be ineffective, therefore we made overlapping bands. We used thicker plastic bands for *magnirostris*.

Fig. 2.16 Predictions of average beak size compared with observations on a one-dimensional adaptive landscape. Histograms of beak depth are for *fuliginosa* (lined) and *fortis* (solid gray) adult males. Arrows pointing up indicate mean beak depths predicted by the food-based expected population density curves (the landscape topography), and arrows pointing down indicate the observed mean beak depths for *fuliginosa* (white) and *fortis* (gray). Adaptive landscapes are usually constructed in three dimensions (Svensson and Calsbeek 2012), but in this case nothing is gained by adding a third dimension such as beak length. From Schluter et al. (1985).

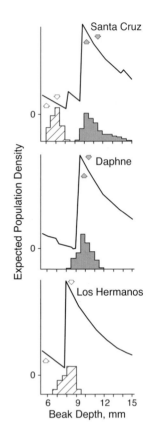

We addressed this task by using seed sampling data from the dry season, finch numbers, and beak sizes on many islands (Schluter and Grant 1984, Schluter et al. 1985). Density curves for a hypothetical solitary population were constructed as a function of average beak size for each island. This involved determining the range of seed sizes available for many populations of ground finch species with known mean beak depths, the density of seeds available within that range, and the number of finches supportable by the seeds. The procedure was repeated for 0.05 mm increments of beak depth across the whole size axis. The curves generated by the procedure can be considered as an adaptive landscape (Svensson and Calsbeek 2012) with peaks and valleys in fitness, unmodified by competition. Peaks and valleys are pronounced (fig. 2.16). Finally the species on each island were placed on the curves at the position(s) of their average beak sizes.

On Daphne there is no competition, and so the average beak size of *fortis* should be closely aligned to the average beak size predicted by the food supply, whereas in the competitive environment of Santa Cruz average beak sizes of *fortis* and *fuliginosa* should be displaced from their expected positions (fig. 2.16). This is what we observe.

On Los Hermanos *fuliginosa* shows an intriguing shift from the normal peak for that species to a second peak that is "occupied" elsewhere by *fortis*. As mentioned above, they consume the moderately large and hard mature seeds of a grass, *Cenchrus platyacanthus* (fig. 2.12), which elsewhere are consumed by *fortis*.

Can *G. scandens* Be Ignored?

The final ecological assumption is that *scandens*, the other finch species on Daphne, can be ignored because it contributes nothing to an explanation of the morphological intermediacy of *fortis*. When we first visited the island we discovered to our surprise that *scandens*, far from being scarce (box 1.3), was quite common. Therefore competition for food with *scandens* is a possible explanation for the relatively small beak size of *fortis* on Daphne. This hypothesis would be supported if *scandens* consumed seeds in the upper size range commonly exploited by the larger members of the *fortis* population on Daphne or other islands. Although *scandens* feed extensively on the seeds of the prickly pear cactus *Opuntia echios* ($DH^{1/2} \approx 3$) (figs. 2.17–2.20), and take a shorter time to crack them open than *fortis* (Grant and Grant 1996b), they do not open the harder, cherry-like, stones of *Bursera mala-*

Fig. 2.17 Exploitation of *Opuntia* cactus flowers. **Upper:** Some pollen has been consumed, but the central style and stigmas remain intact. **Lower:** A finch has removed the stigmas from this flower.

Fig. 2.18 Cactus fruits exploited by Cactus Finches. **Upper:** Opened *Opuntia* fruit on ground; seeds with green arils have been removed and consumed. **Lower:** *Opuntia* seeds have been cracked open, kernels removed, and then discarded. Lower figure from Grant and Grant 2008a.

Fig. 2.19 Cactus feeding. **Upper**: *scandens* pecking at *Opuntia* flower pollen. **Middle**: *scandens* probing an *Opuntia* flower for the basal nectar. **Lower**: *fortis* at an *Opuntia* flower that has been opened by *scandens* and half destroyed.

Fig. 2.20 Cactus seed eating. Cactus Finch cracking an *Opuntia* seed (L. F. Keller).

cophylla (fig. 2.8), nor do they feed on the harder woody fruits of *Tribulus cistoides* (fig. 2.10), unlike *fortis* (Grant et al. 1976, Grant 1981b). In view of dietary overlap *scandens* should not be ignored as a competitive influence on *fortis*, but the morphological shift of *fortis* to intermediate size cannot be attributed to *scandens* depleting the supply of large-hard seeds. The only other granivores on Daphne are doves (fig. 2.21). They may also exert competitive influences on finches, but have been ignored because they feed mainly on the two seed types not eaten by finches (box 2.1), and are generally rare (<20). They occur on all islands (Harris 1973).

Why Is *G. fuliginosa* Absent?

We presume that when Daphne was formed about 15,000 years ago, both *fortis* and *fuliginosa* were present, as they are today on the north shore of Santa Cruz. Subsequently one species became extinct on Daphne: why was it *fuliginosa*? If both species were equally vulnerable to becoming extinct it might have been a matter of chance which particular species did in fact become extinct. Alternatively, *fuliginosa*

Fig. 2.21 Other bird species. **Upper left**: Dove, *Zenaida galapagoensis*. **Upper right**: Yellow Warbler, *Dendroica petechia*. **Lower left**: Galápagos Flycatcher, *Myiarchus magnirostris* (K. T. Grant). **Lower right**: Galápagos Cuckoo, *Coccyyzus melacorhypha* (K. T. Grant). All eat insects, but only doves eat seeds. Doves and warblers are resident, and so are Galápagos Martins (*Progne modesta*), whereas flycatchers and cuckoos are occasional visitors to Daphne, especially after wet years of extensive breeding.

may have been competitively excluded by *fortis*. There is evidence for competitive exclusion in that only one species of granivorous finch is associated with an adaptive peak on all the 15 islands that have been studied. This implies a competitive adjustment of number of species on an island to the number of adaptive peaks (Schluter and Grant 1984). We suggest the reason why *fortis* prevailed on Daphne is that the current peak in the adaptive landscape is closer to the peaks occupied by *fortis* on other islands than to peaks occupied by *fuliginosa* (Schluter and Grant 1984). Extinction of *fuliginosa* rather than *fortis* was therefore not just a matter of chance. The full range of seed sizes on Daphne is now exploited by the *fortis* population but not by *fuliginosa*. In a climatically fluctuating environment with interannual fluctuations in food supply (chapter 4) this puts the small *fuliginosa* at a long-term competitive disadvantage with the larger and more aggressive *fortis* and helps to explain why small numbers of occasional immigrant *fuliginosa* have not established a permanent breeding population.

Discussion

Speciation happened in the past, and we ask how did it happen? This is a historical question, and in the absence of fossils an answer has to be sought with living organisms by looking backward in time: a retrospective analysis. The usual technique is to compare populations at different stages on the way to becoming species, or even closely related species themselves, and to infer the causes of divergence in genetic or phenotypic characters that has already taken place (Coyne and Orr 2004). We adopted the different procedure of using information on contemporary populations to test assumptions of a historical hypothesis, the character release hypothesis of Lack. The conclusion from the tests, in the words of Medawar (1991), is that "the hypothesis has stood up to trial, and remains on 'probation' as before."

For the remainder of the book our approach is prospective. Studying populations through time provides insights into the process of speciation even if the progress toward complete reproductive isolation is small. It provides evidence of evolution, as illustrated in figure 1.6, and also allows us to identify environmental factors that cause it. In addressing the questions raised in chapter 1 we will revisit the competition hypothesis on probation in chapter 7, where we document character displacement in action. In the next chapter we discuss the heritable variation of size traits.

Summary

We chose to study Darwin's finches on the single island of Daphne because of an interesting morphological pattern discovered by David

Lack (1945). The Medium Ground Finch (*Geospiza fortis*) and Small Ground Finch (*G. fuliginosa*) are morphologically distinct where they occur together in the same habitat on 13 islands, whereas *fortis* occupies Daphne in the virtual absence of *fuliginosa* and here it is distinctly smaller, intermediate in body size and beak size between other members of its species elsewhere and *fuliginosa*. David Lack interpreted the pattern as evidence for interspecific competition causing divergence of the two species where they occur together, character displacement, and a release from competition when one species occurs alone. We use quantified information on food availability and diets to test three assumptions of the character release hypothesis, and find support for each. As assumed, food is available for both species on Daphne, beak size influences what size of seeds can be consumed, and a third species, *scandens*, can be ignored as a competitor for large and hard seeds. The evidence from Daphne indirectly supports the Darwinian idea of competitively driven divergence during speciation.

Heritable Variation

Quantitative genetics is concerned with the inheritance of
those differences between individuals that are of degree
rather than of kind, quantitative rather than qualitative.
These are the individual differences which, as Darwin
wrote, "afford materials for natural selection to act on and
accumulate, in the same manner as man accumulates in
any given direction individual differences in his domestic
productions."

(Falconer and Mackay 1995, p. 1)

The journey of populations through genetic space is
intimately linked with the environment around them.

(Coen 2012, p. 278)

Introduction

FOR EVOLUTION TO OCCUR there must be heritable variation, so
one of the first questions that arise from both the small size of
fortis on Daphne and the evolutionary trajectory in figure 1.6 is
how heritable are size traits? We addressed this question by using the
methods of quantitative genetics to estimate heritable variation from
degree of family resemblance. Many nests were found (box 3.1) and
mapped (appendix 3.1), their owners (banded or not banded) were
identified, nestlings were banded when they had reached a posthatch-

Box 3.1 Research Strategy

We studied reproduction in every year for 16 years: 1976–91. We used maps to record nest locations (appendix 3.1). Most finches nest in *Opuntia* cactus bushes, with less than 5 percent nesting in *Bursera* trees or in *Portulaca*, *Chamaesyce*, and *Sesuvium* (perennial) plants close to the ground. Except for a couple of cactus bushes on a sea cliff in which finches may have nested, all nests were accessible at heights of less than 2.5 m. The goal was to determine clutch size, hatching success, and fledging success at every nest throughout each breeding season from 1978 to 1991; the coverage was approximately 50% in 1976, and only a few pairs of *scandens* bred in 1977. Nests were missed very rarely (Boag and Grant 1984b). As the study progressed, the fraction of breeding pairs that had been banded and measured increased to 99.5%; in early 1992 the last *scandens*, a male, was captured and banded, so at that time every adult of both species was identifiable. Nests were checked every 2–4 days, and nestlings were weighed and banded at day 8, counting the day of hatching as day 0. The first two broods produced in 1992 and 1993 were studied with complete coverage. In subsequent years breeding was monitored in an opportunistic way and for a shorter time, except that in 1998, a year of El Niño, breeding was studied again throughout the season (Grant et al. 2000).

Starting in 1988, we took a drop of blood for DNA analysis from the brachial vein of each finch, both adults and nestlings (fig. 2.13). In the absence of a buffer we developed our own method of storage with very good results. Samples were stored on EDTA-soaked filter paper in Drierite® in the field and at –80°C in the laboratory, and they yielded high-quality DNA. A decade later it became possible to use these samples to identify parentage unambiguously with 16 microsatellite DNA markers (Petren 1998). Our identification of mothers by observations at nests was confirmed in all cases, but paternity was sometimes misidentified by observation. The 1988 and subsequent samples were dominated by birds hatched in 1987, but there were some old birds among them, including one that had hatched in 1975! Most offspring of known (genotyped) parents in our samples hatched in 1991.

ing age of 8 days, and, sometime after they had fledged, they were captured in mist nets and measured (box 2.2). By capturing many young birds of different ages, we discovered they ceased growing at an age of about 60 days; therefore only birds of this age or older were included in subsequent analyses (Boag 1984). After the first two years we attempted to find every nest on the island (Millington and Grant 1984, Price 1984a, Grant and Grant 1996c).

This chapter explains how we estimated the genetic component of phenotypic variation in body-size and beak-size traits, and what we found after taking into account possible biases in the estimates that arise from misidentified paternity due to extra-pair mating, maternal effects, and correlations between genotypes and environments. The overall conclusion we reach is that a large fraction of the measurable variation can be attributed to genetic factors: all morphological traits are highly heritable. This applies to both *fortis* and *scandens*, although there are small differences between them. Therefore both species on Daphne possess a strong potential for evolution in all size traits.

Estimating Heritable Variation

A phenotypic trait is heritable to the degree that measured variation in the trait reflects variation in genotype. Family members are genetically more similar to each other than to nonfamily members. This fact is used to estimate the amount of phenotypic variation that can be attributed to the additive effects of genes. In formal terms the heritability (h^2) of a trait in the narrow sense is the additive genetic variance (Va) as a proportion of the phenotypic variance (Vp): $h^2 = Va/Vp$. In practical terms it is estimated most simply by regressing measurements of offspring on the average of the measurements of their two parents. If there is no genetic component to the phenotypic variance, the slope will be 0. The theoretical maximum is 1.0. Most studies of morphological variation in birds find values between 0 and 0.5, sometimes higher (Grant and Grant 2000a, Merilä and Sheldon 2001). There are more statistically refined ways of estimating heritable variation by taking into account genetic effects from relatives other than parents when extensive pedigrees are available, and by estimating and excluding certain sources of environmental variation (reviewed in Kruuk et al. 2008). These give greater precision to the estimates, but even if only one parent is known, an estimate can be obtained by doubling the slope of the regression of offspring on single-parent measurements (Falconer and Mackay 1995, Lynch and Walsh 1998). The main caveat in all these studies of natural populations is the estimates may be biased by uncontrolled factors. For example, heritability estimates may be inflated as a result of parents and offspring growing up in similar environments that affect their final sizes, with different families experiencing different environments (genotype × environment correlations).

Heritable Variation

Our first estimates of heritabilities were remarkably high (fig. 3.1); in fact some of them were statistically indistinguishable from 1.0 (Boag and Grant 1978, Boag 1983). However, the standard errors of the estimates were also large. Sampling variance is large because our samples were relatively small and not because of unreliability in the measurements. In fact measurements of beak traits were highly repeatable (table 3.1, appendix 3.2). We minimized effects of measurement error by averaging repeated measurements of the same bird taken at different times.

The high heritability estimates were confirmed in a follow-up study in 1978 (Boag 1983). The slopes of regressions of offspring on midparent values in 1976 and 1978 were almost identical (Boag 1983), despite a strong difference in rearing conditions. Finches bred at high density and produced one brood in 1976, and bred at low density and produced three broods in 1978. Offspring that hatched in 1978 were closer in size to their parents, on average, than were the 1976 offspring, suggesting density-dependent suppression of growth in 1976: slower, incomplete, or earlier cessation. The fact that the regression slopes were identical in years of different conditions shows that large adult size is not a conspicuous function of the particular rearing environment experienced during growth (genotype × environment interactions). Growth is stunted when conditions are poorer than in 1976, but then the offspring do not survive long enough to be caught and measured as adults (Price 1985) and therefore do not enter the analysis.

We estimated heritabilities from five more years of breeding (1981, 1983, 1984, 1987, and 1991). Number of families ranged from 51 in 1981 to 101 in 1987, and number of offspring varied from 99 in 1981 to 467 in 1987 (Grant and Grant 2000a). The two main results were (a) beak trait heritabilities exceeded those for body-size traits, and (b) for beak depth the estimates were equal to or greater than 0.50 in all seven years (fig. 3.2). Relatively little annual variation in the estimates means relatively little influence of annual variation in growth conditions.

Our overall conclusion is there is substantial genetic variation in all six traits. It supports the hypothesis of evolutionary change of morphological traits on Daphne. However, heritability estimates may be biased, and possible biases need to be examined before the conclusion can be considered secure.

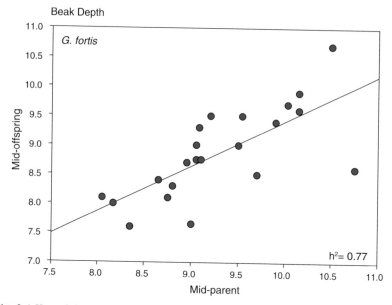

Fig. 3.1 Heritability (h^2) of *fortis* beak depth in 1976, estimated from the slope of the regression of mid-offspring on mid-parent values. From Boag and Grant 1978 and Boag 1983, recalculated with additional measurements.

TABLE 3.1

Repeatabilities: the degree to which repeated measurements of the same individual are the same, as assessed by analysis of variance (Lessells and Boag 1987)

	G. fortis		G. scandens	
	R	s.e.	R	s.e.
Mass	0.829	0.026	0.768	0.055
Wing length	0.783	0.032	0.709	0.067
Tarsus length	0.843	0.024	0.836	0.041
Beak length	0.940	0.010	0.795	0.050
Beak depth	0.938	0.010	0.855	0.037
Beak width	0.931	0.011	0.891	0.028
Sample size (n)	128		46	

Note: Body-size traits have lower repeatabilities (R) than beak-size traits, due to inherently higher measurement error and higher variation within an individual over time; wing feathers abrade and are replaced, and body mass fluctuates daily and seasonally. Measurement error, as a proportion of the phenotypic variance, is 0.06–0.21 for *fortis* traits and 0.11–0.29 for *scandens* traits (Grant and Grant 1994). Abbreviation: s.e. is standard error of the repeatability estimate. Changes from first to second year only are illustrated in fig. A.3.2.

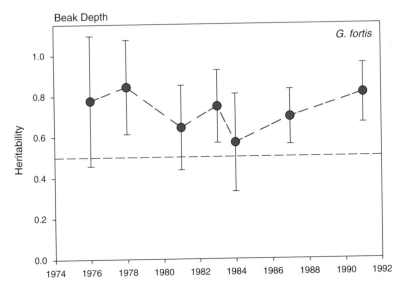

Fig. 3.2 Consistency in the estimates of *fortis* heritability of beak depth. Vertical bars show 95% confidence limits on estimates from offspring mid-parent regressions. The estimates have not been corrected for extra-pair young. From Grant and Grant 2000a.

Potential Biases

High heritabilities may be more apparent than real. Bias can result from similar environments being experienced by parents and offspring during growth, which would increase the estimate, by extra-pair paternity, which would decrease it, and by maternal effects, which could do either. We consider each of these in turn.

EXTRA-PAIR PATERNITY

Initially we believed that pairs were socially monogamous and rarely bigamous, since we saw no extra-pair copulations. Only later, in 1999, did it become possible to identify paternity unambiguously (box 3.1) with the development of 16 microsatellite DNA markers (Petren 1998). And we learned that our initial supposition was wrong! Most offspring of known parents in our samples hatched in 1991. These yielded estimates of 19.7% extra-pair young (EPY) out of a total of 223 genotyped *fortis* offspring, and 35.5% of 93 families with at least one EPY (Keller et al. 2001). Pooling all data from all years gave us an overall estimate of 17.1% EPY (Grant and Grant 2011a). In the total sample there were only two indications of incorrect identification of the mother, and they were ignored because they could have been recording errors.

MISIDENTIFIED PATERNITY

Most offspring and their parents were measured at an age of approximately one year. We calculated heritabilities for the six measured traits after correcting for extra-pair paternity. Because all six traits are positively correlated with each other, we used principal components analysis to find linear combinations of the traits that best summarize their variation and covariation along fewer uncorrelated axes. We did this for the three body-size measures (weight, wing, and tarsus) and separately for the three beak-size measures. In the first analysis PC1-body is a measure of size and PC2-body is a measure of proportions: weight and wing in relation to tarsus. In the second analysis PC1-beak is a measure of beak size, and PC2-beak is a measure of shape: depth and width in relation to length.

The effect of EPY on heritability estimates can be assessed from calculations with and without them. Because family sizes varied from 1 to 9 young, we used a family weighting technique in the calculations and standardized the variables to a mean of 0 and variance of 1.0 prior to analysis to eliminate small differences in either mean or variance between sexes. Excluding all extra-pair young resulted in a 21% increase in heritability of PC1-beak from the mid-parent regression alone, and an average of 21% for the four PC traits together. Effects of EPY on heritability estimates for the individual traits contributing to the PCs were sometimes larger, and varied in proportion to how different social and extra-pair fathers were in those traits (Keller et al. 2001).

MATERNAL EFFECTS

After removing all extra-pair young from the analyses, mother-offspring resemblance still exceeded father-offspring resemblance at PC1-beak and PC1-body by 19% and 23% respectively (Keller et al. 2001). This is one indication of maternal effects on adult size; however, neither difference was statistically significant. By restricting attention to just those offspring produced in 1991 by parents that hatched in 1987, because this was the only cohort of sufficient size for analysis, we found the mother-offspring regression for PC1-beak exceeded (significantly) the father-offspring regression by 41% after we removed extra-pair young. By adopting the same procedure for grandparental regressions we found, after excluding EPY, that maternal grandmother regressions for PC1-body were significantly larger than paternal grandmother regressions. For the other PCs there was no difference, and hence no evidence of maternal effects. Thus there are three indications of maternal effects, but the evidence is inconsistent and their magnitude has not been well estimated. We consider misidentified paternity to be a more serious problem in the estimation of heritable variation than failing to take into account the possibility of maternal effects.

Our overall conclusion is there are substantial extra-pair young effects but inconsistent evidence for maternal effects. After removal of EPY the revised heritabilities become 0.83 for beak size and 0.88 for beak shape. These values are much higher than in other studies (Grant and Grant 2000a, Merilä and Sheldon 2001) but comparable to human height heritability (Visscher 2008). They could result from higher additive genetic variance, lower environmental variance, or some combination. The phenotypic variance is large in this population (Grant et al. 1985); therefore, assuming environmental variances to be similar, elevated genetic variance is the probable cause of the high heritabilities.

GENOTYPE × ENVIRONMENT CORRELATIONS

If one part of the island is richer in food supply and better for rearing young than others, the young there might reach large size, survive well, obtain better territories in the same areas, and rear large offspring. Contrariwise, small parents may raise small offspring on poor territories. Continuation of this tendency from generation to generation would yield a resemblance between parents and offspring at least in part for environmental reasons; they were both raised in food-rich or food-poor territories. Randomizing clutches among foster parents is a technique for breaking any such environment correlations in order to estimate the magnitude of their effects, but experiments like this are understandably not permitted in the Galápagos National Park.

There is some evidence that parents do breed in the same habitat in which they were raised. We know the natal and breeding locations of 962 *fortis* and 304 *scandens*. Most of these birds hatched from nests on the outer slope: 56.4% of the *fortis* sample and 59.5% of the *scandens*. The species were similar in where the rest were hatched: on the inner slope (22.4% and 23.4%) and plateau (21.2% and 17.1%) respectively. These proportions are the basis for expecting in a 2 × 3 contingency χ^2 test what fraction of the samples should breed in each habitat. Males and females were combined because they do not disperse different distances, in either species (Krakower 1996). A disproportionate number of G. *fortis* that hatched on the inner slope bred there (48.9%), twice as many as expected (22.4%; $\chi_1^2 = 30.19$, $p < 0.0001$). Similarly a disproportionate number that hatched on the outer slope bred there (68.9%; $\chi_1^2 = 18.38$, $p < 0.0001$). Males and females contributed roughly equally to these results. There was no tendency for plateau birds to breed on the plateau, and no tendency for G. *scandens* to breed in their natal habitats more often than expected by chance (all p values > 0.1).

These associations could give biased heritability estimates if *fortis* adult size varied systematically across the three habitats (fig. 1.4); they did differ in some size traits according to where they hatched. We tested for differences in size between habitats by two-way analysis of variance (ANOVA), including sex as a factor because males are larger than females on average. After the sex difference was removed, there were no significant differences between habitats in any of the PC indices of size and shape, nor were there any differences within each of the habitats between those that bred in their natal habitat and those that bred in another one (all $p > 0.1$).

CONCLUSIONS ON BIAS

We found substantial extra-pair young effects upon heritability estimates, inconsistent evidence for maternal effects, and no evidence that the estimates are inflated by genotype × environment correlations. Two other pieces of information support the last conclusion. First, regressions of extra-pair young on their apparent, social (cuckolded) fathers were all small and nonsignificant, indicating little or no inflation from common environment effects (Keller et al. 2001). Second, experimental manipulation of clutches in seven studies of birds elsewhere all failed to find an effect of foster-rearing environment on adult size (Grant and Grant 2000a).

Heritabilities: A Comparison of Species

Given these findings, we asked if high heritabilities are solely a feature of *fortis* or if they apply to other species as well. Specifically, can the absence of an evolutionary trend in *scandens* beak depth (fig. 1.7) be

TABLE 3.2

Heritabilities (h^2) with standard errors (s.e.) for *fortis* and *scandens* on
Daphne and *G. conirostris* on Genovesa

	G. fortis		G. scandens		G. conirostris	
	h^2	s.e.	h^2	s.e.	h^2	s.e.
Mass	0.68	0.06	0.50	0.09	0.69	0.20
Wing length	0.64	0.07	0.49	0.09	0.58	0.18
Tarsus length	0.49	0.07	0.43	0.08	0.82	0.16
Beak length	0.85	0.04	0.80	0.07	0.66	0.20
Beak depth	0.75	0.05	0.55	0.10	0.69	0.11
Beak width	0.78	0.05	0.39	0.10	0.77	0.16
PC1	0.76	0.05	0.60	0.09	0.81	0.18
PC2	0.63	0.07	0.52	0.10	0.67	0.17
Families	207		158		57	
Offspring	674		412		81	

Note: PC1 and PC2 refer to the first two components of a principal components analysis of all six measured traits. Separate analyses were performed for each species and are roughly comparable because the loadings of the traits on the components were similar in the species. In each analysis all traits contribute approximately equally to PC1, which is therefore interpreted as a synthetic measure of body size. In all cases PC2 has a beak-shape element, with contributions of beak depth and width being approximately equal and having opposite sign to those from beak length. In all analyses PC1 statistically accounts for 60%–70% of the variance, and PC2 for an additional 11%–12%. Note that other principal components analyses referred to in this chapter were restricted to three body-size traits or three beak-size traits (pp. 64–65). From Grant and Grant 2000a.

explained by lack of genetic variation? The answer is no. *G. scandens* heritabilities were estimated in three years when family sizes exceeded 20 (Grant and Grant 2000a): in 1981 ($n = 37$), 1983 ($n = 68$), and 1987 ($n = 25$) but before we were able to identify EPY. Heritability estimates for the six traits uncorrected for EPY averaged 0.53, which is lower than an average of 0.70 for *fortis* (table 3.2, fig. 3.3). For beak depth, the trait shown in figure 1.7, the heritability estimate is 0.55 (table 3.2). Heritabilities for *scandens* traits corrected for EPY should be almost the same as the uncorrected ones (Keller et al. 2001), because EPY rate for this species is only 8%–10% (Petren et al. 1999a, Grant and Grant 2011a).

G. *scandens* differs from *fortis* in another respect. Heritability of beak length is consistently higher than heritability of depth and width in *scandens*, whereas in *fortis* all three heritabilities are approximately

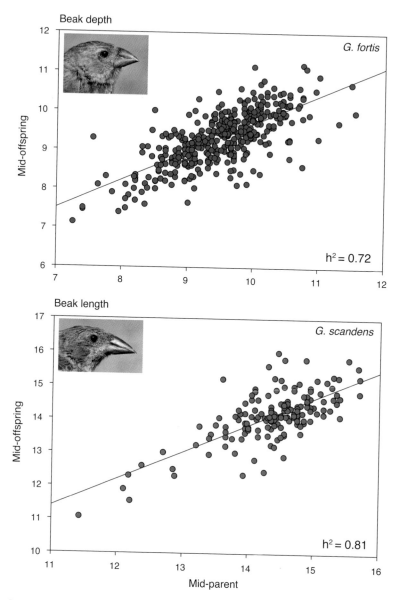

Fig. 3.3 Family resemblance indicates strongly heritable variation in ecologically most important beak dimensions of *fortis* and *scandens* on Daphne; from the combined samples of measurements in all years, and not corrected for extra-pair young or for unequal family sizes. *G. fortis* families (*n* = 413) comprised 1,106 offspring, 1–13 per family, and *scandens* families (*n* = 159) comprised 335 offspring, 1–11 per family. From Grant and Grant 2000a.

TABLE 3.3
Coefficients of phenotypic (CV_P), additive genetic (CV_A) and residual, mainly environmental (CV_E) variation allow comparison of the six measured traits of *scandens* (158 families) and *fortis* (207 families)

	CV_P		CV_A		CV_E	
	scandens	fortis	scandens	fortis	scandens	fortis
Mass	8.41	12.62	5.94	10.38	5.96	7.17
Wing length	3.22	3.54	2.26	2.83	2.89	2.12
Tarsus length	3.78	4.41	2.44	3.07	2.84	3.16
Beak length	6.47	8.13	5.80	7.51	2.86	3.13
Beak depth	4.62	9.89	3.43	8.55	3.10	4.96
Beak width	4.53	7.76	2.82	6.83	3.54	3.67

Note: All coefficients have been multiplied by 100. From Grant and Grant 2000a.

equal in magnitude. Since the two species differ in the average size of most traits, we compared them after standardization by using the coefficient of additive genetic variation, which is the square root of the additive genetic variance divided by the mean (Grant and Price 1981, Houle 1992). It is a comparative measure of "evolvability" (Houle 1992, Garcia-Gonzalez et al. 2012). G. *fortis* is both genetically and phenotypically more variable than *scandens*, especially in mass and beak dimensions, less so in wing and tarsus (table 3.3). Expressed another way, the difference in phenotypic variation between the species reflects an underlying genetic difference (fig. 3.4).

The difference between species in heritabilities is expected from the larger breeding population size of *fortis* than *scandens*, since heritabilities scale positively with effective population size (Lynch and Hill 1986, Houle 1989). However, the magnitudes of the heritabilities are surprisingly high for the small populations of these two species; theoretically they should be in the region of 0.25 when mutation is the sole source of new genetic variation (Houle 1989, Caballero and Keightley 1994). We consider possible explanations in chapter 8.

The combined results stimulated an analysis of heritable variation in a third species, G. *conirostris* on Genovesa Island (Grant and Grant 1989). Heritabilities were found to be high, almost identical to those of *fortis*, and with the same average of 0.70. Therefore all three species of Darwin's finches studied in detail have highly heritable morphological variation and therefore a strong potential for evolutionary change. Evolutionary change as a result of natural selection is the subject of the next chapter.

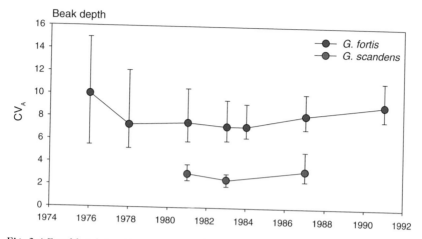

Fig. 3.4 Equilibrial genetic variation in *fortis* and *scandens* beak depth. Estimated coefficients of additive genetic variation (CV_A) are shown with 95% confidence limits. The limits are most asymmetrical where the sample size of families is lowest, as in 1976 for *fortis* (22 families). The estimates for *scandens* (2–4) are consistently lower than the estimates for *fortis* (7–10). Redrawn from Grant and Grant 2000a.

Summary

For evolution to occur there must be heritable variation. This chapter describes our attempts to estimate the magnitude of heritable variation of three body-size traits and three beak dimensions. Heritabilities were estimated from the slope of offspring mid-parent regressions, and found to be uniformly high (>0.50) for *fortis*, especially those of the beak traits, and consistently so in seven cohorts. Misidentified paternity through extra-pair fertilization causes heritability estimates to be underestimated by this method. The frequency of extra-pair paternity was approximately 20%. When corrected for this bias the heritability of beak size is 0.83 and for beak shape it is 0.88. They are exceptionally high values for wild bird populations. The estimates may be inflated to an unknown but small extent by maternal effects on offspring size. We investigated a potential source of bias arising from correlations between genotypes and the environment. Although there is a tendency for *fortis* to breed in the same habitat in which they were raised, the size of adult traits does not vary systematically among habitats; therefore the bias is at most minor. *G. scandens* displays lower heritable variation than *fortis*, due to lower levels of additive genetic variation. The species are similar in one important respect: beak traits are more highly heritable than body-size traits, which is not surprising given the extra sources of environmental variation for body size (daily diet, body fat, etc.). We conclude there is substantial heritable variation in all traits of both species on Daphne, hence a strong potential for evolution.

Natural Selection and Evolution

It may be said that natural selection is daily
and hourly scrutinizing, throughout the world,
every variation, even the slightest.

(Darwin 1859, p. 84)

We now know that species are always evolving.
Natural selection constantly adjusts the traits of
populations generation after generation.

(Thompson 2013, p. vii)

Introduction

G *FORTIS* IS HIGHLY VARIABLE IN SIZE, and the variation is strongly heritable; therefore it has the potential to evolve when the environment changes. This chapter describes what we learned about selection and evolution by following the population through contrasting regimes of wet and dry conditions. Figure 4.1 provides a general framework for the specific observations we describe. It shows how fitness of individuals varies according to the interaction between genotypes, phenotypes, and both social and general aspects of the environment. The center point, indeed the central focus of our field study of evolution, is phenotypic variation. We work back to the genetic variation that generates it, and forward to the fitness variation that ensues. We will refer to this figure several times later in the book.

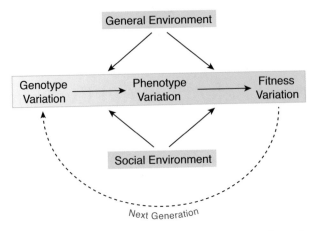

Fig. 4.1 The genotype-phenotype-environment interaction affects fitness. A distinction is made between the social environment of other finches and all other biological and abiotic aspects of the environment. Finches (phenotypes) also affect their environment, for example, when differentially depleting the food supply (this chapter) and destroying the stigmas of cactus flowers (Fig. 2.17; Grant and Grant 1981), and this has feedback effects on fitness.

Expectations

On our first two visits in 1973 the contrast between a wet productive season and a dry nonproductive season was striking (Grant et al. 1976). It led us to speculate that over a long period of time there might be pronounced annual as well as seasonal variation in rain and food production. We reasoned this would have a strong effect upon both the sizes and composition of the population. We then argued that natural selection would fluctuate, favoring different optima in different years, as others have suggested for other systems (Rothstein 1973; reviewed in Bell 2010). Large species would be the most affected, we believed, because fluctuations of the less abundant large seeds would have a potentially greater effect upon birds than fluctuations of the more abundant small seeds (Grant et al. 1976). Hints of natural selection came from two types of observation. First *fortis* individuals with large beaks were found to crack the moderately hard seeds of *Opuntia echios* (fig. 2.18) and the harder, cherry-like "stones" of *Bursera malacophylla* (fig. 2.8; see also fig. 6.10) that others with smaller beaks did not attempt to crack. Those with large beaks also cracked *Opuntia* seeds faster than did others with smaller beaks. Second, between December 1973 and March 1974 *fortis* individuals with long beak tips survived best. Their advantage might have come from an ability to procure seeds in small cracks and depressions in the rocky substrate (Grant et al. 1976).

These observations and extrapolations were a strong motivation to continue the study by returning in both breeding and nonbreeding season to follow the fates of measured and identified birds. Return visits enabled us to estimate heritabilities of morphological traits (chapter 3) and thereby establish a high potential for evolutionary change by natural selection. We did not have to wait long for that potential to be translated into realized evolution.

Natural Selection

Four years after the study began, the archipelago experienced a severe drought (fig. 4.2). Only 24 mm of rain fell on Daphne in the entire wet season of 1977, there was little growth of plants, almost no production of arthropods, and *fortis* did not breed (Boag and Grant 1984b). Almost a thousand birds had been banded and measured, so we were able to follow their fates at intervals up to the end of the drought when the rains returned in January 1978. We found that only 1 of 388 nestlings banded in

1976 survived to 1978 (fig. 4.3, Boag and Grant 1981), and only 15% of adults survived. Importantly, adult survival (or mortality) was not random with respect to size; large birds survived better than small ones (fig. 4.4). Natural selection had occurred. It was surely no coincidence that the single surviving nestling was in the top 1%

Fig. 4.2 Effects of drought on the vegetation. **Upper:** *Chamaesyce amplexicaulis* on the rim of sector 20 (fig. A.3.1), with Daphne Chica in the background. **Lower:** the same view in a drought.

Fig. 4.3 Drought survivors. **Left**: The only *fortis* (number 1929) of the 1976 cohort to survive the 1977 drought. His beak was exceptionally large. **Right**: One of the longest-lived *fortis* (number 2666), hatched in 1978 and lived to 1994. He was genotyped as an F₁ hybrid (*fortis* × *scandens*) and was probably an extra-pair hybrid offspring, since the social parents were both *fortis* and a nest mate, unlike him, had typical *fortis* measurements (Grant and Price 1981).

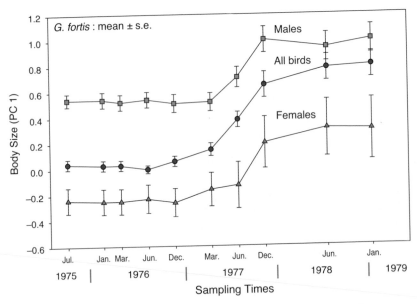

Fig. 4.4 Natural selection on *fortis* body size, indexed by PC1 scores that are explained in the table 3.2 legend. Solid symbols represent means; vertical bars indicate one standard error. Sample sizes of males ($n = 198$) and females ($n = 66$) combined with birds of unknown sex (all birds) varied from 642 in June 1976 to 61 in January 1979. There was no breeding in 1977, and the steady increase throughout the drought that year was the result of selective mortality of small birds. From Boag and Grant 1981.

of the sample of measured birds alive in 1976 ($n = 932$) in weight, beak depth, and beak width.

Birds that disappeared might have flown to other islands; however, most appear to have died on the island, to judge from those we found dead. Measurements of 38 individuals banded before June 1976 and found dead on Daphne in 1977 and early 1978 were statistically indistinguishable from measurements of the other birds missing from the postselection population but not found. Furthermore, like those missing birds, the dead were significantly smaller than the survivors. The entire sample and males and females treated separately showed the same pattern. Thus the cause of disappearance was predominantly if not entirely death on the island.

CAUSES OF SELECTIVE MORTALITY

In the absence of any indication of emigration, disease, substantial predation, or abrupt temperature change being the cause of disappearance, the principal cause must have been death due to starvation. This interpretation is supported by the parallel declines in population size and seed abundance in 1977 (fig. 4.5), coupled with the fact that small seeds declined in abundance faster than large seeds, and as a result the average size and hardness of available seeds sharply increased. New seeds were not produced at this time; therefore the finches themselves brought about the change in composition by consuming seeds differentially.

Initially they depleted the supply of small and soft seeds, and as these seeds became increasingly difficult to find, the finches had to turn progressively more to the large and hard seeds such as those produced by *Opuntia echios* and *Tribulus cistoides*. For example in May 1976 only 17% of feeding was on medium and large seeds, while one year later the figure had risen to 49%. Only large *fortis* with large beaks are able to crack hard seeds, and only the largest can do so quickly and efficiently (Grant 1981b, Boag and Grant 1984a, Price 1987), so they survived best by exploiting these valuable foods. They may have gained two extra advantages, first in being able to dissipate heat loss without water from the base of their beaks better than smaller birds (Greenberg et al. 2012), and second in being socially dominant at contested food sources (Boag and Grant 1981).

This was the most intense selection recorded for continuously varying traits in a natural population at that time (Boag and Grant 1981). It was especially intense for females. They are smaller than males by about 4% on average, and proportionately more of them died. As a re-

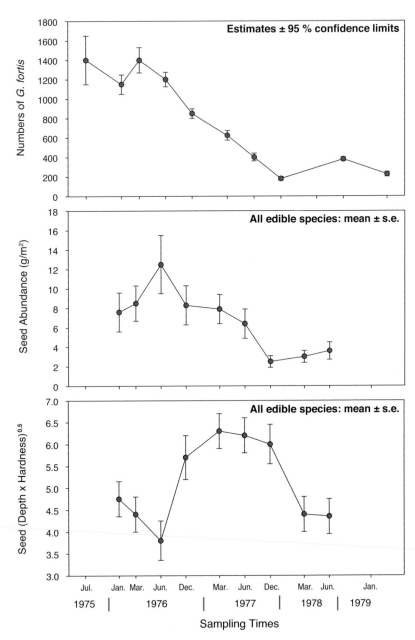

Fig. 4.5 Temporal changes in *fortis* numbers and their principal dry-season food in the drought of 1977. **Upper**: Finch population estimates and 95% confidence limits derived from a Lincoln index based on regular visual censuses of marked

sult of differential and size-selective mortality a roughly equal sex ratio in 1976 became skewed to as much as 6:1 before falling to 3:1 at the time of breeding in 1978 (Boag and Grant 1984b). Size-selective mortality and the dependence of finches on a declining supply of seeds ceased a little earlier, at the end of 1977, when *Opuntia* cactus began flowering. All birds fed heavily on its pollen and nectar (figs. 2.17 and 2.19), and their body condition, which had been gradually deteriorating, noticeably improved. Molting resumed then after a hiatus of more than a year for most birds.

THE TARGETS OF SELECTION

What exactly were the targets of selection, the traits that made a difference between life and death? Two discriminant function analyses pinpointed beak depth as a key variable (Boag and Grant 1981). In the first analysis, birds that were observed to feed solely on small seeds were compared with those seen to feed on medium and large seeds. The standardized coefficients of the discriminant function that separated the two groups with different diets gave greater weight to beak depth than to the six other morphological variables (wing length, beak length, etc). In the second analysis, males that survived and died were compared. Beak depth was the strongest contributor (highest coefficient) to the discrimination of the two groups. This analysis was repeated with females, and beak depth was found to be the third-strongest variable in the discriminant function after beak length and an index of beak pointedness. Thus *fortis* with deep beaks fed on large seeds and survived best.

The importance of beak depth for survival was confirmed by selection analysis (Price et al. 1984b). For a cluster of intercorrelated variables the partial regression method developed by Lande and Arnold (1983) allows identification of the direct relationship between any one of them and fitness (survival) while holding constant the statistical effects of the other variables on fitness (box 4.1). Success of this method is in part dependent on large samples of measurements (Hersch and Phillips 2004), which we have. Survivors were larger than nonsurvivors

Fig. 4.5 (*continued*) birds throughout the island. **Middle:** Estimates of mean seed abundance and one standard error. **Lower:** Estimates of the average size-hardness index ($DH^{1/2}$) of edible seeds and one standard error. D is the depth of seed, its second-longest dimension, and H is its hardness, measured as the force in newtons needed to crack it open (box 2.1). From Boag and Grant 1981 and Grant 1986.

Box 4.1 Selection Analysis

We have used the following procedure to estimate selection (Lande and Arnold 1983). Before each analysis we ln-transformed morphological data and then standardized them by dividing by the standard deviation so that they had zero mean and unit variance. Relative fitnesses were obtained by dividing absolute fitness (0 or 1) by the absolute mean fitness, that is, the proportion surviving. The net effect of selection on a trait, the selection differential (s), combines the direct effect with indirect effects arising from correlations with other measured traits. It is simply measured as the difference between the mean value of the trait before and after selection. Statistical significance of the selection coefficient was assessed with t-tests comparing survivors and nonsurvivors. Multiple regression analysis was used to separate the direct effects of selection from indirect effects. The direct effect of selection on each trait, or selection gradient (β), was estimated by the partial regression coefficient of relative fitness (survival) on each trait. Significance of the gradients was checked with t-tests of the coefficients from a logistic regression analysis.

in all dimensions, as shown by the selection differentials (s) in table 4.1. Nevertheless beak depth and body size were most strongly associated directly with fitness, as shown by the entries (β) in the selection gradient—the partial regression coefficients—and therefore these were the primary "targets" of selection.

Evolution in Response to Selection

Given the high heritability of beak and other dimensions of *fortis* (chapter 3), an evolutionary response to strong selection is to be expected in the next generation, and was in fact observed in 1978 (fig. 4.6). As a result of the 1977 selection episode the mean value of all six traits increased in the next generation (1978) to a varying extent (0.21–1.07 standard deviations [s.d.]; Grant and Grant 1995a).

Observed responses can be compared with responses that are predicted by taking into account the genetic correlations between characters (Lande 1979), as explained in box 4.2. Actual evolutionary re-

TABLE 4.1
Standardized selection differentials (s) and entries in a standardized selection gradient (β) with standard errors (s.e.) for two periods of heavy mortality associated with dry conditions

	1976–77		1981–82	
	s	β ± s.e.	s	β ± s.e.
Weight	+0.62	**+0.51 ± 0.14**	+0.15	+0.13 ± 0.08
Beak length	+0.49	+0.17 ± 0.18	+0.13	+0.06 ± 0.09
Beak depth	+0.60	**+0.79 ± 0.23**	+0.12	+0.17 ± 0.12
Beak width	+0.49	**−0.47 ± 0.21**	+0.08	−0.20 ± 0.12
Sample size (n)	640		197	
Proportion surviving	0.15		0.65	

Note: Coefficients in **boldface** are significantly different from zero at $p < 0.05$. Variances remained unchanged by selection. From Price et al. 1984a.

sponses to selection closely matched the predictions (fig. 4.6; Grant and Grant 1995a), so if any targets were omitted from our analysis, they must have been either highly correlated with the included ones or entirely independent of them. The offspring were large, though smaller on average than their parents as is to be expected from heritabilities that are less than 1.0. Large size of the offspring could be attributed in part to the highly favorable growth conditions that year (Boag 1983), since the breeding density was low in 1978 and food relatively plentiful. The effect, if present, seems to have been minor because parents that bred in the same pairs in 1978 and again in 1981 under more crowded conditions ($n = 7$ families) produced offspring of the same size in the two years (Grant and Grant 1995a).

Selection Occurs Repeatedly

Selection occurs repeatedly and in the same direction when the same or similar environmental conditions are repeated. We observed this in two dry periods that followed in the half-dozen years after the drought of 1976–77. In the first, from the middle of 1979 to the end of 1980, 53 mm of rain fell, 22% of *fortis* adults died, and although there was no significant selection, the signs of the selection coefficients were the same as in an equivalent 18-month period in 1976–77 (Price et al.

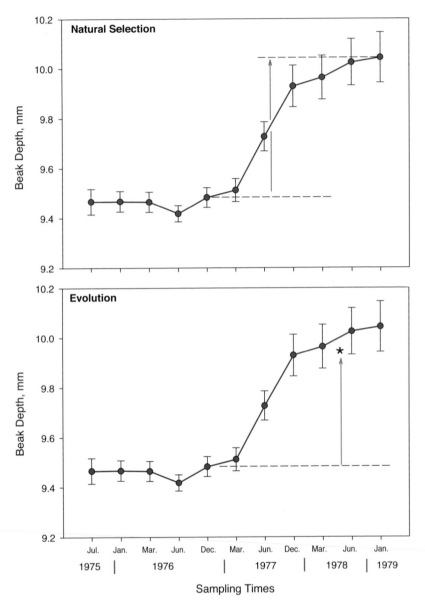

Fig. 4.6 Natural selection within one generation (above) followed by evolutionary change between generations (below). The asterisk (below) indicates the average beak depth of the next generation (9.94 mm), which is the observed evolutionary response to natural selection. It is well within sampling error of the predicted average (9.83 mm) (Grant and Grant 1995a). From Grant and Grant 2010a.

Box 4.2 Prediction of Evolution

Evolution occurs when the effects of selection on a heritable trait in one generation are transmitted to the next generation. An evolutionary response to selection is calculated as the difference between the mean of a trait before selection and the mean of the same trait in the offspring. The observed response may then be compared with the response predicted from the net selection (box 4.1) and the inheritance of the trait. In the simplest case a single heritable trait is subject to directional natural selection. The response (R) to selection is then estimated by the product of the heritability of the trait (h^2) and the strength of selection, which is measured by the selection differential (s). Thus $R = h^2 s$. This is the breeders' equation (Falconer and Mackay 1995). The breeders' equation does not work well in several situations where either heritabilities or selection coefficients are small and imprecisely estimated, and when traits are phenotypically plastic (Kruuk et al. 2008). When two (or more) traits (i and j) are subject to selection, an evolutionary response is predicted on the basis of the strength of selection on each trait (β_i, β_j), the genetic variance of each trait (G_{ii}, G_{jj}), and the genetic covariance between the traits (G_{ij}) (Lande 1979); see box 4.1 for the β coefficients. This is usually represented in matrix form, $\Delta \bar{z} = GP^{-1}S$, where G and P are genetic and phenotypic variance-covariance matrices, S is a vector of selection differentials, and $\Delta \bar{z}$ is a vector of changes in morphological trait means. In the case of two traits, such as beak length and beak width (Grant and Grant 1993), the analysis is reduced to

$$\Delta \bar{z}_1 = \beta_1 G_{11} + \beta_2 G_{12}$$

$$\Delta \bar{z}_2 = \beta_2 G_{22} + \beta_1 G_{21}$$

The first term on the right-hand side of each equation represents the direct response to selection on the trait, and the second term represents the indirect response arising from the direct effect of selection on a genetically correlated trait. Note that when several correlated traits are in the analysis, their indirect effects may cancel if the β coefficients are of opposite sign, even when all genetic correlations are positive, as in the finches (Grant and Grant 1994). The cumulative indirect effects may be so large as to overwhelm the direct effects. For possible biases in analyses and interpretation see Grant and Grant (1995a).

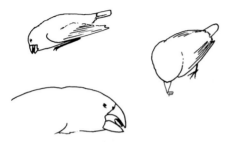

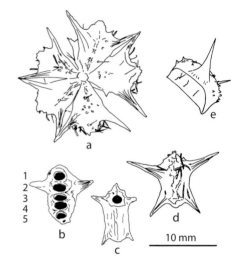

Fig. 4.7 *G. fortis* use different maneuvers to open *Tribulus* mericarps and extract the seeds: either by crushing or by twisting and biting at the corners. Exploited mericarps are shown below. Small individuals sometimes feed parasitically by waiting for a large finch to split a mericarp into two by crushing it, and then seizing an ejected seed. From Grant 1981b.

1984b). In the second period, in 1981–82, 51 mm of rain fell, 35% of adults died, and beak depth and weight were once again selected traits and in the same direction as in 1976–77 (table 4.1).

Surprisingly, fitness was negatively associated with beak width after other traits were controlled, as in 1976–77 (table 4.1). This may be biologically significant: some birds, especially the smaller ones, twist the end of a *Tribulus* fruit to extract a seed, and a relatively narrow beak may facilitate this maneuver (fig. 4.7; Grant 1981b, Price 1987). However, there is some doubt about this because the beak width result might be partly a statistical artifact arising from the strong correlation between beak depth and width. This is the problem of multicolinearity; there is so little independent variation in either one of two strongly correlated characters that variation in each of them cannot be reliably associated with a third variable such as fitness. The phenotypic correlation between beak depth and beak width is 0.89 (Grant and Grant 1994), and since $R^2 = 0.89^2 \approx 0.80$, only 20% $(1 - R^2)$ of one dimension varies independently of the other. In contrast to this pair of variables, separate correlations between each of them and beak length are less than 0.75, and therefore a large amount of variation in beak length, approximately half, is independent of variation in beak depth and beak width. Notice in table 4.1 that the net effect (s) of selection on beak width in 1977 was positive, even though the direct effect (β) was negative, because the direct effect was overwhelmed by positive selection on all the other, positively correlated, characters.

Selection Oscillates In Direction

By extending the study beyond the first few years, we learned that selection changes direction. Changes in the terrestrial environment are driven by interannual variation in rainfall (fig. 4.8). The environment changed profoundly in 1982–83 (figs. 4.9–4.11) with abundant rain (1,359 mm; fig. 4.12) and a prolonged wet season (eight months) associated with an exceptional El Niño event (Gibbs and Grant 1987a).

This transformed the composition of the seed supply from one dominated by large and hard seeds to one dominated by small and soft seeds (fig. 4.13). There was no immediate selective effect on *fortis*; survival was high, and effects were delayed. In the following year 53 mm of rain fell, and finches bred once or twice, but then a drought began, and the island received a mere 4 mm of rain in two days throughout the whole of 1985. The drought ended in 1986 with the return of rain (49 mm) and a partial resumption of breeding, which was almost entirely unsuccessful.

Fig. 4.8 Weather. **Left**: Typical cumulus clouds in the wet season over Santa Cruz. **Right**: Rainbow, same view.

Fig. 4.9 Effects of El Niño. **Right**: Breeding of Blue-footed Boobies, *Sula nebouxii*, in March 1976. **Left**: Same view in the El Niño year of 1983, showing the effects of abundant rain on the crater and surrounding vegetation, and the absence of boobies (N. Grant).

Fig. 4.10 Rampant growth. **Right**: The only place to put a tent is on the outer slope in sector 20 (fig. A.3.1), shown in 1975. **Left**: Same view in 1983, an El Niño year, after abundant rain has promoted prolific growth and flowering of *Cacabus miersii*.

Fig. 4.11 Seasonal and annual variation in vegetation. **Upper left**: Dry season (*Bursera malacophylla* trees are leafless). **Upper middle**: Wet season (trees are in leaf). **Upper right**: Extensive growth of annual plants in early stages of El Niño development. **Lower left**: Late stage of El Niño. **Lower middle**: Cactus bush almost smothered by *Merremia aegyptica* vines. **Lower right**: Aftermath of El Niño (dead vines drape cactus bushes and trees). From Grant and Grant 2008a.

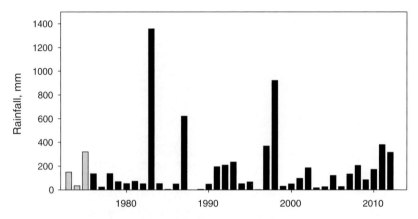

Fig. 4.12 Variation in total annual rainfall on Daphne. The unknown rainfall in 1973–75 (gray bars) has been estimated from a relationship between annual rainfall on Daphne and at the Charles Darwin Research Station on Santa Cruz Island ($r = 0.93$).

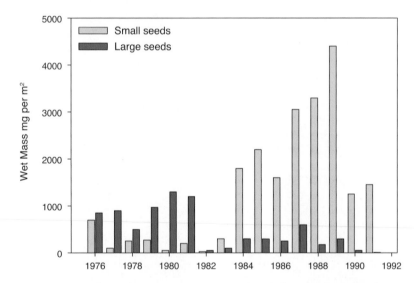

Fig. 4.13 Changes in seed composition associated with the El Niño event of 1982–83. Values are mean wet mass in milligrams per square meter. The seed supply was sampled twice a year from 1976 to 1991 and averaged (chapter 2). Annual totals for each seed species were converted to biomass by multiplying by their mean wet mass. Individual seeds fall into three discrete size-hardness classes. More than 20 species contribute to the small-soft category (size-hardness index $DH^{1/2} < 2$), and of these only 14 are common (appendix 1.1). *Tribulus cistoides* is the only species in

TABLE 4.2

Standardized selection differentials (s) and entries in a standardized selection gradient (β) with standard errors (s.e.) for two periods of heavy mortality associated with dry conditions before (1976–77) and after (1984–86) the El Niño perturbation of 1982–83

	1976–77		1984–86	
	s	$\beta \pm$ s.e.	s	$\beta \pm$ s.e.
Weight	+0.74	+0.48 ± 0.15	−0.11	−0.04 ± 0.10
Wing length	+0.72	+0.44 ± 0.13	−0.08	−0.01 ± 0.08
Tarsus length	+0.43	+0.01 ± 0.11	−0.09	−0.05 ± 0.08
Beak length	+0.54	−0.14 ± 0.17	−0.03	+0.24 ± 0.09
Beak depth	+0.63	+0.53 ± 0.21	−0.16	−0.13 ± 0.14
Beak width	+0.53	−0.45 ± 0.20	−0.17	−0.15 ± 0.12
Sample size (n)	634		549	
Proportion surviving	0.15		0.32	

Note: Coefficients in **boldface** are significantly different from zero at $p < 0.05$. From Grant and Grant (1995a). Small corrections to the sample for 1976–77 in table 4.1 and inclusion of wing length and tarsus length changed coefficients for the other variables to a minor extent.

The critical years were 1984 to 1987, when adults survived poorly (32%; Gibbs and Grant 1987a). Although the seed supply declined during 1984 and 1985, it remained higher than in 1977 and 1982. More important than numbers was the composition; the proportion of the total seed biomass made up by small soft seeds in 1984 and 1985 was 21%–80%, which is 2–10 times greater than the previous maximum of 8% following a normal wet season. Correspondingly, proportions of *Tribulus* seeds and *Opuntia* seeds in the diet declined (Grant and Grant 1993). In this altered environment small birds with small beaks had a strong selective advantage over the rest (table 4.2; Gibbs and Grant 1987b, Grant and Grant 1995a).

Fig. 4.13 (*continued*) the large-hard category ($DH^{1/2} \approx 10$). The third class ($DH^{1/2} = 2$–7), not shown, comprises seeds of *Opuntia echios*, *Desmodium glabrum*, *Tephrosia decumbens*, *Cenchrus platyacanthus*, and *Bursera malacophylla*; the last two are consumed as they ripen, and only *Opuntia echios* contributes substantially to the regular dry-season food supply. Mean biomass of small-soft seeds was 206.14 ± 76.74 mg (s.e.) per square meter in the seven years before the 1982–83 El Niño event, and 2441.0 ± 361.49 mg in the eight following years. Mean biomass of large-hard seeds was 843.89 ± 165.07 mg before 1983 and 244.66 ± 50.96 mg afterward. From Grant and Grant 1993.

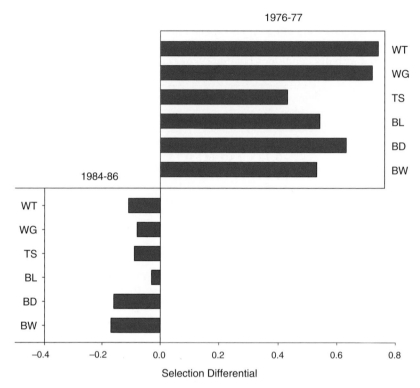

Fig. 4.14 Contrasting selection differentials for *G. fortis* in two droughts before (1976–77) and after (1984–86) the El Niño event of 1982–83 that transformed the vegetation on the island. Abbreviations: WT weight; WG wing length; TS tarsus length; BL beak length; BD beak depth; BW beak width. Modified from Gibbs and Grant 1987b.

Two differences from selection in the preceding years stand out. First, ignoring statistical significance, all selection differentials (s) were negative in 1984–85, whereas they were all positive in 1976–77 (fig. 4.14). The only (significant) target of selection in 1984–85 was beak length (table 4.2), which was selected in the opposite direction (increase) to all other traits (decrease). The indirect effects of the other traits, even though individually none were significant, apparently outweighed the direct effect on beak length because beaks became more pointed as a result of a reduction in average beak depth and beak width without a net change in beak length. Second, males experienced stronger selection than females. For example, selection differentials for overall body size (PC1) were almost twice as large for males (–0.41,

$p < 0.001$) as for females (-0.22, $p < 0.05$). Despite that, males and females survived equally well, unlike in 1977 (Boag and Grant 1981, Gibbs and Grant 1987b).

Thus small individuals with relatively pointed beaks survived the stressful period of 1984–85 best (Gibbs and Grant 1987b, Grant and Grant 1993). Large birds may have been at a disadvantage because (a) they have a higher resting metabolic rate (e.g., Nilsson et al. 2009), (b) they suffer from relatively high water loss associated with higher metabolic rates (MacMillen 1990, Williams 1996, Williams and Tieleman 2005; but see Greenberg et al. 2012), (c) associated with their greater daily energy needs they may have had difficulty in finding enough large seeds, and (d) they may not be as skillful as small birds in acquiring small seeds. As in previous years, birds feeding on large seeds in 1984 and 1985 ($n = 108$–124 birds) were larger in beak dimensions, especially depth and width, than those feeding on small soft seeds (Gibbs and Grant 1987b). Birds with relatively pointed beaks may have gained an advantage in being able to pick up small seeds with their forceps-like beaks, and to process them rapidly and in sufficient numbers to maintain a positive energy balance.

Evolutionary Response

All evolutionary responses to the 1984–86 selection were significant except for tarsus and beak length. They agreed in direction with the signs of the β coefficients, in that offspring were smaller on average than the adults before selection in all traits except for beak length. With heritabilities of less than 1.0 they should have been more similar to the population average before selection and hence larger than their parents, but in fact they were even smaller than their parents, except in beak length. Differences from predicted means in three traits were statistically significant: weight, wing length, and beak depth (Grant and Grant 1995a). The net result was that observed decreases (mean standard deviation shift = 0.24) were twice as large as those predicted (mean = 0.12 s.d.). Why?

We considered several possible biases in the analysis that might have led to the unexpectedly small size—incorrect estimation of heritabilities and selection coefficients, incorrect identification of fathers (chapter 3), missing variables, selection against large offspring in 1987 before they were measured—and found evidence for only one, a distortion due to sustained effects of poor nutrition during growth (Grant and Grant 1995a). Offspring that hatched in the second half of the breeding season and survived to be measured were significantly smaller than the early ones in weight, wing length, beak depth, and, almost significantly, beak width. This is understandable because breeding density was high (100 pairs), unlike in 1978 (~50 pairs), and food supply declined seasonally. Exclusion of the late offspring resulted in a much better fit to predicted values, and now only wing length was significantly smaller than predicted, probably through abrasion during extensive feeding on the ground in the ensuing drought when most of the birds were measured. Consistent with this explanation, late offspring measured by one of us (PRG) in 1988 and again in 1989 generally increased in wing length as a result of molting (15 of 22), whereas early ones did not (6 of 16). The difference is close to statistical significance ($\chi_1^2 = 3.527$, $p = 0.06$).

These analyses illustrate the important point that selection analysis assumes the parental and offspring generations experience similar environmental conditions during growth to maturity. If they are not similar, environmental differences may cause generations to differ in average morphology, thereby giving a misleading estimate of the evolutionary response to selection.

Selection in Opposite Directions

Correlations between characters introduce uncertainty into the direction of evolution, especially when characters are selected in different directions. Directions may differ in two ways: two correlated traits are selected in opposite directions at the same time, or the same trait is selected in opposite directions at different times. A possible example of the first is the contrast between directions of selection on beak length and beak depth described above. This is rare in *fortis*. More often one trait is selected and another is not selected. An example of the second is the contrast between directions of selection on beak depth in 1976–77 and 1984–85.

Antagonistic selection may take the form of higher survival of small juveniles and large adults (Price and Grant 1984). Selection against large juveniles was observed in 1978 and again in 1979, but not in 1984 and 1985 (Gibbs and Grant 1987b). We do not know how often such selection takes place or its magnitude because usually we are not present on the island when fully grown juveniles can be measured and then followed in their first year. Another potential form of antagonistic selection is bidirectional selection on males and females (Price 1984b). Again it is hard to judge how common this is. Typically selection is in the same direction on both sexes, and if it is strong, it is unequal and most severe for the sex furthest from the selected joint mean. Antagonistic selection dampens the net effect of directional selection and hence the strength of evolutionary change.

Conclusions

The trajectory shown in figure 1.6 is not a straight line through time but rises and falls. In this chapter we have shown that natural selection is the agency and evolution is the outcome because the traits subject to selection are highly heritable. The environment changed, and selection occurred in different directions tracking the change. The two contrasting selection episodes illustrate the value of considering selection separately from evolution. Traits subjected to selection underwent evolutionary change, but did not do so always because correlated characters were selected as a set, and the genetic correlations affected the evolutionary consequences. Beak length was not a target of selection in 1976–77; nevertheless it evolved as a correlated response to selection on other traits. In contrast, beak length was selected in the 1984–86 episode but did not evolve, partly because the effect was nullified by selection in the opposite direction on positively correlated traits. Evolutionary change in this dimension, or lack of it, could not be understood without the correlation structure. The fact that sets of correlated traits were selected in different ways in the two episodes, and also in a parallel study of *G. conirostris* (Large Cactus Finch) on Isla Genovesa (Grant 1985, Grant and Grant 1989), raises the interesting possibility that genetic correlations, which can constrain evolution (Agrawal and Stinchcombe 2009, Walsh and Blows 2009), may be modified (weakened) more easily by oscillating than by unidirectional selection.

Before discussing events in the second half of the study we will first consider how differential reproduction contributes to overall fitness, in both *fortis* and *scandens* (next chapter).

Summary

Natural selection occurs most strongly when the environment changes. We began the research speculating that natural selection fluctuates in direction, with different optima in years of contrasting food supply. This was confirmed with an example in the first twelve years of the study. The strongest selection on *fortis* occurred under severe drought conditions in 1976–77 and 1984–85, when large fractions of the population died. In the first episode large birds, especially those with deep beaks, survived best, apparently as a result of their ability to crack open the large and hard woody fruits of *Tribulus cistoides* and the hard seeds of *Opuntia echios* that became relatively abundant after finches had depleted much of the supply of small and soft seeds. Evolutionary responses of morphological traits were well predicted from estimates of selection strengths and genetic variation. The second selection episode followed an exceptionally strong and prolonged El Niño event. The abundant rain altered the composition of the food supply, which became dominated by an abundance of small and soft seeds. A drought occurred in 1984–85, and under the changed conditions of the food environment small birds with pointed beaks had a selective advantage over the rest. Evolutionary responses to selection were not predicted as well as in the first episode, apparently because offspring suffered effects of poor nutrition during growth as the breeding season progressed. Thus selection oscillates in direction in different droughts depending on the preceding conditions during the growing season. The net effect of selection is influenced by antagonistic selection on the same trait in opposite directions at different times and on different, correlated traits in opposite directions at the same time. In later chapters (8 and 12) we will extend the discussion of selection to other species and later years.

Breeding Ecology and Fitness

Natural selection works not only through mortality but
also through fertility.

(Cavalli-Sforza and Cavalli-Sforza 1995, p. 247)

If you can look into the seeds of time,
And say which grain will grow and which will not. . . .

(Shakespeare, *Macbeth*, 1.3)

Introduction

THERE IS MORE TO FITNESS than surviving dry seasons (fig. 4.1). The genetic fitness of an individual is its ability to survive *and* reproduce (Charlesworth 1980, De Jong 1994). Whether its potential is realized or not is determined by its behavior, physiology, and morphology, as well as by chance. At certain critical times in dry seasons, survival is dependent on beak and body size (chapter 4). It is possible that breeding success is also dependent on beak size and body size in wet seasons. Some birds may be better than others at gaining a territory and a mate, and converting eggs into breeding members of the next generation (Reid et al. 2003, Jensen et al. 2004, Brouwhuis et al. 2010). If so there could be a trade-off between morphological factors influencing survival and those influencing reproduction. For example, some individuals at a disadvantage under stressful conditions of a dry season may survive and then perform particularly well when breeding. Alternatively, survival and reproduction components of fitness may be uncoupled (Siepielski et al. 2011), such that a particular beak morphology or body size enhances dry season survival but has no influence on reproductive success (Jensen et al. 2004).

Thus finches can be thought of as occupying two niches separated in time, a wet-season and a dry-season niche, just as migrant birds occupy two habitats separated in time. This means that two components of biological success, in addition to chance, have a bearing on life history patterns (box 5.1). The first component is an ability to find foods (seeds) in dry years and dry seasons when food is scarce and there is no breeding, and the second is an ability to find insects and spiders for food and avoid interference at the nest from intruders during breeding. Finches have the same beaks in both niches but the fitness consequences may be different because they use them differently to exploit a different food supply. Their lifetime fitness is determined by how well they perform in both niches.

This chapter introduces the basic features of finch breeding, compares reproduction in years of different rainfall, and discusses the way in which morphology affects reproductive success. Immediate success is raising offspring to independence. Ultimate success is producing offspring that become breeders (recruits). The chapter is not restricted to morphology but considers more broadly the causes of variation among individuals in their reproductive success and their fitness in terms of the recruits they produce. Daphne is unusually suitable for this type of study because the island is small and the small populations of finches are resident and nonmigratory, and banded birds can be recognized as individuals. A few other insular populations have similar advantages (e.g., Gustafsson and Merilä 1994, Komdeur 2003, Smith et al. 2004, Postma and van Noordwijk 2005, Jensen et al. 2004, Schoener et al. 2004, Wheelwright et al. 2007, Calsbeek and Smith 2007, Pelletier et al. 2007), and so do some semi-insular ones (Garant et al. 2005, Price et al. 2008, Wesolowski 2011), though in general it is rarely possible to follow known individuals from birth to reproduction to death, estimate their contributions to the next generation, and identify the important factors determining their success.

Basic Breeding Biology

Rainfall is the driver of productivity in the seasonally arid terrestrial environment (figs. 4.2 and 4.8). Finches are in a state of reproductive readiness when rain arrives sometime in the first two months of a typical year (Grant and Boag 1980, Hau et al. 2004). The first heavy rain

Box 5.1 The Unusual Combination of Large Clutches and Long Lives

Darwin's finches on Daphne, Genovesa (Grant and Grant 1989), and elsewhere in lowland Galápagos habitats deviate from the standard tropical pattern of a slow pace of life (Wikelski et al. 2003) by combining tropical (long life span) and temperate (large clutch size) life histories. It seems an anomalous combination when the standard patterns are long life and small clutches (humid tropics) and short life and large clutches (temperate regions). The explanation lies in opportunities and constraints. Their pace of life is adjustable.

The opportunities for repeated breeding with large clutches arise from strong pulses of food production in some very wet years, coupled with low breeding density following heavy mortality in the preceding dry season. In this respect the finches on Daphne resemble birds in temperate regions (Ashmole 1963, Ricklefs 1980, Grant and Grant 1980b) and in the seasonally arid coastal region of Ecuador (Marchant 1958), in marked contrast to the stable populations under density-dependent competition for food and territories in the relatively unvarying tropical humid environments, as well as on many islands elsewhere in the world (Covas 2012). Significant constraints on reproduction and survival from predators and parasites are present in continental regions generally (Marchant 1960, Martin 2004) but lacking in Galápagos. Snakes and mammals that prey on nestlings in the tropical mainland are absent, and the only conspicuous disease is avipox, which can be debilitating but is rarely fatal for finches (Huber et al. 2010). In the absence of these nest predators, clutches are larger and nestling periods longer than in birds of the coastal region of mainland Ecuador (Marchant 1958, 1959, Grant and Grant 1989). Darwin's finches can live a long time because they do not have to face the hazards of migration, and they do not breed in every year owing to the failure of sufficient rain in some years: in a quarter of the years either no finches bred or just a very few adults did.

Darwin's finches lay smaller clutches in less seasonal habitats. The single species of finch in rain-forested Cocos Island lays clutches of two eggs (Slud 1960, Grant 1986), like unrelated species in similar, rain forest, habitat on the continent. In the absence of nest predators on Cocos Island, food supply and its variation in time is probably the most important factor affecting clutch size. In the mesic habitat of Galápagos highlands finches lay smaller clutches than in the lowlands (Kleindorfer 2007). Their food supply may vary less seasonally than in the lowlands. This has not been studied.

Fig. 5.1 Reproduction. **Upper left**: Nest of *fortis*. **Upper right**: Clutch of 4 *fortis* eggs. **Lower left**: Single egg of each of 4 species, from left to right *magnirostris, scandens, fortis,* and *fuliginosa*. **Lower right**: Fledgling *fortis*.

stimulates growth of leaves and flowers, and induces singing, chasing, courtship, and pair formation (box 3.1). Nests are built in cactus bushes, rarely in trees (*Bursera*) or on the ground at the base of grasses (*Cloris*) and herbs (*Chamaesyce* and *Sesuvium*). By the end of the second week after the first heavy rain caterpillars appear on the flowers of *Bursera, Croton,* and some of the annual plants, and finches begin to lay eggs. Clutches of 2-4 eggs (fig. 5.1) are laid in a typical year, incubated by the female, and hatch 12 days later. Both parents feed the nestlings for 12–14 days, and the fledglings for the next 2–4 weeks, mainly on caterpillars (fig. 2.6). If sufficient rain has fallen, the female starts a second clutch in a new nest within 2 weeks of the fledging of the young from the previous nest, usually with the same male but sometimes with a different one. Occasionally the second clutch is started before the nestlings from the first nest have fledged. This summary captures the essence of all breeding seasons (Boag and Grant 1984b), except for one additional feature. In several years a few *scandens* pairs start to breed in December or January before rain arrives.

They do so by intensively exploiting pollen and nectar from *Opuntia* flowers (figs. 2.17 and 2.19; Grant 1996, Grant and Grant 1981). The longest-lived bird, male *scandens* 18111, hatched in early 1993 before the rains arrived three weeks later, and lived for 17 years.

ANNUAL VARIATION IN REPRODUCTION IN RELATION TO RAIN

Sea temperatures rise in January of a typical year, causing air temperatures to rise, cumulus clouds to develop, and, following the cooling of moisture-laden air, rain to fall (fig. 5.2). The timing and amount of rain and the duration of the wet season vary greatly from year to year under the influence of the El Niño–Southern Oscillation phenomenon. At the extremes are droughts with no breeding (La Niña) and years of abundant rain (El Niño). The extremes of annual rain we have recorded on Daphne are 1 mm in 1988 and 1,359 mm in 1983 (which includes December 1982, when the rains began). In a quarter of the years (10/40) little or no breeding took place, and in 44% of the years almost no offspring survived to the following year. In contrast a wet season with abundant rain can last up to eight months (1982–83), and finches breed at the rate of one brood per month for a potential maximum of eight broods.

The fluctuating climate and somewhat erratic breeding schedules have three main demographic consequences. First, population sizes of breeding finches differ in years of plenty and scarcity by an order of magnitude or more (fig. 5.3). Second, the age structure of populations undergoes substantial change, reflecting the contrasting years of high production and low survival (fig. 5.4). This is more typical of marine fish such as herring and cod, whose population dynamics are dominated by strong cohort effects, than most bird populations that have been studied in detail. Third, the timing of recruitment varies greatly. Entry into the breeding population occurred gradually over five years for members of the 1978 cohorts, abruptly after two years for 1981 cohorts, and after as little as three months at rapidly rising density for 1983 cohorts (Gibbs et al. 1984).

In general the more rain that falls the longer is the breeding season, the more clutches are laid, and the more broods are produced (fig. 5.5). Clutches vary from 2 to 6 eggs and again, in general, the larger the clutch the more fledglings are produced and survive to the next year (Gibbs 1988). Clutches are larger on average in El Niño years than in other years (fig. 5.6), but hatching success and fledging success are not enhanced, in fact they may even be reduced (fig. 5.7), by disturbances caused by prolonged bouts of rain and density-dependent interactions.

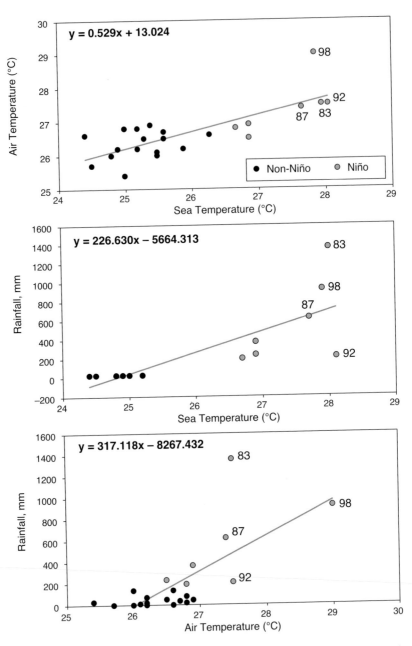

Fig. 5.2 Average daily maximum air temperatures of the warmest two successive months of a year on Santa Cruz Island (1965–98) is a function of the average daily sea temperature of those months at the same location (upper). Annual rainfall on Daphne (1976–98) is a function of the average daily sea temperature (middle) and of the average daily maximum air temperature (lower) recorded on Santa Cruz Island in the warmest two successive months of the year. Seven years of El Niño conditions are shown by green circles. Conditions in four of those years labeled

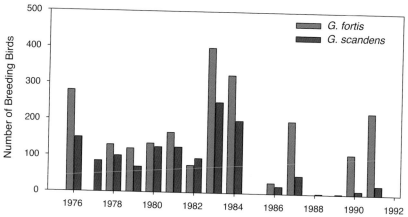

Fig. 5.3 Number of breeding individuals in 1976–91. Not all pairs were studied in 1976, and the numbers of breeders were estimated. There was no breeding in 1985, and only two pairs of *scandens* attempted breeding in 1988 and 1989. Harmonic mean breeding population sizes were 94 *scandens* and 197 *fortis*. Genetically effective population sizes were a quarter to one-half of these values principally as a result of a large variance in the production of recruits per parent (see also Nunney 1993, Engen et al. 2011). From Grant and Grant 1992a.

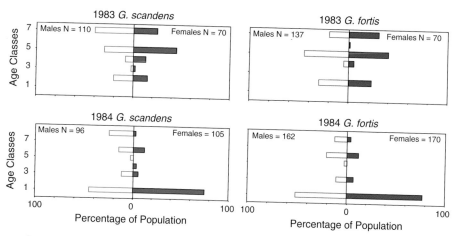

Fig. 5.4 Age structure of *fortis* and *scandens* at the time of breeding in 1983 and 1984. Not included are the individuals that hatched in 1983 and bred later in the same breeding season. From Grant and Grant 1992a.

Fig. 5.2 (*continued*) with two digits were clearly extreme, whereas conditions in three others (1991, 1993, and 1997, unlabeled green circles) were not markedly different from the warmest and wettest of the non–El Niño years (black circles) except in rainfall (middle). From Grant et al. 2000.

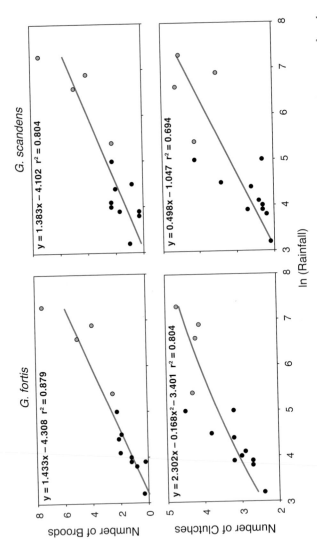

Fig. 5.5 Number of broods per female per year (upper) and the largest average clutch size per female among broods in a year (lower), as functions of rainfall measured in millimeters. Excluded are birds that bred in the year of hatching. All linear relationships, and the squared term in the bottom left relationship, are statistically significant ($p < 0.05$). Green circles identify El Niño years; 1983, 1987, 1991, and 1998. From Grant et al. 2000.

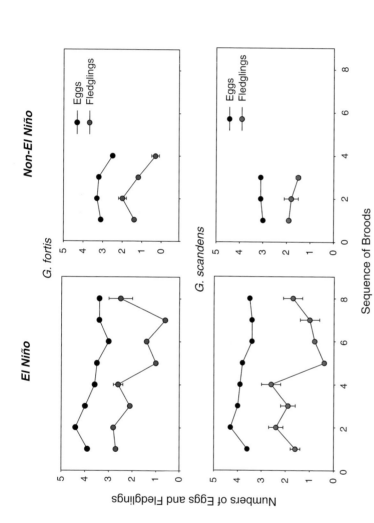

Fig. 5.6 Seasonal variation in mean number of eggs and fledglings produced by identified females in El Niño and non–El Niño years (1976–91, 1998) Brood means for sample sizes of fewer than 10 females have been omitted. Vertical bars indicate 1 s.e. above and below the mean; these are sample-size weighted averages of the means of individual years. Number of broods refers to their sequence. From Grant et al. 2000.

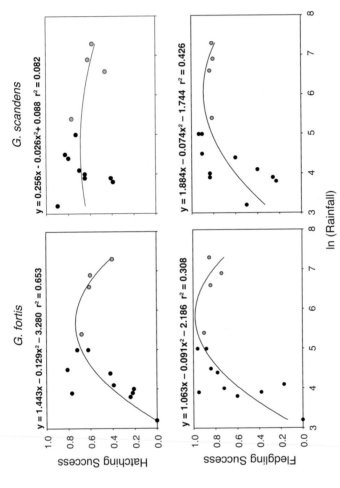

Fig. 5.7 Annual variation in hatching and fledging success in relation to rainfall (with years of El Niño conditions shown by green circles as in fig. 5.5). Both linear and squared terms are significant in the *fortis* relationship between hatching success and rainfall. Of the remainder the only significant (linear) relationship is *scandens* fledging success in relation to rainfall; nevertheless curvilinear lines of best fit are shown. Neither hatching success nor fledging success is enhanced in years of extensive rain. From Grant et al. 2000.

Despite this possible reduction, the increased number of broods means that the greatest potential reproductive fitness is to be gained in years of El Niño conditions (fig. 5.8). These occur at intervals of approximately four to seven years, as do droughts.

It follows that the fittest individuals are those that can survive the droughts and then take advantage of breeding repeatedly in El Niños. An extreme example is provided by the longest-lived female *fortis*, 5960. She hatched in the El Niño of 1983 and died after another in 1998, having experienced four others, 1987 and 1991–93, and several droughts. Starting in 1984 she produced a minimum of 23 clutches of 72 eggs with 8 mates, yielding a minimum of 45 nestlings, 33 fledglings, and 5 recruits. These large numbers are minima because we were not present on the island for all of her breeding. Extremes like this are valuable because they reveal what is possible (table 5.1); see box 5.1 for a brief discussion of the unusual combination of prolific reproduction and long lives. Extreme successes are put in perspective by the numerous failures. As is general with bird populations (Newton 1989), most fledglings do not survive their first year, and many breeders fail to fledge any young (Grant and Grant 1992a, 1996c).

Predicting Reproductive Success

The longer a bird lives, the more opportunities it has to breed and thereby gain high fitness. On this simple basis we used multiple regression analysis to predict the number of fledglings an individual produces in its lifetime and the number of recruits it contributes to the next generation, which vary considerably (fig. 5.9). The variables employed to do this and the procedures are explained in the legend of figure 5.10.

THE COHORTS OF 1975

We begin with the cohorts produced in 1975. The advantage of focusing on a cohort is that all members are the same age and experience the same environment. This eliminates some uncontrolled variables that could complicate interpretation of the predictions. Members of the 1975 cohorts bred for the first time in 1976 and then suffered heavy mortality, and as a result our *fortis* samples are small (20 males and 21 females) and *scandens* samples are too small for analysis. Nonetheless, despite this restriction, we found strong predictors of fitness (fig. 5.10). Longevity predicts number of clutches, number of clutches predicts

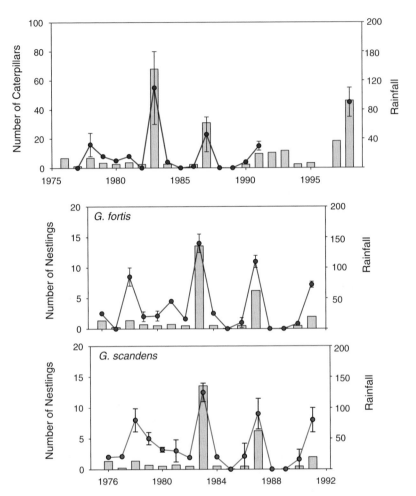

Fig. 5.8 Upper: Annual variation in caterpillar numbers (mean ± s.e. of weekly or biweekly samples) in relation to rainfall shown by histogram bars (from Grant et al. 2000). El Niño years were 1983, 1987, 1991–93, and 1997–98. **Middle and lower:** Annual variation in the mean number of nestlings per female in relation to rainfall. Vertical bars are 95% confidence intervals on the estimates of the means. Only those females breeding in the first month of the breeding season after the first rains are included. Others, usually young, began breeding later in some years, and in a few years there was no breeding at all. Pre-rains breeding by a few *scandens* females have been ignored. For each species mean number of nestlings is highly correlated with rainfall, and so is the mean number of clutches per female (from Grant and Grant 1996b).

TABLE **5.1**
Maximum lifetime values of reproductive traits of females

	G. fortis	G. scandens
Eggs	97	99
Clutches (total)	29	27
Clutches (successful)	15	17
Nestlings	49	60
Fledglings	39	46
Recruits	15	10
Life span	16	17
Number of mates	8	11

Note: Successful clutches are those that yielded at least one fledgling. The total biomass of eggs, each about 2 g (Grant 1982), is approximately 10 times the body weight of the mothers, which is far greater than the ~1.4 theoretically expected from metabolic life history theory (Charnov et al. 2007). Note that high numbers of mates imply high genetic diversity of offspring.

number of fledglings, and number of fledglings predicts number of recruits. There is strong statistical support at each step in the linear flow of causality. Morphological traits have little influence on any of the predictions of reproductive fitness, in contrast to their importance in dry season survival (chapter 4), partly as a result of a restriction of the analysis to those that bred successfully at least once by excluding all the failures.

One relationship deserves comment. Beak size of males was negatively correlated with number of fledglings produced over their lifetime after removal of the substantial statistical effects of numbers of clutches on fledglings. This is a small support ($p = 0.025$) for the idea of a trade-off between reproduction and survival; large males survived better than small males (chapter 4), but small males were relatively successful in producing fledglings, though not recruits.

FOUR LATER COHORTS

The same, strong, links in the linear chain of causality were shown by the larger cohorts produced in 1978, 1981, 1983, and 1987 (Grant and Grant 2000b, 2011a). These results are summarized in figure 5.11, where the numbers of statistically significant relationships out of a total of 15 groups are shown above the arrows: 4 cohorts, 2 species, 2 sexes, minus one group with an insufficient sample size. The trends

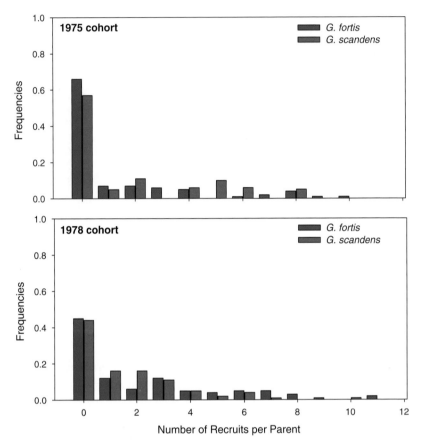

Fig. 5.9 Variation in lifetime recruitment of offspring produced by members of two cohorts of *fortis* and *scandens*. Based on Grant and Grant 1992a.

strongly apply to females, to the males paired with them, to both *fortis* and *scandens*, and to all cohorts despite marked differences in environmental conditions experienced in their starting years (Grant and Grant 2011a). Other factors predict fitness in some years, but to a minor extent.

THE CONTRIBUTION OF MORPHOLOGY TO FITNESS

Morphological variables are rarely significant predictors of longevity or fitness beyond what is expected from occasional size-selective mortality in dry seasons. To make this point, we summarize the results of

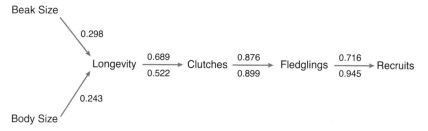

Fig. 5.10 A causal flow diagram for the 1975 cohorts of male and female *fortis*, showing how lifetime production of recruits is determined by lifetime production of clutches and fledglings and ultimately by lifespan (longevity). Body size and two beak variables (size and shape) were used to predict each of the independent variables to the right, and additionally longevity was used to predict clutches, fledglings and recruits, clutches were used to predict fledglings, and fledglings were used to predict recruits. A step-down procedure was used to eliminate variables that were not significant (Grant and Grant 2011a). R^2 values of the single best predictor are shown above the arrow lines for males and below them for females. They give the proportion of variation statistically explained by the predictor variables. Two minor indirect predictors have been omitted. Statistical significance of each of the predictions of longevity is $p < 0.05$. All other predictions are significant at $p < 0.0001$. Sample sizes are 20 males and 21 females.

conducting a total of 68 multiple regression analyses of all five cohorts. Body size or the beak traits were significant predictors of longevity, reproduction, or recruits in only 18 (26.5%). In 14 of the 18 their signs, positive or negative, were consistent with the direction of dry-season selective mortalities (e.g., fig. 11.1) and therefore explicable solely by differential mortality. Moreover, with one exception all morphological variables were minor predictors in analyses of lifetime production of clutches, fledglings, and recruits. Regression slopes were relatively low, and so were R^2 values. Significance levels ranged from 0.038 to 0.003 and generally exceeded 0.01 (12 of the 18), whereas those of the major predictors were usually less than 0.0001. Finally, when cohorts experienced no selection, as was the case for the 1987 cohort of *fortis* (beak shape) and *scandens* (beak size and shape), morphological variables did not predict longevity or any of the reproductive fitness measures.

The ultimate measure of fitness is lifetime number of recruits. In three of the regressions morphological variables added significantly to the prediction of recruits by number of fledglings, but two of the partial regression coefficients were marginally significant ($p \sim 0.04$) and therefore of doubtful reliability. Simple regression analyses predicting number of recruits give the same result: three significant predictions, one

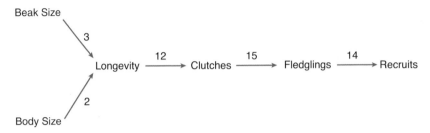

Fig. 5.11 A causal flow diagram, similar to figure 5.10, for the 1978, 1981, 1983, and 1987 cohorts of both *fortis* and *scandens*: four cohorts, two species, two sexes. Males and females of each species of the four cohorts were analyzed separately by the same procedure described in figure 5.10. In total 15 analyses were performed for each variable in the chain; the sample size of female *scandens* in the 1981 cohort was insufficient for analysis. The numbers of partial regression coefficients out of the total that were significant at $p < 0.05$ are shown above the arrows. For the prediction of clutches, fledglings, and recruits along the chain there are 45 partial regression coefficients, and 35 of them are significant at $p < 0.0001$. In contrast none of the other coefficients are significant at this level. The maximum number is five for the prediction of fledglings by longevity $p = 0.0334–0.0001$). These and other secondary predictors of lesser explanatory value have been omitted for simplicity. Sample sizes are approximately the same as in Grant and Grant 2011a. Paternity was checked in the 1987 cohort with microsatellite markers (chapter 3). Taking extra-pair paternity into account does not change the predictions of fitness and fledglings for either species of this cohort; therefore analyses of other cohorts are not likely to be distorted by ignoring extra-pair paternity (appendix 5.1). The probability of a *fortis* male gaining paternity from extra-pair mating is close to 0 until age 5–6 yr, and gains exceed losses on average only when males reach age 7 or 8 yr (Grant and Grant 2011a). This is an additional advantage of a long life span.

body size, the others beak size, and in all three instances the significance level is between 0.05 and 0.01.

We conclude that morphology makes little difference to whether a finch breeds successfully or not. It may affect mating success to a small extent in some years, as happened in the four years following the 1977 drought, when the *fortis* sex ratio was strongly male-biased (Price 1984a, 1984b, Price and Grant 1984), but the effects of morphology on breeding success are minor. From the cohort analyses there is little evidence of a trade-off between wet-season and dry-season performance of birds with a particular beak morphology or body size. Kingsolver and Diamond (2011) concluded from a large meta-analysis that trade-offs among fitness components are generally rare.

THE CONTRIBUTION OF OFFSPRING
TO PARENTAL FITNESS

The close association between recruits and fledglings (figs. 5.10 and 5.11) means that the number of recruits per parent can be considered a random sample of the offspring they produce. Given this, there is little additional scope for recruitment to be influenced by the particular properties of recruited offspring. We compared offspring that became recruits with those that did not and found few, and inconsistent, differences in morphology between them (Grant and Grant 2000b). Female recruits from the 1981 cohort of *fortis* were larger than those that failed to become recruits, whereas male recruits from the 1983 cohort of *fortis* were smaller than the failures. Female recruits from the 1978 cohort of *scandens* hatched earlier than the failures, whereas female recruits from the 1981 cohort of *fortis* hatched later than the failures. Inconsistencies are explained by the fact that different cohorts experience different conditions, the most important being the conditions in the first year of life, when mortality is highest.

LONGEVITY

The longer a bird lives the more offspring it produces. In the preceding statistical relationships longevity has a positive influence on number of fledglings of five groups out of the fifteen , and on recruits of two more, over and above its influence on number of clutches (fig. 5.9): any bias due to unrecorded extrapair paternity (appendix 5.1) should be small (fig. 5.11). Longevity could influence reproductive output in various ways. For example, long-lived birds may gain an advantage when old and experienced in producing healthy and vigorous offspring by breeding early in a season and feeding them a rich supply of insect food (Grant and Grant 2000b).

Another possibility is that long-lived birds gain a reproductive advantage in acquiring different mates in successive breeding seasons while being socially monogamous at any one time. There is statistical evidence for this in the production of fledglings by females of the 1983 cohorts of *fortis* and *scandens*: those females with most mates in their lifetime produced the most fledglings (Grant and Grant 2011a). The causes of this relationship could lie in the social environment, such as behavioral compatibility of particular combinations of parents, or it could be genetic, such as compatibility or diversity of the genetic contributions of the parents to their offspring (e.g., see fig. 4.1). However, in only one case is there evidence of a positive influence of lifetime

Fig. 5.12 Predators of finches. **Upper left**: Short-eared Owl, *Asio flammeus*. **Upper right**: Galápagos Hawk, *Buteo galapagoensis*. **Lower left**: Great Blue Heron, *Ardea herodias*. **Lower right**: Great Egret, *Casmerodius albus*. Owls breed on other islands and visit Daphne for long periods. Analysis of pellets shows they sometimes feed on other islands and roost on Daphne (appendix 5.2). They also die there. A pair of hawks bred on the island in the 1920s (Beebe 1924). Individual hawks, herons, and egrets are rare visitors. All four predators feed on finches but not on nestlings.

number of mates on lifetime production of recruits, and that is from the 1983 cohort of *scandens* females (Grant and Grant 2011a). All these statistical indications of influences from partial regression coefficients are relatively weak and account for little additional variation in fledgling or recruit production after controlling for the number of clutches.

The question left unanswered by these analyses is what governs longevity? The short answer is dry-season survival, which is sometimes dependent on morphology, and random environmental effects. It may be largely a matter of chance which birds survive a long time and which die young. Those that survive long enough to breed may be more or less

Fig. 5.13 Debility and mortality. **Upper left**: Avipox on leg of immigrant *fuliginosa*. **Upper right**: Banded finch in owl pellet. **Lower left**: Egg and killed nestling in *magnirostris* nest (two other nestlings were killed, one with its head bitten off). **Lower right**: Two dead *fortis* with claws apparently entangled. The lower two figures illustrate the fact that finches are each others' own enemies in the breeding season.

equally equipped to continue living a long time, while subject to the random vicissitudes of territory quality, mate quality, susceptibility to predation by short-eared owls and egrets (fig. 5.12 and appendix 5.2), and other mortality factors (fig. 5.13), as well as accidents. Alternatively there may be genetic factors influencing survival, and also behavioral, ecological, and physiological factors that we have failed to detect. Whatever the causes, the consequence of a long life is high reproductive fitness.

These remarks apply to variation in longevity among members of the same cohort. There are differences among cohorts in average longevity caused by enduring effects of different environmental conditions (food supply, population density) experienced by finches early in life (fig. 5.14), as has been observed in other species elsewhere (e.g., Reid et al. 2003).

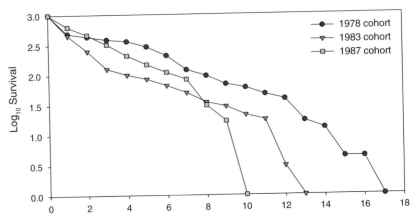

Fig. 5.14 Differences in survivorship among cohorts of *fortis* males. Numbers have been standardized to 1,000 alive in year of hatching prior to log-transformation. Actual numbers are 223, 1,019, and 955 for cohorts produced in 1978, 1983, and 1987 respectively. Differences in survivorship reflect differences in population density and food supply. Survival was greater in the cohort produced in a non–El Niño year (1978) than in the two El Niño years owing to low density in early life following the drought of 1977. *G. scandens* survival varies among cohorts in a similar way.

INBREEDING

Inbreeding makes a small contribution to variation in longevity. Inbreeding is associated with reduced fitness in both species, especially in the rare instances when close relatives breed with each other. We estimated a reduction of 21% in first year survival of a *scandens* offspring with an inbreeding coefficient (f) of 0.25, and 45% reduction in annual probability of survival of adults with $f = 0.25$ (Keller et al. 2002). The magnitude of inbreeding depression in both juveniles and adults is most severe in dry years of low food supply and high number of competitors. For example, under conditions of low rainfall the survival of a juvenile *scandens* with $f = 0.25$ was reduced by 65% compared with a noninbred juvenile, but under wetter conditions there was no difference. Thus inbreeding depression was revealed only under stressful conditions. In adults inbreeding depression was five times more severe in years of low food availability and high population density. Very few inbred *scandens* succeeded in breeding, and none produced grandoffspring. *G. fortis* experienced inbreeding depression only as a reduced probability of recruitment (Keller et al. 2002, Markert et al. 2004). Despite the adverse effects of inbreeding there is no statistical evidence of inbreeding avoidance (Gibbs and Grant 1989), nor is inbreeding minimized by differential dispersal of the sexes from birth to breeding territory (Krakower 1996). The same, a lack of inbreeding avoidance, was found in small insular populations of Song Sparrows, *Melospiza melodia* (Keller and Arcese 1998), and House Sparrows, *Passer domesticus* (Billing et al. 2012).

Variation in Fitness

The fitness of an individual is governed by its ability to survive and reproduce, and by the ability of its offspring to survive to breed. Fitness varies substantially among breeders (fig. 5.9). If same-sex members of a breeding population have equal abilities to survive, reproduce, and produce recruits, and they and their offspring are subject to random hazards, the expected variation in fitness of the breeders is a Poisson distribution, characterized by a variance that is equal to the mean. However, the variance is greater than the mean in both *fortis* and *scandens*; moreover the variance increases in relation to the mean during the life span of cohorts as a consequence of a few individuals living long enough to take advantage of exceptionally favorable conditions for breeding in El Niño years (fig. 5.15). The increase in fitness with age demonstrates both deterministic and stochastic factors: the increase is governed deterministically, but who survives is partly stochastic.

Evolutionary consequences of fitness variation depend on whether it is heritable or not (Charlesworth 1987). If it is heritable, the lifetime number of recruits (generation 2) produced by 1978 males or females (parental generation 1) should predict the mean lifetime number of *their* recruits (generation 3). We used regression analysis to explore this possibility and found no evidence of heritable variation in clutch size (Gibbs 1988), number of fledglings, number of recruits, or longevity (Grant and Grant 2000b). The lack of heritable variation in fitness was also discovered at about the same time in red deer (Kruuk et al. 2000), although Merilä and Sheldon (2000) found significant but low heritability for male flycatcher life span and female recruitment success. Lack of heritable variation in fitness is expected in varying environments (Charlesworth 1987).

However, there is a subtle but important distinction to be made between average and total lifetime number of recruits to the grandparental generation 3 in the finches. The number of recruits (generation 2) produced by members of the 1978 cohorts (generation 1) does predict the total number of recruits that they contribute to the next generation (generation 3) in simple regressions (Grant and Grant

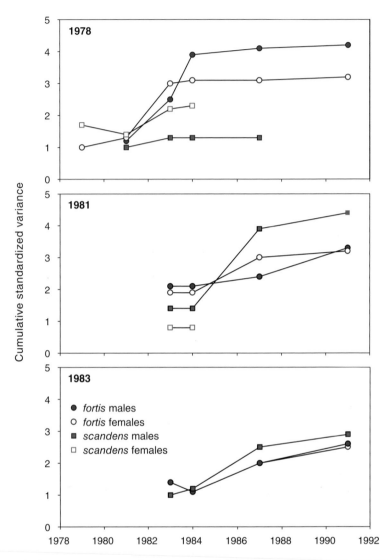

Fig. 5.15 Increase in nonrandom recruit production per parent (males and females averaged) over the lifetimes of three sets of cohorts: 1978, 1981, and 1983. The standardized variance (variance/mean) of recruits per parent is expected theoretically to remain at 1.0 in an idealized randomly breeding population, but instead increases systematically with the age of the breeders as a result of some gaining disproportionate recruitment success (adapted from Grant and Grant 2000b). Cohorts vary in recruit productions according to an unpredictable temporal pattern in the environmental fluctuations: at what ages they experience times of plenty and times of scarcity. For example, on average a female *fortis* hatched in 1983 had to live for 6 yr and produce 10 fledglings to replace herself (genetically) with 2 recruits, whereas a female of the 1978 cohort achieved replacement in 2.5 yr with 5 fledglings.

2000b): *fortis* males ($p = 0.0001$), females ($p = 0.0004$), *scandens* females ($p = 0.0072$) but not males ($p > 0.1$). Thus absence of detectable heritable variation does not mean the absence of predictability of fitness between parents and offspring. In general more recruits beget more recruits, with the result that some lineages proliferate and others do not.

Discussion

The principal contributors to lifetime fitness variation are morphology-dependent survival in the dry-season niche and morphology-independent reproduction in the wet-season niche. The lack of seasonal coupling means there are no specific demographic consequences of morphological evolution such as higher or lower reproductive rates, unlike some other species elsewhere (Ozgul et al. 2009, Schoener 2011, Miner et al. 2012): the feedback from ecological to evolutionary change is unidirectional and not bidirectional except in long-term coevolution of beaks and seeds or the woody tissues that contain them (Grant and Grant 1982, Benkman 1999).

Morphological and reproductive traits are sensitive to environmental change in fundamentally different ways. Morphology is almost fixed after growth has ceased in the first year (Boag 1984) and is highly heritable (chapter 3), whereas clutch size is phenotypically plastic, adjusted by females to the prevailing feeding conditions, and its heritability is close to zero (Gibbs 1988). Finches are at the mercy of the environment in the sense that being large, for example, may be selectively favored in one drought and disfavored in another (chapter 4): a finch may or may not be lucky to have the appropriate beak and body size for the circumstances it encounters. In contrast finches do control their reproduction, and adjust their reproductive effort and output in response to food supply and population density. Their behavioral flexibility extends to skill in rearing the offspring, to a degree dependent on age and experience, by among other things choosing defendable and cryptic nest sites and regulating interference from other finches there (Grant and Grant 1996c). This is easier when populations are low and food is abundant. In these respects both the general and the social environments influence fitness (fig. 4.1).

The contrast between morphology and reproductive traits has left its mark on the finch radiation. Species have diversified in morphology yet have largely retained uniformity and flexibility in breeding characteristics.

Summary

The chapter considers the causes of variation among individuals in reproductive success and overall fitness in terms of the number of recruits they produce in their lifetime. Reproduction is dependent on the amount of rain and the duration of the wet season, which vary enormously from year to year under the influence of the El Niño–Southern Oscillation phenomenon. Living a long life is key to an individual's fitness because the longer a bird lives the more opportunities it has to breed. Longevity is determined by an ability to survive droughts, which is sometimes dependent on morphology (*fortis*), but beyond that it may be largely a matter of chance which birds survive a long time and which die young. Those that survive long enough to breed may be more or less equally equipped to continue living a long time, while subject to random environmental effects such as interactions with aggressive neighbors, susceptibility to predation, and accidents. Longevity predicts total number of clutches, number of clutches predicts number of fledglings, and number of fledglings predicts number of recruits.

Morphological variables are rarely significant predictors of longevity or fitness over and above what is expected from occasional size-selective mortality in dry seasons. There is little evidence of a trade-off between wet-season and dry-season performance of birds with a particular morphology. When measured as lifetime production of recruits, fitness varies from 0 to 15 offspring per breeder. Variation in fitness is not heritable. A few individuals live for an exceptionally long time, up to 17 years, and breed many times. The variation in fitness has environmental causes and evolutionary (genetic) consequences, despite the lack of heritability, because in general more recruits beget more recruits, with the result that some lineages proliferate and others do not.

PART 2

Developing a Long-Term Perspective

A Potential Competitor Arrives
on Daphne

The really surprising fact in this case of the Galapagos
Archipelago, and to a lesser extent in some analogous
instances, is that the new species formed in the separate
islands have not quickly spread to the other islands.
(Darwin 1859, p. 401)

There is a large element of chance in
successful island colonizations.
(MacArthur 1972, p. 84)

Introduction

D ARWIN ANSWERED THE DILEMMA above in two ways. First, he
suggested that deep channels and rapid currents would be
strong barriers to the dispersal of sedentary organisms, such as
snails and centipedes, although not for mobile animals like birds. Second, he pointed out that potential colonists arriving at an island would
have to compete with the local inhabitants. Darwin's second answer
remained in the realm of conjecture until an unexpected colonization
event occurred in the extraordinary El Niño year of 1982–83. The
Large Ground Finch *Geospiza magnirostris* (fig. P.1) established a
breeding population on Daphne. We were exceptionally lucky to be in
the right place at the right time, because natural colonization is rarely

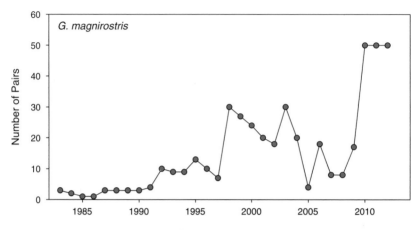

Fig. 6.1 Changes in the size of the breeding population of *magnirostris*.

observed (Sprunt 1953, Bock and Lepthien 1976, Larsson et al. 1988). Instead it is usually reconstructed and interpreted from evidence obtained after the event, often several years after the event (Diamond 1974, Conant 1988, Mathys and Lockwood 2009), or by statistical inference from genetic differences between populations (e.g., Tonnis et al. 2005, Farrington and Petren 2011, Päckert et al. 2013). The *magnirostris* population increased slowly over the first decade and then accelerated, with the result that by 2012 there were 50 pairs breeding on the island (fig. 6.1). This chapter describes the founding of the population and the genetic and morphological changes it underwent in the first three generations (~15 years). The next chapter discusses competition for food with the resident finch species and the influence on the *fortis* trajectory depicted in figure 1.6.

The first few generations after colonization are crucial to the outcome. There is a high risk of failure (extinction) for a variety of reasons associated with the small number of founders involved. First, the habitat may be unsuitable for sustaining a breeding population. Second, inbreeding occurs, with possibly deleterious fitness effects and a loss of genetic variation, and variation may also be lost through random drift. Therefore the population must have sufficient genetic variation to survive a period of inbreeding when it is most vulnerable to extinction through chance demographic and environmental factors (Lande 1993). Despite the likelihood of extinction the population may nonetheless thrive. The external (ecological) environment may be highly favorable because resources are plentiful and enemies are scarce or absent, and the internal (genetic) environment may be highly favorable because rare alleles may rise in frequency and come together in new, favorable combinations that enhance evolutionary divergence (Mayr 1954, Barton and Charlesworth 1984, Carson and Templeton 1984).

Here we discuss causes and consequences of the colonization, and reflect on how the information illuminates or qualifies theory. The addition of a closely related species to the community is relevant to ideas about speciation (fig. 1.2), so in the following chapters we discuss ecological and evolutionary effects of *magnirostris* on the other species of finches (chapter 7), behavioral interactions with the other species in the breeding season (chapter 8), and long-term morphological and genetic changes in the *magnirostris* population (chapter 11).

Founder Event

At least one pair of *magnirostris* bred on Daphne in 1923 (Beebe 1924). Sometime later the population became extinct (Lack 1945, 1947, Harris 1973); therefore we were surprised to find a male on our first visit in April 1973. In the next 10 years others in immature plumage were occasionally seen and captured in the nonbreeding season. None of these immigrants stayed to breed, and almost all disappeared soon after or at the beginning of the following breeding season, possibly returning to their island of origin. A few *magnirostris* were present when El Niño began in late November 1982, but this time they did not all disappear: five of them, two females and three males, stayed and bred (Gibbs and Grant 1987a, Grant and Grant 1995b). Starting with the first egg laid on December 15 they bred in three combinations and produced a total of 17 fledglings from eight broods. Evidently it took a rare event to nudge a species into joining a community whose changed environment was able to accept it (Grant and Grant 1996c).

CAUSES

What triggered breeding? *G. magnirostris* do not have unique requirements for breeding. There was no obvious reason why they could not have bred on the island before 1982–83, and indeed a male was seen building a nest in 1974 and others were heard singing in 1979–81. It seems certain that the unusually early onset of El Niño was responsible for the founding of the new population, but in what way?

Unfortunately neither we, nor our assistants, were on the island when the first rains fell at the end of November; assistants arrived in December. One probable factor inducing the immigrants to stay was a rapid response of the vegetation to an unusually large input of rain in the early stages of El Niño. A key component of the vegetation was *Croton scouleri*, a shrub or small tree that responds rapidly to rain by leafing and flowering, and produces an abundance of large and easily accessible, energy-rich seeds (fig. 2.8) that are a major source of food for nestlings and fledglings, as well as adults. A second factor may have been the breeding of *fortis* and *scandens*, which started soon after the first heavy rains fell. We speculate that social stimulation by them combined with a rapidly growing food supply helped to bring the *magnirostris* into breeding condition. Perhaps *magnirostris* needed to build energy reserves before leaving, and as they did so they also developed reproductively. Some emigrated while others stayed and bred. The social stimulation from the song and chasing of resident finches was perhaps enough to tip the hormonal balance of five birds in the direction of staying, and a rapid increase in numbers of *Croton* seeds and arthropods then facilitated their breeding.

In retrospect this was a colonization waiting to happen. It just needed a particularly favorable wet season. Studies of birds in Europe and elsewhere have shown that colonization of a new area has often been preceded by short-term visits of nonbreeding immigrants (O'Connor et al. 1986, Garnett et al. 1992). Sol et al. (2005) surveyed more than six hundred species worldwide and showed that successful colonizers had large brains relative to body size. Perhaps *magnirostris* do also. It is easy to believe by watching them that they have enhanced cognition. For example, when approached by us they are more likely than *fortis* and *scandens* to stand and stare at us before flying away.

First Few Generations

All the original founders disappeared and were replaced by the next generation, a sister and her two brothers. The offspring of one of the brothers did not survive long enough to breed; therefore the population was essentially started by one, presumably unrelated, pair in the F_0 generation, and a single son and daughter in the F_1 generation. The son and daughter, in turn, contributed seven individuals to the next generation (F_2). The small number of founders and the inbreeding that followed conform to the general understanding of the structure of founder populations, and the likelihood of inbreeding (Crow and Kimura 1970, Charlesworth and Charlesworth 1987).

INBREEDING

We followed the population demographically; established the pedigree by observation, banding, and DNA analysis; and found, as expected theoretically, inbreeding was frequent in the first few generations (fig. 6.2). A high frequency of inbreeding may be due to the high probability of mating with a relative through random mate choice early in the establishment of a small breeding population, or to active mate choice. Using the coefficient of coancestry (ø; Falconer and Mackay 1995) as a measure of the relatedness of breeding pairs, we found no difference between observed and randomly expected average coefficients in the years 1987–92, with one exception: close inbreeding occurred more frequently than expected by chance in 1992. This resulted from closely related birds that formed pairs early in the establishment of the population surviving well and remaining in pairs, even when a few unrelated potential mates became available through immigration. The coefficient of coancestry was 17 times greater in *magnirostris* than in the much larger and long-established *fortis* population (Grant and Grant 1995b).

FITNESS COSTS OF INBREEDING

The population was supplemented by a trickle of immigrants, and at least one of them bred in each year of breeding from 1987 to 1993. This allowed us to compare an inbred group with an outbred group. As expected, inbreeding was found to carry a cost. In the well-sampled 1991 cohort, longevity (maximum = 7 years) varied inversely with inbreeding coefficient (Grant et al. 2001). Inbred birds survived in their first year much less well on average than noninbred birds (fig. 6.3). As far as we could tell, inbreeding depression was restricted to this part of the life history. There was no further inbreeding depression manifested as low clutch size or poor hatching success. Their clutches of 2–5 eggs were typical of the other species. The proportion of eggs that yielded fledglings was lower (50%) than the other species (>60%) but the difference was not statistically significant (Grant and Grant 1995b). Nor was it much lower than on Genovesa, where breeding success of *magnirostris* was estimated to be 60% (Grant and Grant 1989).

The cost of inbreeding was estimated to be 4.5 lethal equivalents (Keller et al. 2002). This is defined as the number of deleterious genes whose cumulative effects equal that of one recessive lethal (Morton et

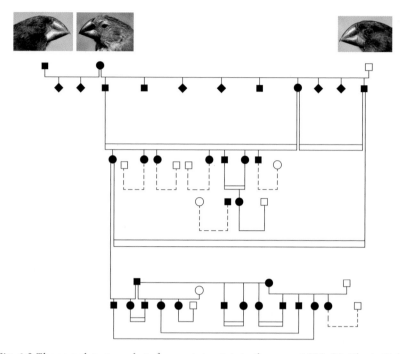

Fig. 6.2 The complete genealogy for *magnirostris* in the years 1983–92. The initial breeders are shown at the top, together with all the fledglings they produced; three of the fledglings became recruits. Below that, only the fledglings that became recruits are shown. Squares, males; circles, females; diamonds, birds of unknown sex. Solid symbols represent banded birds; open symbols, birds without bands. Double horizontal lines indicate breeding pairs of relatives. Broken lines indicate new pairings in 1993 with unknown reproductive outcome. From Grant and Grant 1995b with corrected orientation.

al. 1956), and is measured by the slope of the logistic regression of fitness (0 or 1) on relatedness (f value). For comparison the estimated number for a small population of Song Sparrows (*Melospiza melodia*) in Canada was 2.7 (Keller et al. 2006), whereas larger values have been calculated for island populations of flycatchers (*Ficedula albicollis*) in Sweden (7.5; Kruuk et al. 2002) and Stitchbirds (*Notiomyotis cincta*) in New Zealand (6.9; Brekke et al. 2010).

IMMIGRATION

Immigrants arrived in pulses after the breeding season; most disappeared, but at least one bred in each year of breeding from 1987 to 1993. Immigrants could be identified unambiguously up to March 1992,

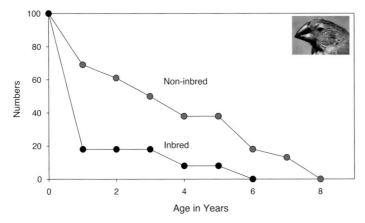

Fig. 6.3 Inbreeding depression in first-year survival but not later. Survival of inbred ($f \geq 0.125$, $n = 17$) and noninbred ($f \leq 0.124$, $n = 14$) members of the 1991 cohort of *magnirostris* are compared. From Grant et al. 2001. Numbers are scaled to an initial 100 of each. Cuckoldry, at 6% in the population ($n_{families} = 15$, $n_{offspring} = 86$), is not large enough to be a biasing factor. The severity of inbreeding, as measured by the number of lethal equivalents, is 4.47. For comparison it is close to 0 for *fortis* but 4.27 for *scandens* (Keller et al. 2002).

when continuous and complete monitoring of breeding of all finches on the island ceased, and with a high degree of confidence in the next few years when most breeding was monitored. Captures of immigrants varied annually in the years 1973 to 1995 (fig. 6.4), and were positively correlated with rainfall in the preceding year when presumably they hatched (Grant and Grant 1995b).

By breeding with residents, immigrants had two effects on the population. First, they caused the mean coefficient of inbreeding to decline from 0.25 in 1986, when the brother-sister pair formed, to 0.11 in 1992, 0.03 in 1993, and 0.01 in 1995. Second, as a result, deleterious effects of inbreeding declined.

SELECTION

Up to 2000 there was a trend toward more robust beaks in the *magnirostris* population, but no trend in beak size or body size. Changes in beak shape (table 6.1) are better described as a couple of step-function changes, occurring first after the initial breeding and second from 1991 to 1992. The second transition from relatively pointed beaks to relatively robust beaks was sustained for almost a decade, with all mean values after 1991 being larger than all means up to and

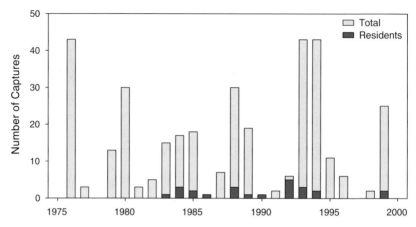

Fig. 6.4 Mist-net captures of *magnirostris* in nonbreeding seasons. Captures were dominated by immigrants after 1983, when breeding began. Netting effort varied among years; standardized captures per unit time are given in Grant and Grant 1995b. From Grant et al. 2001.

TABLE 6.1

The meaning of principal components of morphological variation

	G. fortis	G. scandens	G. magnirostris
PC1-body			
% variance	69.74	70.31	65.33
Weight	0.594	0.604	0.588
Wing length	0.585	0.575	0.584
Tarsus length	0.552	0.552	0.560
PC1-beak			
% variance	86.45	68.29	79.42
Beak length	0.557	0.443	0.550
Beak depth	0.592	0.649	0.596
Beak width	0.584	0.619	0.584
PC2-beak			
% variance	9.57	25.23	13.44
Beak length	0.825	0.887	0.825
Beak depth	−0.313	−0.213	−0.283
Beak width	−0.662	−0.411	−0.488
Sample size (n)	3,868	1,419	632

Note: Each analysis was performed on only one species. Values in the body of the table show the magnitude and sign of the loadings of each trait on the principal components. PC1-body and PC1-beak are positively correlated in *fortis* ($r = 0.743$, $p < 0.0001$), *scandens* ($r = 0.701$, $p < 0.0001$), and *magnirostris* ($r = 0.627$, $p < 0.0001$). Beak shape (PC2-beak) varies independently of body size in *fortis* ($r = 0.000$) and *magnirostris* ($r = 0.067$, $p = 0.0918$) but not in *scandens* ($r = 0.253$, $p < 0.0001$).

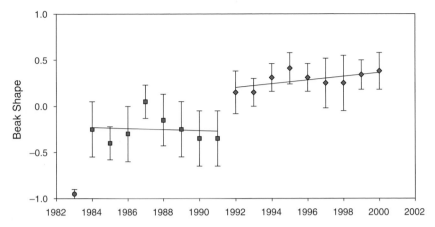

Fig. 6.5 Changes in beak shape, separated into two step-transitions. Beak shape is the second component of a principal components analysis of beak length, depth, and width (table 6.1). Beaks became less pointed and more robust. Lines are least-squares best fits from linear regressions, and are used heuristically to show apparent parallel and flat but displaced morphological trajectories. Annual means and one standard error are shown. After 1983 ($n = 6$), sample sizes varied from 4 (1986) to 29 (1994). From Grant et al. 2001.

including 1991 (fig. 6.5). The pattern is striking, but the interpretation is not straightforward. The transition from 1991 to 1992 could have been the result of chance, selection, or both. It involved the loss of two birds and the addition in 1992 of one immigrant and 10 offspring hatched in 1991.

Genetic Diversity

The founding of a new population is understood better with genetic data. In 1988 we began taking samples of blood for microsatellite DNA analysis, and continued every year thereafter. Ken Petren (1998) developed 16 microsatellite markers for characterizing individuals genetically. Fourteen of the loci were autosomal, and 2 were Z-linked. Some of the birds banded before 1988 were captured and sampled, and the pedigrees of these birds enabled us to reconstruct genetic characteristics of the breeding population before the blood sampling began. None of the original colonists was genotyped, but genetic inferences about them are possible from their offspring. The brother (6501) and sister (6509) of the F_1 generation were each heterozygous at 10 of the 14 autosomal loci, but different at 6 of them. Either the parents were similarly heterozygous or genetically very different from each other.

LOSSES AND GAINS OF ALLELES

As expected theoretically, the population lost alleles in the first few generations. The brother (6501) and sister (6509) possessed a combined total of 34 alleles at the 14 autosomal loci. Their daughter and son that formed a pair had 27 alleles, and the two offspring of this pair that survived to breed had a total of 22 alleles. Whereas 6501 and 6509 were each heterozygous at 10 loci, their two grandchildren were heterozygous at only 4 and 5 loci respectively. These numbers illustrate the expected pattern of random loss and reduction. The other brother (6505) was not genotyped, nor were his (nonsurviving) offspring.

As the population increased in size, the number of alleles in the population increased, and average heterozygosity, that is, the proportion of loci at which an individual is heterozygous, remained undiminished for many years (fig. 6.6). The standing genetic (allelic) variation doubled in the space of 10 years as a result of input from immigrants. A few individuals contributed disproportionately to the increase. While most immigrants contributed 1 or 2 new alleles to the population, one exceptional male (6102) contributed 11 new autosomal alleles in 1990 and 1991, and two others each contributed 8 new alleles in 1995. Even as late as 1998 one male contributed an additional 6 new alleles (Grant et al. 2001).

THE SOURCE OF IMMIGRANTS

Where did the immigrants come from? At the outset the large neighboring island of Santa Cruz, at a distance of 8 km, seemed the most likely source, but we also knew that the founders were unusually variable in beak morphology (Grant and Grant 1995b), which was a hint that they may have come from more than one island. Two of the five founders were captured and measured. The beak depth of one was 18.6 mm, and that of the other was 15.0 mm. Their measurements did not differ from those of four others that did not stay to breed; nevertheless the larger of the two measured founders was exceptional, possessing the second-largest beak among the total sample of 243 individuals measured up to 1993.

We used genetic information to answer the question of origin by comparing genotypes of birds on Daphne with genotypes of samples

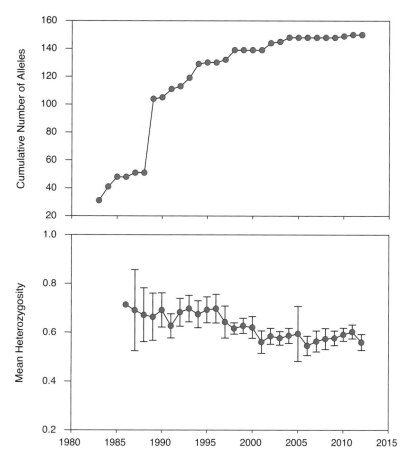

Fig. 6.6 Cumulative numbers of alleles (above) and average observed heterozygosity (below) through time ($n = 588$ individuals). The cumulative numbers increased toward an apparent plateau; 95% of the total 150 alleles had been recorded by 2002. The first point in the allele figure is the number estimated for the five founders from the minimum number known for the two parents of the second generation (Grant et al. 2001). The introduction of new alleles from immigrants far exceeded subsequent losses and the loss of two of the original alleles. As a result of these dynamics the number of alleles in the breeding population had approximately doubled seven years after the population was founded. For the lower figure an index of heterozygosity was calculated for each individual as the proportion of 14 autosomal loci that were heterozygous. Means with 95% confidence limits are shown. There are no bars on the first point, because the two birds then had an identical heterozygosity index. Mean heterozygosity remained stable until about 1996, after which it declined to a lower level in the next five years.

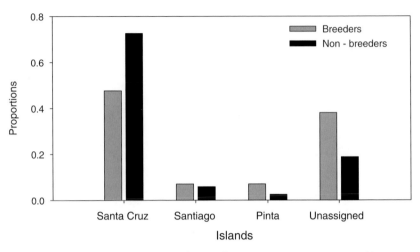

Fig. 6.7 Proportions of immigrants from different islands, as determined by using microsatellite DNA markers. Up to 1998, 117 birds without bands were captured on Daphne. We used a Bayesian modeling approach to assign them to potential source populations (Pritchard et al. 2000), and the following samples to genetically characterize the source populations: Santa Cruz (12), Santiago (10), Rábida (5), Isabela (6), Marchena (10), Pinta (7), Genovesa (32), and Fernandina (9). Three islands were identified as sources of birds on Daphne by the majority rule: Santa Cruz (85), Santiago (7), and Pinta (3). Twenty-two could not be assigned to any island. Of the 22 birds that bred on Daphne, 18 were from Santa Cruz, 1 was from Santiago, and 3 were unassigned. The analysis was repeated with 159 birds captured in the years 1998–2008 with almost identical results (Grant and Grant 2010b), and with 127 additional birds captured in 2009–12, again with almost identical results (two came from Marchena, and one bred). These results indicate continuing, heterogeneous immigration. It is striking that none were assigned to the genetically well-characterized Genovesa population.

from populations on other islands. Multiple sources were confirmed: immigrants ($n = 117$) came from not one but three islands (fig. 6.7). The majority came from Santa Cruz (72.6%), as suspected. The others came from Santiago, Pinta, and possibly other islands or other locations within the large islands of Isabela and Santa Cruz because a significant minority (18.8%) could not be assigned to a single island (Grant and Grant 2010b). One bird from Santiago and one from Marchena bred on Daphne, and the remaining breeding birds all originated from Santa Cruz. This heterogeneity demonstrates much more inter-island movement of *magnirostris* than had been suspected.

NONRANDOM COLONIZATION

It is generally assumed in models of gene flow that colonization is random with respect to genotype and phenotype (Rannala 1996, Slatkin 1996, Johnson et al. 2000). However, there are genetic and phenotypic indications of nonrandom colonization of *magnirostris*. For those immigrants genotyped at 12–14 autosomal loci, the mean heterozygosity of the 22 breeders (0.707 ± 0.125 s.d.) was nearly 10% higher than the mean heterozygosity of 91 nonbreeders (0.639 ± 0.114). With regard to morphology, the immigrants that bred (26 measured) were larger in beak size than 174 contemporary immigrants that did not breed on Daphne (Grant and Grant 1995b). The morphological difference may reflect an underlying genetic difference because beak size is a heritable trait (Grant and Grant 1994; also chapter 3). These differences demonstrate that colonization was selective and not random.

Song

Song is a nongenetic, culturally transmitted trait (Bowman 1983, Grant and Grant 1996d). Only males sing, and they learn their single song type as juveniles, usually from their fathers when dependent on parents for food. A species often sings different songs on different islands (Bowman 1983, Grant et al. 2000b); therefore song is a cultural signature of an immigrant's island of origin. Three *magnirostris* song types have been recorded on Daphne (fig. 6.8). Song A has been recorded only on Santa Cruz, type B has been recorded on Santa Cruz, Santiago, Pinta, and, rarely, Genovesa, and type C has not been recorded anywhere except on Daphne. Thus at least one source of the founders was Santa Cruz, as shown by both song and genetic data. Furthermore, song may influence the decision of immigrants to stay on Daphne and join a breeding population. The evidence for this is that immigrants came from three of the islands where song type B has been recorded (fig. 6.7).

The initial immigrant breeders brought two song types to the island in 1982: A and B (fig. 6.8), rendered onomatopoeically as *pee-oo* and *chi-urrrrr* respectively. The frequency of the three types subsequently changed radically (fig. 6.9). Type B was not present when the third, type C (*zee-urr*), was first heard in 1986. The first breeder (6102) with type C song bred in 1990. Remarkably, from that year on type C rapidly became the majority type, and in some years the only song type, largely owing to 6102's reproductive success. There is no evidence to suggest

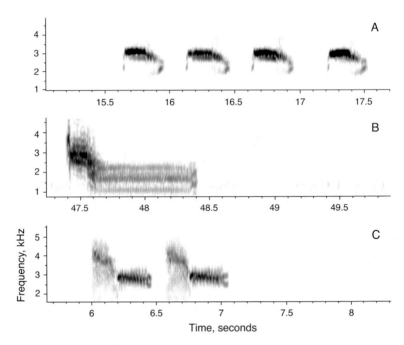

Fig. 6.8 Sonograms of the three types of song sung by *magnirostris* males on Daphne. Note the difference between B and C in which part has maximum amplitude (blackness). From Grant and Grant 1995b.

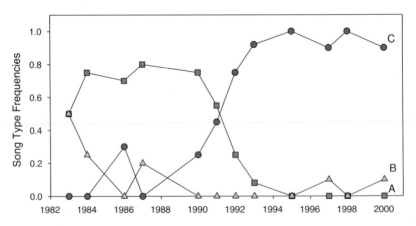

Fig. 6.9 Changes in the frequencies of the three song types. The song type of one of the three original colonists is not known. Up to 1998 all but one of 49 sons of known paternal song type sang the type of their (social) father, be it A or C. In dry years there was no singing or breeding. From Grant et al. 2001.

that song type C increased in frequency so dramatically because it was an intrinsically more effective territorial and mating signal than the others. Rather it was brought into the population by a robust and large male, who carried 12 new alleles at the 16 loci assayed, at a time when the population was experiencing inbreeding depression. His lineage had better fledging and recruitment success than the others. Changes in song type frequencies appear to be a consequence of his reproduction, aided by small population size and chance events. By 1995 types A and B were extinct, although type B reappeared several times, presumably as a result of recurrent immigration, yet remained rare.

Colonization Success

What determines colonization success? A broad study of more than 2,700 avian introductions found that living long was a stronger influence than reproducing fast (Sol et al. 2012). Although ours is only a single study, several additional lessons can be learned from our experience with *magnirostris* about the role of chance and deterministic factors (ecological and genetic) in the founding of a new population. We believe they apply to similar environments elsewhere.

At the outset the odds were against colonization success, for three reasons. First, the expected population size is low, in fact so low there is no detectable peak in the finch density distribution that is determined by food supply and depicted as an adaptive landscape in figure 2.16. Second, if the density distribution remains unchanged, the *magnirostris* should be especially vulnerable to extinction because the strongly fluctuating climate periodically reduces both food supply and finch numbers to low levels (chapters 4 and 5). Third, ignoring the very rare *fuliginosa*, Daphne was already occupied by as many species (two) as are predicted from the relationship between species number and island area (Grant and Grant 1996c).

Nevertheless, in spite of the odds against success, four factors aided colonization: a favorable environment, genetic diversity of the colonists, recurrent immigration, and competitive dominance. First, the abundant rain and vegetation response that lasted for eight months in 1982–83 allowed a breeding population to become established and produce many young. Subsequent El Niños gave additional boosts to population growth. Second, the population started genetically diverse at microsatellite loci and possibly fitness-related loci as well. The brother and sister that constituted the F_1 generation were highly heterozygous, implying genetic diversity in the founder parents. Third, immigration was recurrent, and by introducing new alleles immigrants helped the population to emerge from a demographic and genetic bottleneck at the end of the first decade. This had the important effect of decreasing the chances of inbreeding through random pairing, and helped to end the period of time when the population sustained a genetic load through close inbreeding. Repeated addition of immigrants to the breeding population was possible because the habitat was suitable and the breeding population remained small for several years.

The original pair (F_0 generation) and most of their offspring died before 1985, which shows how precarious population establishment can be. However their son (6501) and daughter (6509) survived for 10 and 11 years. To illustrate how unusual this is, only one bird in our total sample of 398 *magnirostris* banded before the year 2000 lived for 12 years! The 6501 × 6509 pair also produced more fledglings (27) than any other pair, and seven out of the eight that survived for at least one year became breeders. It appears anomalous that these highly inbred birds ($f = 0.25$) experienced the highest success. However, they were unusually heterozygous, and they had a demographic advantage in entering the population at extremely low density. Their diets, typical of the species, were dominated by the large and hard seeds of *Tribulus*, *Opuntia*, and *Cacabus* (fig. 6.10). Although the diet of *magnirostris* overlaps the diets of *fortis* and *scandens* (fig. 6.11), *magnirostris* has two advantages over these species: being larger they are socially dominant at individual food sources, and having larger beaks they can crack the large seeds faster (Grant 1981b). Moreover *magnirostris* are presumably able to dissipate heat with relatively little water loss as a result of having large beaks (Greenberg et al. 2012). So in addition to genetic factors that contributed to their robustness, these two individuals started long lives under favorable environmental circumstances.

Another lesson is that with recurrent immigration important events with long-term consequences may occur a few generations after the founding of a population (see also chapters 7 and 11). For example, a single immigrant male, 6102, had a disproportionate effect on the genetic (microsatellite) and song characteristics of the population from 1991 onward, a little more than eight years after a breeding population was founded. Genetic intervention and augmentation by this robust male helped the population to emerge from a genetic bottleneck and to increase substantially in number of breeders (fig. 6.1). Genetic rescue has been observed elsewhere. An example occurred coincidentally in the same year in Sweden, when a single immigrant wolf (*Canis lupus*) rescued a severely bottlenecked population in Scandinavia, with almost identical results. It introduced 10 new microsatellite alleles, elevated heterozygosity, and eliminated inbreeding depression (Vilà et al.

Fig. 6.10 Large Ground Finches feeding. **Upper left:** *Croton scouleri* fruits. **Upper right:** Red aril of *Bursera graveolens* (on Genovesa) removed before the fruit is cracked (G. B. Estes). **Lower left:** *Tribulus cistoides* mericarp (K. T. Grant). **Lower right:** *Tribulus cistoides* mericarp cracked with beak braced against a rock.

2003; Adams et al. 2011 give a similar example). Short-term genetic rescue was documented in a study of Song Sparrows *Melospiza melodia* on Mandarte Island in Canada (Marr et al. 2002).

The fourth factor, competitive dominance, forms the subject of the next chapter.

Summary

G. magnirostris became a member of the community of finches breeding on Daphne when, in 1982–83, two female and three male immigrants stayed to breed. They produced 17 fledglings, among which only a sister and two brothers survived and bred. This illustrates the precarious nature of colonization at the earliest stages. We used observations of pairs, diets, morphological measurements, and microsatellite DNA

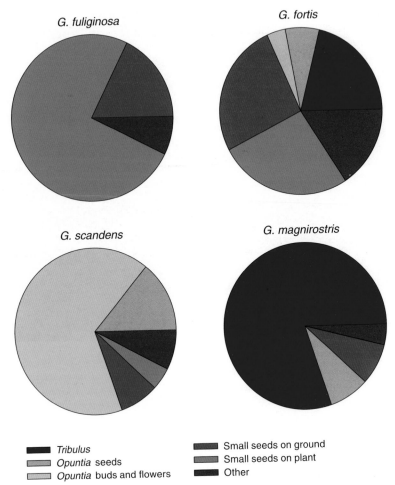

Fig. 6.11 Dry-season diets of the four *Geospiza* species on Daphne. Feeding observations were made on large samples of birds in January to March 1985 and 1986 spanning a drought: 164 *fortis*, 116 *scandens*, 40 *fuliginosa*, and 15 *magnirostris*. The food eaten by an identified bird was recorded only once, when it was first seen feeding. Dietary categories are *Tribulus* seeds, *Opuntia* seeds, *Opuntia* flowers or buds, small seeds on plants, small seeds on the ground, and others, including insects and nectar from *Bursera* and *Portulaca* flowers. All *fuliginosa* and some *magnirostris* were immigrants. Feeding is illustrated in figures 2.19, 2.20, B.2.2, and 6.10.

markers to interpret the initial colonization and subsequent population development. Some expectations from theory were confirmed. The environment was favorable, founders were genetically variable, inbreeding occurred at a high frequency as a result of random mating when numbers remained low, inbreeding depression in survival was strong, and some microsatellite alleles were lost. Unexpectedly colonization was not random with respect to genotype, as often assumed by theory: individuals that immigrated and bred differed from those that disappeared. Immigration occurred repeatedly, from more than one island, with the result that introgression of alleles from later colonists increased genetic variation and alleviated the effects of inbreeding depression. One exceptional individual had a disproportionate effect on the genetic and song characteristics of the population when, eight years after the breeding population was founded, he introduced 11 new autosomal alleles and a new song type, which rapidly became the most abundant type. The arrival of this particularly robust individual at a time when the population was showing signs of inbreeding depression illustrates the role of chance in colonization. Thus the characteristics and fate of the population were molded by an interplay of stochastic and deterministic processes (fig. 4.1) that affected genotype, phenotype, and a culturally inherited trait, song. The interplay governed which particular individuals stayed to breed and, as modulated by the environment, their subsequent success.

Competition and Character Displacement

No scientific theory is worth anything unless it enables us
to predict something which is actually going on. Until that
is done, theories are a mere game with words, and not
such a good game as poetry.

(Haldane 1937, p. 7)

For empirical research, every island is a cause for
celebration, a natural laboratory.

(Schalansky 2009, p. 17)

Introduction

COLONIZATION OF DAPHNE BY *MAGNIROSTRIS* had a profound
effect on the community of finches, after a lag of 22 years. *G.
magnirostris* apparently caused an evolutionary change in *fortis* morphology as a result of competing with it for food (character displacement; Grant and Grant 2006, Pfennig and Pfennig 2010, 2012).
An evolutionary effect of competition in the present is the complement
of character release, evolution of *fortis* morphology in the absence of
competitors in the past (chapter 2). *G. magnirostris* is a potential competitor with the two resident species, especially *fortis*, because their
diets overlap (fig. 6.11). The potential for competition could scarcely
be realized in the first decade, when *magnirostris* numbers were so low
(fig. 6.1), even though nonbreeders occasionally immigrated, reduced

the supply of large seeds to some extent (Boag and Grant 1984b), and might have had short-term effects on the *fortis* population. The potential was realized much later, during a severe drought.

The chapter is structured to show how a competitive influence of one species on another can take three forms: diets diverge in the food-limited dry season, mortality rates increase, and survival depends on phenotype. These competitive effects are respectively behavioral, ecological, and evolutionary. The first two are explored with *fortis* and *scandens*. Before the mid-1990s, at a time when *magnirostris* were scarce, population sizes of *fortis* and *scandens* were large enough that occasional competition between them could be expected in the dry season. Food availability varies among dry seasons as a result of strong interannual variation in seed production in preceding wet seasons (e.g., fig. 7.1). We investigated the possibility of competition by (1) comparing diets with food availability, and (2) examining mortality in relation to food availability and the numbers of potential competitors.

Competition and Diet Overlap

Smith et al. (1978) suggested that if species competed for food, their diets would diverge as food supply declined from wet season to dry season. The predicted divergence was observed in a comparison of wet- and dry-season diets of different ground finch species on several islands in 1973, including *fortis* and *scandens* on Daphne. The comparison of dry-season diets was extended for another 12 years on Daphne (Boag and Grant 1984b, Grant and Grant 1996c), and the predicted divergence was observed in 10 of them, more frequently than expected by chance (Grant and Grant 1996b). The two exceptional years (and one other) were unusual in one respect that could justify their exclusion from the comparison. *Opuntia* flowers were abundant, and almost certainly this alleviated competition between the species because they were too numerous to be defended by *scandens*. The flowers are always an important component of the feeding niche of *scandens*, and important to *fortis* in proportion to their (flower) abundance.

Effects of Competition on Survival

By depleting some parts of shared resources, one species could increase the mortality rate of another through starvation. If such a competitive process is going on, we expect a negative relationship between the numbers of a pair of interacting species (Gibbs and Grant 1987c,

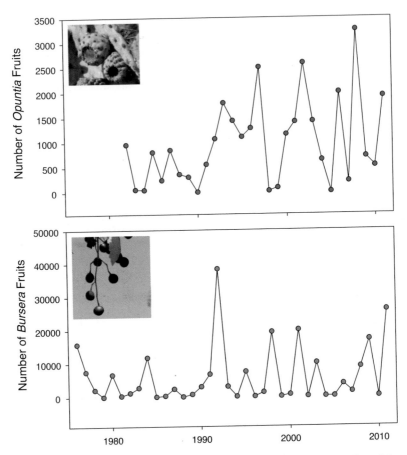

Fig. 7.1. Two indices of annual variation in plant reproduction: *Opuntia echios* (above) and *Bursera malacophylla* (below). Each year all fruits were counted on the same six *Opuntia* bushes on the plateau in or close to the first week of February, and all berries were counted or estimated on the same 79 *Bursera* trees on the plateau in years of production (21 others were males). Annual production is influenced by rain, but numbers of fruits or berries are not correlated with total rain in either *Opuntia* ($r = 0.104$, $p = 0.5837$, $n = 29$) or *Bursera* ($r = 0.182$, $p = 0.2873$, $n = 35$). Rain at the end of the dry season initiates flowering of *Bursera*, but heavy rain may destroy many flowers, as happened in the El Niño years of 1983 and 1987. Numbers of *Opuntia* fruits overestimate seed production in some years when alternative foods are scarce because *scandens* destroys the stigmas in reaching the pollen and nectar (fig. 2.17; Grant and Grant 1981); for example, almost all stigmas were destroyed in this way in 2004. Almost all *Bursera* berries are consumed in the first month after they mature, whereas *Opuntia* seeds and fruits remain available for the whole year. *Bursera* trees live for up to 200 years, extrapolated from a 25-year growth study of 100 trees (1976–2001) on the plateau. An *Opuntia* bush identified in a 1923 photograph (Beebe 1924) lived until 1989, for more than 66 years and an estimated minimum 80 years. Hamann (2001) estimated maximum life expectancies of 200 years for *Bursera* and 150 years for *Opuntia* from growth studies on three islands.

Grant and Grant 1996c): as one species becomes more common, the other becomes less common. For testing the competition hypothesis it is important to statistically control the influence of food supply on finch numbers because population sizes of *fortis* and *scandens* fluctuate together: high food production in some years and scarcity in others affect the two species similarly. When experimental control of food supply is not available for a test of competition (Martin and Martin 2001, Dhondt 2012), as here, statistical control is achieved by regression analysis. We regressed mortality of birds in their first year of life on abundance of small, medium, and large seeds, and on the number of adults of each of the species. Data were transformed appropriately (Grant and Grant 1996c).

Young birds of both species feed predominantly on small seeds in their first dry season because they are relatively common and easy to handle. Not surprisingly, therefore, first-year mortality of each species was found to vary negatively with the abundance of small seeds, and not with the abundance of medium and large ones (Grant and Grant 1996c). After the effects of small seeds were controlled statistically, the mortality of each species was found to vary positively with the abundance of adult *fortis* but not with the abundance of *scandens*. Small-seed abundance is more important (47.8% variance explained) than adult *fortis* numbers (18.7%) in determining survival of young *fortis*, whereas *fortis* numbers (38.9%) account for more of the variation in young *scandens* mortality than do small seeds (25.7%).

These results suggest a competitive asymmetry: intraspecific competition contributes to the mortality of young *fortis*, whereas interspecific competition with the more abundant *fortis* contributes to *scandens* mortality. Even though not demonstrated by the results, *scandens* probably does have some effects on juvenile *fortis* survival because they physically exclude them from feeding sites, especially *Opuntia* flowers and fruits.

Adult *fortis* mortality is influenced, like juveniles, by intraspecific and not interspecific competition. Their annual mortality is strongly predicted by their numbers. *G. scandens* adults are less dependent than the juveniles upon seeds, and their diet is dominated by various products of *Opuntia* cactus at which they, and especially the males, are socially dominant to *fortis*.

Character Displacement

The third type of interspecific influence is evolutionary, causing a shift in phenotype in the next generation. Some phenotypes might be at a selective disadvantage compared with others in the same population when competing with another species. Phenotypes best adapted at exploiting that part of the diet spectrum shared by another species would survive less well than the remainder of the population. The result would be character displacement, a shift in the phenotypic mean (Schluter 1994), and, to the extent that the phenotype is heritable, an evolutionary change in the next generation (Grant and Grant 2006).

G. magnirostris and fortis are potential competitors as a result of an overlap in their diets (figs. B.2.1 and 6.10), especially in the dry season (table 7.1 and fig. 6.11), when food supply is likely to limit population sizes (Grant 1986). The principal shared food is Tribulus cistoides (fig. 2.10). Seeds of this plant are contained within a hard mericarp and exposed when a finch cracks or tears away the woody outer covering (fig. 7.2). Only the large-beaked members of the fortis population are capable of removing seeds (chapter 4), and on average they take three times longer than magnirostris to gain a seed reward (Boag and Grant 1984a, Grant 1981b). G. magnirostris compete with fortis in two ways, by interference and by exploitation. They physically exclude them from Tribulus feeding sites, and they reduce the density of Tribulus

TABLE 7.1

Proportions of seeds in the diets of three finch species

Species	Year	N	Small	Medium	Large
G. fortis	1977	216	0.731	0.102	0.167
	1985	205	0.805	0.000	0.195
	1989	628	0.771	0.051	0.162
	2004	97	0.804	0.113	0.082
G. magnirostris	1985	27	0.185	0.000	0.815
	1989	68	0.059	0.118	0.823
	2004	110	0.045	0.264	0.691
G. scandens	1977	115	0.852	0.148	0.000
	1985	96	0.771	0.219	0.000
	1989	145	0.234	0.697	0.000
	2004	98	0.174	0.826	0.000

Note: See also fig. 6.11. Small seeds are a composite group of 22 species, medium seeds are Opuntia echios, and large seeds are Tribulus cistoides. N is the number of observations. There is strong heterogeneity in the fortis feeding data ($\chi_6^2 = 30.979$, $p < 0.0001$). The reduction in G. fortis feeding on Tribulus in 2004 makes a significant contribution ($\chi_1^2 = 3.912$, $p < 0.05$). Data were obtained by observations in the first three months of each year. In 1977 (only), when fortis with large beaks had a selective advantage, the proportion of large seeds in the diet rose to 0.304 (June) and 0.294 (December). From Grant and Grant 2006.

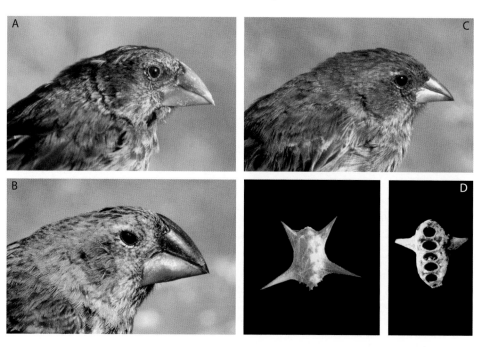

Fig. **7.2** Character displacement. Large members of the Medium Ground Finch (*fortis*) population (A) died at a higher rate than small members (C) in 2004 owing to depletion by Large Ground Finches (*magnirostris*) (B) of the large and hard fruits of *Tribulus cistoides* (D). Five mericarps constitute a single fruit. The left-hand mericarp is intact. The right-hand mericarp, viewed from the other (mesial) side, has been exploited by a finch, exposing five locules from which seeds have been extracted. Mericarps are about 0.8 mm long. From Grant and Grant 2006.

fruits to the point at which they are probably no longer profitable for *fortis* to exploit, owing to handling inefficiencies in relation to search and metabolic costs (Schluter et al. 1985, Boag and Grant 1984a, Grant 1981b, Price 1987). Thus competition is mutual but not equal. By depleting the supply of *Tribulus* fruits, *magnirostris* is predicted to cause a selective shift in *fortis* in the direction of small beak size.

The predicted shift occurred 22 years after the founding of the *magnirostris* population (fig. 7.3; Grant and Grant 2006). By 2003 the combined numbers of residents and immigrants reached a maximum of 354 ± 47 (s.e.). Scarcely any rain fell in 2003 (16 mm) or 2004 (25 mm), there was no breeding in either year, and numbers of both species declined drastically from 2004 to 2005. During the last phase of the decline *fortis* with large beaks were at a strong selective disadvantage, and the average beak size fell to an unprecedentedly low level

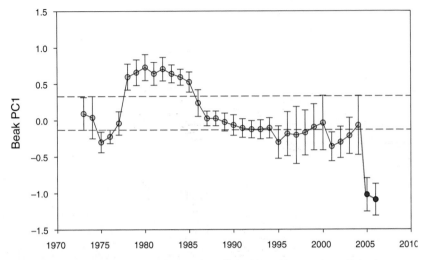

Fig. 7.3 Mean beak size (PC1-beak) of adult *fortis* (sexes combined) in the years 1973–2006. Vertical lines show 95% confidence intervals for the estimates of the mean. Horizontal, parallel dashed lines mark the 95% confidence limits on the estimate of the mean in 1973 to illustrate subsequent changes in the mean. Sample sizes vary from 29 (2005) to 950 (1987). The effect of strong selection in 2004 is highlighted in dark blue. Adapted from Grant and Grant 2006.

(fig. 7.3). Two factors that might have contributed to the selective advantage of *fortis* with large beaks in 1977–78—social dominance of large birds and the higher rate of heat dissipation without water loss from the base of large beaks (chapter 4) (Greenberg et al. 2012)—could not have contributed to selection in 2003–5, despite the same climatic conditions.

STRENGTH OF SELECTION

Coefficients of selection on *fortis* were uniformly large and negative, whether males and females were treated separately as in table 7.2 or combined. Average selection differentials for the six traits in standard deviation units (males 0.522, females 0.972) are unusually large compared with values reported in studies of other organisms (e.g., Hoekstra et al. 2001, Kingsolver et al. 2001, Kingsolver and Diamond 2011). All traits are strongly and positively correlated, so the question arises as to which, if any, was the most important factor that distinguishes survivors from nonsurvivors. To answer this we used principal components analysis to reduce the six intercorrelated traits to three independent

TABLE 7.2

Standardized selection differentials (*s*) for *fortis* in the presence (2004) and absence (1977) of *magnirostris*

	2004		1977	
	Males	*Females*	*Males*	*Females*
Weight	–0.36	–0.93*	+0.88****	+0.90****
Wing length	–0.42	–0.69	+0.43***	+0.75***
Tarsus length	–0.44	0.23	+0.29	+0.22
Beak length	–0.78***	–1.50****	+0.73****	+0.76***
Beak depth	–0.49	–1.39****	+0.81****	+0.73***
Beak width	–0.54	–1.09***	+0.70****	+0.62*
PC1-body	–0.46	–0.79	+0.68****	+0.75***
PC1-beak	–0.64*	–1.39****	+0.79****	+0.74***
PC2-beak	–0.63*	–0.19	+0.20	+0.24
Overall mean	–0.55	–0.96	0.61	0.63
Sample size (*n*)	49	23	171	58
Proportion surviving	0.41	0.61	0.47	0.45

Note: Statistical significance at $p < 0.05, < 0.01, < 0.005$, and < 0.001 is indicated by *, **, ***, and **** respectively. Not included in the 1977 calculations are 135 birds of unknown sex (chapter 4; cf. tables 4.1 and 4.2); the sex was known for all birds in the 2004 sample. Despite the small proportion surviving (0.10), their selection differentials are almost identical (average 0.642) to an average of the male and female differentials. Also excluded were birds captured in 1978 (one) and 2006 (two) that may have been alive in 1977 and 2004 respectively but could have hatched later. Variances did not differ significantly between survivors and nonsurvivors in each selection event.

ones: body size, beak size, and beak shape. The answer is overall beak size; PC1-beak (beak size) was a strongly selected trait in females and in both sexes combined, whereas PC2-beak (beak shape) was not selected in males or when sexes were combined. Multiple regression analysis confirmed the primacy of beak size. Moreover there was little effect on body size, unlike the 1977 episode. In contrast to *fortis*, *magnirostris* experienced nonselective heavy mortality, although the small samples do not permit a sensitive test: 5 surviving males did not differ from 28 nonsurvivors in any of the six measured traits (all $p > 0.1$), nor did the 2 surviving females of the 32 alive in 2004 (Grant and Grant 2006, with minor changes).

THE CAUSAL ROLE OF *G. MAGNIROSTRIS*

There are four reasons for believing that *magnirostris* played a causal role in the character displacement of *fortis*.

HIGH IMPACT ON FOOD SUPPLY. The potential impact of *magnirostris* was greatest at the beginning of 2004, when their numbers (150 ± 19) were closer to those of *fortis* (235 ± 46) than at any other time. Daily energy intake of the two populations and hence impact on the food supply was approximately the same at this time since *magnirostris* individuals (~30 g) are approximately twice as heavy as *fortis* (~17 g).

SUPERIOR FEEDING EFFICIENCY ON SHARED COMPONENT OF THE DIET. *G. magnirostris* are largely dependent on *Tribulus* seeds (fig. 6.11), which are depleted but not renewed during droughts. A much higher fraction of *magnirostris* than *fortis* feed on *Tribulus* (table 7.1), and they deplete the supply of seeds faster than *fortis*. We have calculated that seeds consumed by a *magnirostris* individual each day are sufficient for two *fortis* individuals to meet their energy needs if they feed on nothing else (Boag and Grant 1984a). As a result of their joint reduction in the number of seeds, *fortis* fed on *Tribulus* in 2004 only half as frequently as in other years (table 7.1). Feeding rates of *magnirostris* were exceptionally low, implying food scarcity. For example, in 2004 at least 90 individuals were observed foraging for *Tribulus* mericarps for 200–300 s, and none obtained seeds from more than two mericarps. Under the more typical conditions prevailing in the 1970s a total of eight birds observed for the same length of time fed on 9–22 mericarps, and the average interval between successive mericarps was only 5.5 ± 0.5 s (s.e.) (Grant 1981b).

PARALLEL DECLINE DUE TO STARVATION. Numbers of *fortis* declined to a lower level in 2005, estimated by observations to be 83 (Grant and Grant 2006), than at any time since the study began in 1973. Numbers of *magnirostris* declined so strongly from the 2003 maximum through 2004 that by 2005 only 4 females and 9 males were left. The population was on the brink of extinction as a result of exhaustion of the standing crop of large seeds and starvation. Of 137 *magnirostris* that disappeared in 2004–5, 13.0% were found dead, and so were 21.7% of 152 *fortis*. They starved to death (fig. 7.4), as was evident from their empty crops and stomachs (23 *magnirostris* and 45 *fortis*). Absence of alternative large seeds was a contributing factor to the high mortality of both species. *Opuntia* cactus production in 2004 (fig. 7.1) was the fourth lowest since records were first kept systematically in 1982. Cac-

---

Fig. 7.4 (*facing page*) Starvation during droughts. **Upper:** Dead *magnirostris*, skeletal, bands showing, 1990. **Middle:** *scandens* (above), *fortis*, and *magnirostris*, 2004. **Lower:** about 50 individuals of all species in 2004. Most were found in the small craterlet above the main crater illustrated in figures 1.4 and 1.5. From Grant and Grant 2008a.

tus seeds were insufficient for the two granivore species to escape the dilemma of a diminishing supply of their preferred foods, and insufficient for the cactus specialist *scandens* (~20 g). Their numbers, like those of *fortis*, fell lower (to 50) than in any of the preceding 32 years, without there being any connection between survival and beak size.

KEY DIFFERENCE BETWEEN 1977 AND 2004. In 1977, a year of only 24 mm of rain and no breeding, body size and beak size of both male and female *fortis* considered separately were subject to selection (table 7.2). Average selection differentials were 0.611 for males and 0.634 for females, and they were uniformly positive. In 2004, with the same amount of rain, selection differentials were similar in magnitude to those in 1977 but uniformly negative. Selection occurred in the intervening years (chapter 4), but the events of 1977 and 2004 stand out against a background of relative morphological stability (Grant and Grant 2002a) (fig. 7.3). Immediately prior to 2004 there was no unusual rainfall to cause a change in composition of the food supply or plant community, and no other unusual environmental factor such as temperature extremes or an invasion of predators such as owls or egrets, yet large finches survived at a low frequency in 2004 and at a high frequency in 1977. The conspicuous difference between these years was the number of *magnirostris*: 2–14 occasional visitors in 1977 (Boag and Grant 1984b), in contrast to 150 ±19 residents at the beginning of 2004.

EVOLUTION OF A DISPLACED CHARACTER

An evolutionary response to strong directional selection against large size is to be expected from the high heritability of beak size of *fortis* (Boag 1983, Grant and Grant 1995a, Keller et al. 2001). This was in fact observed. The mean beak size (PC1-beak) of the generation produced in 2005 and measured in 2006–08 was significantly smaller than the 2004 sample of the parental generation before selection ($t_{184} = 4.55$, $p < 0.0001$). The difference between generations is 0.65 standard deviations, which is exceptionally large (Kingsolver et al. 2001, Grant and Grant 2002a). It may be compared with the range of values predicted from the breeders' equation (Falconer and Mackay 1995, Lynch and Walsh 1998), namely, the product of the average selection differentials of the two sexes and the 95% confidence intervals of the heritability estimate. We used the heritability estimated from the father-offspring regression, having excluded offspring of extra-pair paternity (chapter 3), and thereby avoided inflation due to maternal effects of unknown magnitude. The observed value of 0.65 s.d. falls within the predicted

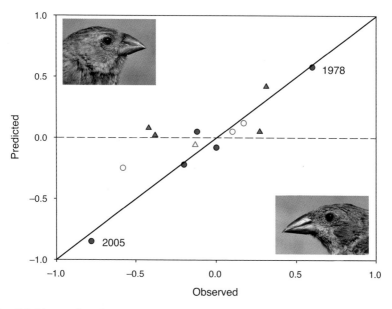

Fig. 7.5 Observed evolutionary responses to natural selection on beak size (filled symbols) and shape (open symbols) in *fortis* (circles) and in *scandens* (triangles) compared with predicted values. Note the response of *fortis* in 2005 is in the bottom left-hand corner. The reference line has a slope of 1.0: observed = predicted. Predictions are the product of the selection differential and heritability in standard deviation units at different times in the study. All selection differentials were significant at $p < 0.01$. Responses are positive for size increases and negative for decreases. Equations that take genetic correlations between traits into account (box 4.2) produce similar results because the correlations are all strong and positive (Grant and Grant 1994, 1995a).

range of 0.48–1.06 s.d. While a small component of the response is possibly attributable to environmental factors (Boag 1983, Grant and Grant 1995b; also chapter 4), the major component is genetic. There was plenty of food and low finch density after selection over the next few years, yet average beak size remained small, reflecting high heritability and no further selection. This was the strongest evolutionary change in the whole study (fig. 7.5).

G. *FORTIS* AND *SCANDENS* COMPARED

The difference in diet between *fortis* and *scandens* explains why *fortis* underwent evolutionary change in beak depth (fig. 1.6) whereas *scandens* did not (fig. 1.7). An ability to find and extract seeds of *Tribulus cistoides* made the difference between survival and starvation in the *fortis* population at two times under contrasting conditions. *Tribulus* seeds are not in the *scandens* diet (fig. 6.11).

Some Implications

We began by considering the absence of *fuliginosa* as a key factor in permitting an evolutionary shift of *fortis* to small size (chapter 2). We concluded that the particular food supply on Daphne and freedom from interspecific competition jointly explained why *fortis* were small: the hypothesis of character release was upheld. We now see that directional evolution of *fortis* in the past may have been partly driven by *magnirostris* and not just permitted by the absence of *fuliginosa*. From this we gain an insight into the early insular history of the Daphne finch community discussed in chapter 2. A possible reason why the *fuliginosa* population became extinct is that it suffered from competition with *fortis*, which were small as a result of competition with *magnirostris*.

Whether this speculation is correct or not, the evolutionary changes we observed are much more complex than those envisaged by Lack (1947), and by us. Composition of the food supply changes across years; so does the composition of the finch community. The average beak size of *fortis* does not occupy a fixed position on an adaptive landscape as implied by figure 2.16 but changes because the landscape, dominated by *Tribulus* seeds, is dynamic on a scale of years (fig. 7.6), and even months within years (fig. 4.5). The average beak size of *fortis* was closely aligned to the peak in the absence of a competitor, *magnirostris* (fig. 2.16). But as the present chapter shows, it was displaced from the peak by *magnirostris* to the left, toward smaller size, as predicted. The strong decline in *fortis* numbers further indicates a strong reduction in the height of the peak, although this was not measured. Recognizing the lack of constancy in an adaptive landscape such as this, Merrell (1994) has recommended "adaptive seascape" as a better term!

These observations have some interesting implications for evolution on Daphne on a scale longer than a few decades. *G. magnirostris* is likely to have been a resident breeder in the past, perhaps many times, and to have become extinct, perhaps many times, which implies a long-term, changing composition of the finch community and its evolutionary dynamics. The long-term evolution of *fortis* is therefore best viewed as determined by the presence or absence of two species of competitors and not just one: a seesaw dynamic of oscillation in eco-morphological

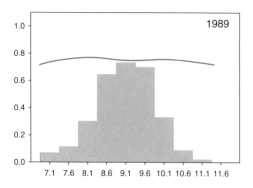

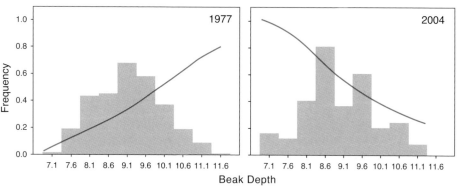

Fig. 7.6 *G. fortis* selection functions in three contrasting years. Selection functions are cubic splines (smoothing function 10) fitted to survival data. They are superimposed on beak-depth frequency distributions. Relative fitness does not vary significantly in most years; the 1989 sample ($n = 554$, omitting two outliers) is the largest.

space defined by reciprocal interactions with the other species and each other, and a changing food supply (fig. 7.6). The result is a nondirectional cycle of expansion and contraction of the community about the core central species, *fortis*, driven by competition, evolution, extinction, and reinvasion. In other systems, including island populations of lizards (Roughgarden 1995), cycles of changing community membership are directional through sequential replacement of one species by another that is larger or smaller.

A second implication follows from the long delay before character displacement occurred; it requires special circumstances. Those circumstances were an initially high density of consumers (finches) and a food supply that is strongly depleted and not replenished for many months, resulting in unusually heavy mortality of the consumers. Thus, although the species were in potential competition at all times, there was in fact little or no competition, and no microevolutionary effects, until those special circumstances arose.

A third implication is that current morphology may reflect past evolutionary forces to some extent, as well as contemporary ones. After the rains of 2005 ended the two-year drought, the supply of *Tribulus* seeds quickly regenerated because *Tribulus* plants are perennial. *G. fortis* did not return to its pre-2005 morphological state of large size when *magnirostris* density stayed relatively low in 2006–8, but remained morphologically frozen in the new state of small size. There were no droughts, and dry-season survival was greater than 60%. The *fortis* population may have experienced stabilizing selection around a new optimum, but if so it was too weak to be detected, and we saw no signs of it in finch measurements. We prefer an alternative interpretation to stabilizing selection: no selection. *G. fortis* were small in 2012 because of selection in 2004–5 and not because of any selection in 2012. In 2008 the *magnirostris* population size rapidly rose to an unprecedented high level (fig. 6.1), so competition with *fortis* for large seeds may have resumed at that time.

Summary

In this chapter we return to the question of competition raised in chapter 2. Species may compete for food, with behavioral, ecological, and evolutionary consequences. By studying populations for many years, we observed three kinds of competitive influences on Daphne: behavioral, ecological, and evolutionary. Diets of *fortis* and *scandens* overlap, and from wet season to dry season their diets diverge in most years, which is the expected consequence of competition. Survival of each species was assessed by multiple regression analysis of first-year survival in relation to seed supply and numbers of adults. Interspecific competition with the more abundant *fortis* contributes to *scandens* mortality, whereas mortality of young *fortis* is affected only by intraspecific competition. Competition with *magnirostris* caused an evolutionary change in *fortis*. This happened toward the end of a two-year drought (2003–5) as a result of their joint and severe depletion of the supply of *Tribulus* seeds. *G. fortis* with large beaks were at a strong selective disadvantage; average beak size of the population fell to an unprecedentedly low level, and remained there for the next few years because no further selection occurred. This is an example of character displacement. The observed evolutionary response to natural selection was the strongest recorded in the study, and exceptionally strong in comparison with other organisms elsewhere. The long delay before character displacement occurred shows it requires special circumstances: large numbers of competitors and severe competition for food. *G. magnirostris* may have been an important influence on *fortis* morphology in the past, even if it underwent repeated cycles of colonization, numerical increases, decreases, and extinction. Therefore current morphology may be partly a reflection of past evolutionary forces as well as contemporary ones.

Hybridization

The extraordinary variants that crop up in a series
[of museum specimens] give an impression of a process
of change and experiment going on.

(Swarth 1934, p. 231)

It would be of great interest to determine the
critical factors controlling the variability of
each species, and to know why some species are
so much more variable than others.

(Lack 1947, p. 94)

Introduction

OUR STUDY ON DAPHNE BEGAN as a small part of an archipelago-wide attempt to understand the evolutionary diversification of the finches. Since the strength of the evolutionary response to natural selection is governed by the amount of genetic variation (chapter 3), morphological averages should not be considered in isolation from the variances. Average beak size differs between species as well as between populations of the same species (fig. B.1.2), and so do phenotypic variances (Lack 1947). For example, in chapter 3 we showed that *fortis* vary more than *scandens* phenotypically and genetically. These contrasts raise the question of why some populations vary in morphological traits such as beak size much more than others and hence have greater potential for evolutionary change. Up to now we have discussed factors that are relevant to the average size and shape of beaks. In this chapter and the next we explain how addressing questions of population variation, that is, variation among members of the same population, led to a study of hybridization.

Background

Daphne is a good choice for the variation question, even if it is not ideal. The coefficient of variation for beak width of *fortis* males in museum collections is 6.48 ($n = 31$) (Grant et al. 1985). On other islands it varies from 3.63 (Fernandina, $n = 14$) to 10.98 (Floreana, $n = 276$) with an average of 7.08 for 12 populations with sample sizes greater than 10 individuals. The coefficient of variation for *scandens* males in museum collections is distinctly lower, and varies from 4.25 (Floreana, $n = 141$) to 6.44 (Marchena, $n = 10$) with an average of 5.78 for 10 populations. There are too few specimens from Daphne (box 2.2) for calculating the coefficient, but a large sample of live birds (both sexes) from Daphne ($n = 570$) has a much smaller coefficient (4.53) than the comparable one for *fortis* (7.76, $n = 881$; Grant and Grant 2000a). Thus *fortis* is a more variable species than *scandens*. Why is this?

When we started in 1973, there were two major ecological ideas to explain differences in variation between natural populations in traits that vary continuously. One hypothesized a role for diversifying selection in heterogeneous environments, and was explicitly adaptive (Van Valen 1965, Roughgarden 1972; more recently Bolnick et al. 2003, Araújo et al. 2011): the more heterogeneous the environment, the more variable the population as a result of advantages experienced by extreme individuals relative to those with average characteristics (fig. 8.1). The other suggested that gene exchange between slightly differentiated populations could elevate the variance in each one without the variation in either being adaptive; populations that exchange genes vary more than those that don't (Soulé 1971, 1972).

Hybridization, which is one form of gene exchange, was not part of mainstream discussions of population variation at that time. Introgressive hybridization, that is, hybridization followed by backcrossing to a parental species, was considered to be generally rare and inconsequential in animals (Mayr 1963), although common in plants. Where it was observed, it gave rise to concerns about the loss of genetic identity and uniqueness of a species. This was appropriate in situations where invasive species hybridized with local residents in human-altered habitats (Cade 1983, Rhymer and Simberloff 1996). Lack (1945, 1947) looked

Fig. 8.1 **Upper**: Graphical models of the way resources are exploited by members of a species. A broad range of resources (A–C) is exploited by populations 1a and 1b; in the first, individuals are generalists; in the second, they are specialists. A narrow range of resources (A–B) is exploited by all members of the specialist species 2. Overlap of individuals in 1b is shown to be symmetrical, for simplicity. In fact overlaps are likely to be asymmetrical. From Grant et al. 1976, based on the adaptive variation arguments of Van Valen 1965 and Van Valen and Grant 1970. Alternatively specialist species might be divided into different specialist individuals (Bolnick et al. 2010).

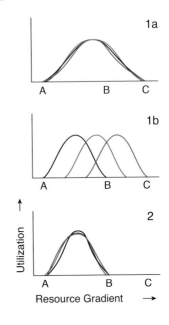

for field evidence of hybridization in the Galápagos, failed to find it, and rejected it as a cause of the small size of *fortis* on Daphne (chapter 2). Although hybrid finches were produced in captivity, they died within a week of hatching (Orr 1945). Nevertheless, we paid attention to the natural breeding of finches at the start of our study because neither of the two ecological explanations seemed fully satisfactory. Our reasons were, first, Daphne appeared to be too small and homogeneous for a variable population to be maintained by diversifying selection, and second, the island appeared to be too isolated (by 8 km) from the nearest island to receive many, if any, immigrants.

Our first study of breeding in 1976 yielded evidence of hybridization (Grant and Price 1981, Boag and Grant 1984b). In addition to *fortis* and *scandens* there were 14 *fuliginosa* individuals breeding. This was a little surprising because, although very small numbers of this species had been reported on the island on previous visits (Harris 1973; box 1.3), it was not known to breed there. What was even more surprising was that eight *fuliginosa* females bred with *fortis* males. Furthermore their eggs hatched, and the young fledged. This was the first demonstration that the reproductive barrier between finch species in nature was not absolute. It raised a set of questions concerning the circumstances and consequences of interbreeding that are relevant to the potential role of hybridization in causing an elevation of morphological variances. This chapter, which focuses on the causes and fitness consequences of hybridization, leads into the topic of morphological consequences in the next chapter. Our study shows that a low level of hybridization and introgression occurs naturally in pristine, undisturbed, habitats. It gives insights into the degree to which genetic variation of a population is enhanced, particularly when species of different genetic correlations and allometries hybridize. It also gives insights into the origin and dynamics of reproductive isolation, whether partial or complete.

Frequency of Hybridization

Hybridization occurred in most years of breeding between 1976 and 1995, and several later years, but was always rare (Grant and Grant 1992b, 1996d, 1997a). The numbers of interspecific pairs across those 20 years were 45 *fuliginosa × fortis* and 8 *scandens × fortis*. These numbers contrast with numerous intraspecific pairs; 23 *fuliginosa*, 1,268 *fortis*, and 562 *scandens*. Thus 2.8% of *fortis* males and 2.3% of females paired with *fuliginosa*, and an additional 0.1% of males and 0.8% of females paired with *scandens*. The *fortis × fuliginosa* pairs comprise 3.4% of pairs involving *fortis* but 65.2% of pairs involving *fuliginosa*. Despite that, the proportion of *fuliginosa* pairs is much higher than would be expected if *fortis* and *fuliginosa* were to mate at random (Grant and Grant 1997a). Males and females of *fortis* and *fuliginosa* hybridized at approximately equal frequencies, on average, whereas *scandens* males hybridized consistently more frequently than *scandens* females.

These estimates come from observations of adult males and females attending nests. They rest on the assumptions that we identified the species correctly by morphology (box 8.1), and that we identified paternity correctly (box 3.1).

Causes of Hybridization

Why do species hybridize? Two plausible explanations for Darwin's finches are, first, a scarcity of mates, and second, perturbation of the process of sexual imprinting, when offspring learn characteristics of parental phenotypes early in life and select mates with those characteristics. The two possibilities are considered in turn.

A SCARCITY OF CONSPECIFIC MATES

A relative scarcity of mates could explain why a rare species (*fuliginosa*) of either sex hybridizes with a common species (*fortis*). This is sometimes referred to as the Hubbs principle (Mayr 1963, Grant and

Box 8.1. Identifying Species

At the beginning of the study *fortis* and *scandens* clearly differed in morphology. *G. scandens* could be recognized unambiguously by long beaks, both absolutely and relative to beak depth or width (fig. 8.7). The males of each species could also be identified unambiguously by their songs (figs. 8.2 and 8.5). On the other hand identifying *fuliginosa* gave us a problem in that there were no discrete differences in morphology between *fortis* and *fuliginosa*; the frequency distributions ran into each other without a break, with the result that the combined frequency distribution had a long tail at the lower end of the axis of beak or body size measurements (fig. 9.6). Nor are the songs diagnostically different.

To identify *fuliginosa* individuals, we therefore adopted an operational definition by using the frequency distributions of *fuliginosa* and *fortis* measurements on the northern coast of Santa Cruz Island at Borrero Bay, where the distributions discretely differ (Grant 1993). The upper measurements of *fuliginosa* beak dimensions on Santa Cruz were used as markers for classifying individuals on Daphne as *fuliginosa* or not *fuliginosa*. Those individuals classified as not *fuliginosa* were assumed to be either *fortis* or *fortis* × *fuliginosa* hybrids. Pedigrees were subsequently used to assign individuals to species and hybrids. F_1 hybrids and backcrosses were identified when the parents were known from observations of adults that were incubating eggs (females only) or feeding nestlings and fledglings (both social parents). All birds—F_1 hybrids and backcrosses as well as nonhybrids—mate according to paternal song type (box 8.2); therefore F_1 hybrids and backcrosses, including those that failed to breed, could be assigned to one population or another. We considered the 1983 populations as the baseline for studies of hybrids because we did not know if hybridizing individuals were in fact pure species, F_1 hybrids, or backcrosses. In the next two chapters (9 and 10) we discuss tests of hybrid identification with genetic data, and identification of island sources of *fuliginosa* immigrants that sustain the population on Daphne.

Grant 1997a, Randler 2002), because it was invoked by Carl Hubbs to explain hybridization in fish species in terms of encounter rates (Hubbs 1955). Hybridization of *scandens* is also consistent with the principle. Even though *scandens* were not rare, males outnumbered females at all times, many failed to obtain mates, and male *scandens* (seven) hybridized more than females (one). But why did the common species, *fortis*, hybridize? The sex ratio of *fortis* was usually close to 1:1, except after the drought of 1977 (chapter 4). One possible answer is a scarcity of available mates at the end of the pairing period. As the supply of potential mates declines, the last few individuals remaining unmated might lower a critical threshold of mate acceptability and accept a heterospecific mate. Moreover the last few individuals could be young and inexperienced, with incompletely developed mate recognition systems. These plausible expectations were tested with observations on the temporal pattern of pairing, but not supported (Grant and Grant 1997a). Interspecific pairs formed throughout the pairing formation period, and therefore a scarcity of mates does not explain why *fortis* hybridized. Possibly female *fortis* chose heterospecific males holding high-quality territories, but we have no measure of territory quality to assess this. For example, territory quality confers direct benefits to hybridizing flycatchers in Europe (Wiley et al. 2007).

IMPRINTING

Playback experiments with song and mount experiments showed that finches on Daphne can discriminate between their own and another species using either visual or acoustic cues alone (Ratcliffe and Grant 1983, 1985). Furthermore, experiments with captive finches have demonstrated that males learn their short adult song early in life from a male they interact with socially (Bowman 1983). Mates are chosen according to species-specific song and morphology, which are learned early in life though imprinting on parents and occasionally others (box 8.2). Hybridization can occur after young birds of either sex imprint on an adult of a different species and use the learned cues in choosing a heterospecific mate. Heterospecific imprinting is sometimes referred to as misimprinting (Grant and Grant 2008a). Song plays a significant role. Song is probably an important factor in the interbreeding of *fortis* and *fuliginosa* because their songs are similar. However, *fuliginosa* immigrate to Daphne, so we do not know the origin of their songs and the learning environment.

SONG INHERITANCE

Males, only, sing a single, structurally simple song, which is retained unaltered for life (fig. 8.2). Both sons and daughters learn songs, mainly from their fathers, according to (1) laboratory studies, (2) father-son resemblances (fig. 8.3), and (3) pairing patterns in nature (box 8.2). Learning takes place during a short sensitive period extending from day

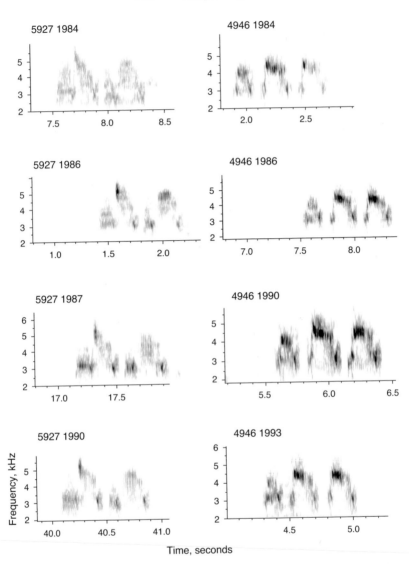

Fig. 8.2 (*above and facing page*) Four song variants sung by *fortis*. Repeated recording across years shows that an individual's song is retained for life. Adapted from Grant and Grant 1996d, 2008a.

10 to day 40 after hatching. On a coarse scale songs of *fortis* can be classified into four types; 72% of 263 sons sang the same type as their fathers (Grant and Grant 1996d). However, there is variation within song types, and the song types are not discretely different, so we measured eight variables to quantify temporal and frequency features of

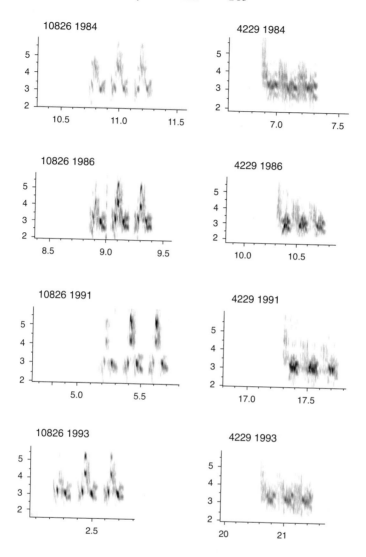

songs in order to analyze variation with principal components analysis (PCA).

Family resemblance could be due to transmission of genetic factors in addition to learning. If so we would expect the songs of sons to resemble not only the songs of their fathers but also the songs of both paternal and maternal grandfathers, whereas with learning there should be no relationship between the songs of sons and maternal grandfathers. We used PCA to discriminate between these alternatives. The resulting regressions supported the hypothesis of learned cultural

Generation

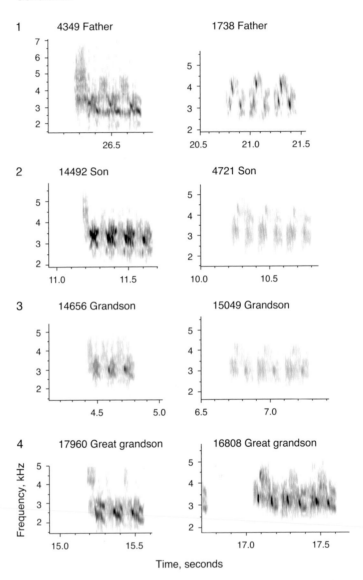

Fig. 8.3 (*above and facing page*) Songs of *fortis* sons usually resemble the songs of their fathers. From Grant and Grant 1996d, 2008a.

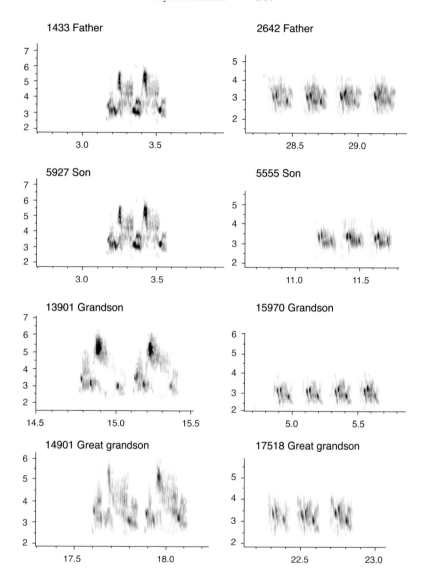

transmission from father to son and gave no support to the alternative of genetic transmission (fig. 8.4). Variation in song among *scandens* individuals is not so striking; nevertheless the father's song does predict the length of the first note sung by the sons (Grant and Grant 1996d) as well as trill rate (Grant and Grant 2010c). *G. magnirostris* sons sing the same song type as their fathers (chapter 6). We know of only a single exception.

Box 8.2 Cues Used in the Choice of Mates

The six species of ground finches have similar plumage, and do not differ consistently in patterns of courtship behavior, but do differ conspicuously in size and shape and in song. They have apparently identical mate recognition systems, yet pair conspecifically solely as a result of learning the specific features of their parents or similar models (Grant and Grant 1997a, 1998, Podos 2010). Experiments with stuffed museum specimens showed that individuals discriminate between their own and another species on the basis of beak size and body size in the absence of movement and acoustic cues (Ratcliffe and Grant 1985). A similar set of experiments showed that they also discriminate between species on the basis of song alone in the absence of movement and visual cues (Ratcliffe and Grant 1983).

Visual cues of appearance (beaks) are inherited genetically (chapter 3), whereas acoustic cues are inherited through learning, that is, culturally. Only males sing. By raising finches in soundproof chambers and playing tape-recorded conspecific or heterospecific song to them, Bowman (1983) found that song is learned during a short sensitive period early in life between day 10 and day 40 post-hatching, in other words for a few days in the nest and then for the first month as fledglings when dependent on parents for food. The song is learned mainly from the father, and once learned it is retained for life. We presume the appearance of the parents is also learned at this early time of life in association with species-specific song. This is an imprinting process that influences mating behavior later in life (the evolutionary significance is reviewed in Irwin and Price 1999, ten Cate and Vos 1999, Slabbekoorn and Smith 2002, and Price 2008). An important part of the process is learning that song and acoustic cues of similar sympatric species are different from their own species, as shown experimentally with the Blackcap (*Sylvia atricapilla*), a European warbler, learning the features of the Garden Warbler (*S. borin*) (Matyjasiak 2005).

There is some evidence of nonrandom mating within species with respect to size, but it is not strong (Grant and Grant 1997a, Grant and Grant 2008c). The evidence is correlations between the body size of parents and the mates of either their sons or daughters (Grant and Grant 2008c). The strongest associations are between fathers and the mates of their sons. In some years males change mates and the second mate is similar in size to the son's father whereas the first is not. With regard to song, the mating rule for females is that they pair only with a male of the same or different species that sings the same species song as their father. For example, of 90 *scandens* females for which both the father and the mate were recorded, 86 mated with *scandens* males that sang *scandens* songs, while the remaining 4, all daughters of a *scandens* male that sang a *fortis* song, mated with *fortis* males. Rare exceptions to the rule, detectable only when the song of mate and father are known, amount to 2 of 482 *fortis* females (Grant and Grant 1996d). For non-random mating in the *fortis* population on Santa Cruz see Podos (2010), and for a general review of learning, sexual selection, and speciation see Verzijden et al. (2012).

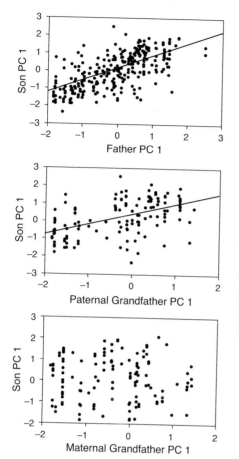

Fig. 8.4 Songs of *fortis* sons resemble the songs of their fathers and paternal grandfathers but not maternal grandfathers, indicating cultural but not genetic inheritance. The slope of the son-father relationship is 0.725 ± 0.047 s.e. ($r = 0.671$, $p < 0.0001$), based on 273 father-son pairs involving 133 different fathers. The slope of the son–paternal grandfather relationship is 0.540 ± 0.076 ($r = 0.523$, $p < 0.0001$), based on 136 son–paternal grandfather pairs involving 47 different paternal grandfathers. The slope of the son–maternal grandfather relationship is 0.036 ± 0.101 and not significantly different from zero ($r = 0.032$, $p > 0.1$), based on 124 son–maternal grandfather pairs involving 45 different maternal grandfathers. The slope of the paternal grandfather regression (0.540) is close to the expected value (0.525) from the square of the son-father regression (0.725). From Grant and Grant 1996d.

PERTURBATIONS OF IMPRINTING

Species differ in song characteristics (compare figs. 8.2 and 8.5), so under the normal imprinting process unambiguous song cues to species identity are learned in association with morphology and later used at the time of courtship and mate choice. Imprinting in effect sets up a premating barrier between species. However, the barrier is not impermeable; it occasionally leaks.

The normal learning process is vulnerable to perturbation if a young bird hears and learns the song of another species during its sensitive period. This we have observed happening in three ways, each rarely. First, a nest with one or more eggs is taken over by another species, and the young that hatch learn the foster father's song. We know of one case where a *fortis* nest with an egg was taken over by a *scandens* and a

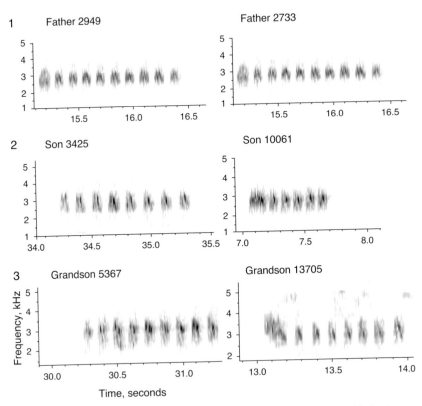

Fig. 8.5 (*above and facing page*) Songs of *scandens*. Close copying of father's song by sons is a rule that is occasionally broken. Songs of males 15901 and 13705 deviated from their respective fathers' songs. From Grant and Grant 1996d.

young *fortis* was raised to fledging and sang the song of its foster father. Second, if the father dies or disappears toward the end of the nestling period, the young birds may imprint on their natal neighbor, which can be a member of another species: we know of three such cases. Third, a dominant heterospecific neighbor may repeatedly chase the father away from its nest and then sing, with the result that the young imprint on that song. We know of one case where a *magnirostris* influenced the subsequent song of two juvenile males out of three from two *fortis* nests (fig. 8.6): the third sang the same song as his father's. Thus perturbation of the normal learning process can result in the son singing a heterospecific song, even when the father sings a normal song. However,

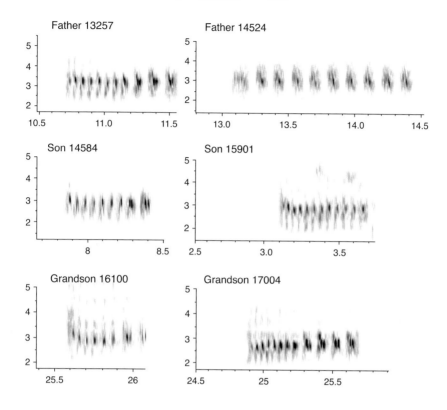

for most cases of heterospecific singing we do not know how the imprinting occurred (box 8.3).

Imprinting on another species can lead to hybridization (fig. 8.7) but does not always do so. Heterospecifically imprinted males transmit ambiguous signals, in having the song of one species and the morphology of the other. As a result not all hybridize: some mate conspecifically. Over a period of more than 20 years we noted 11 instances of *fortis* and *scandens* males singing each other's (heterospecific) song, and three of them paired heterospecifically (Grant and Grant 1996d, 1998). Apparently some females are more influenced by morphology than by song when the two sets of cues are in conflict, and for others the reverse is true. An example is given in figure 8.8. Two *fortis* sisters paired with different heterospecific mates even though the father sang a normal *fortis* song. We presume the normal imprinting process was perturbed after fledging. In both cases the morphology of the mates was just beyond the normal intraspecific range, and in one case the song was clearly beyond the normal conspecific range.

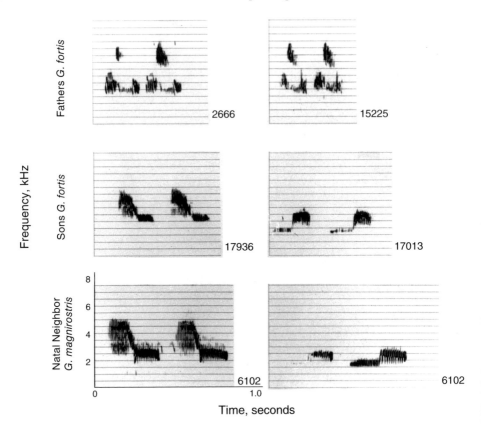

Fig. 8.6 Heterospecific copying of song. Two *fortis* (17936 and 17013) copied the two songs sung by a neighboring *magnirostris* male (6102) and not the songs sung by their respective fathers (2666 and 15225). Their fathers were adjacent neighbors, and the *magnirostris* territory overlapped both *fortis* territories. Male 15225 was the son of 2666—a rare case of a son occupying a territory next to that of his father. The *magnirostris* is also a rare case of an individual singing two songs. In neither case did the imprinting lead to hybridization. In contrast to 17013, a brother (17011) raised in the same nest was not imprinted on the *magnirostris* song. He sang the same song as his father and paired with a *fortis* female. From Grant and Grant 1996d.

Such variation in the pattern of mating gives no indication of genetically based mate preferences. Controlled experiments would be needed to expose them if they exist. Experimental studies of imprinting on other species by cross-fostering have given similar, nonuniform, results in the pattern of mating (Clayton 1990, Slagsvold et al. 2002).

Box 8.3 Heterospecific Imprinting as a Cause of Hybridization

The general impression is one of heterogeneity, and at best only partial predictability, in the pattern of hybridization. This could be a consequence of the fact that heterospecifically imprinted birds, like F_1 hybrids themselves, have a broader morphological range of potential mates from which to choose (Gee 2003) as a result of early experience in or out of the nest. Two examples illustrate this possibility. First, two sisters from the same nest paired with males of different species (fig. 8.8). Second, finches that pair heterospecifically may change mates and pair conspecifically. For example, a female *fortis* paired first with a *scandens* male, then a *fortis* male, returned to the first *scandens* male, and following his death paired with a new *fortis* male (Grant and Grant 1997a). Experience after leaving the nest, or possibly even off the natal territory, may occasionally modify the canalizing influence of early experience. For example, sexual and aggressive interactions may influence choice of a mate (Bischoff and Clayton 1991).

G. MAGNIROSTRIS

G. magnirostris did not hybridize with the other species at any time throughout the study, in spite of nine *fortis* learning and singing a *magnirostris* song (fig. 8.6). Female *magnirostris*, being twice their size, may not have recognized them as potential mates, but often they were not given a chance to explore courtship possibilities, because whenever a *fortis* sang a *magnirostris* song it was immediately and repeatedly chased and harassed by *magnirostris* males. The only one of these *fortis* males that acquired a mate did so after reducing song output and singing at low volume and low in the bushes: the mate was a *fortis*. A single *scandens* male imprinted on a *magnirostris* song (fig. 8.9) received the same harassment, responded in the same quiet and cryptic way, and also acquired a mate, but in this case it was a *fortis*! Her history was not known. Song copying must have been precise to elicit such strong responses from *magnirostris* males; usually each species ignores songs of the others.

Importantly the mating patterns show that acoustic cues and visual cues complement each other (Baker and Baker 1990). When morphological differences between species are small, as between *fortis* on the

Fig. 8.7 Birds imprinted on another species and a hybrid. **Upper:** *scandens* 2054 sang a *fortis* song. **Middle:** *fortis* 18459 sang a *scandens* song. **Lower:** *scandens* 4053 sang a *fortis* song. The top two birds have measurements typical of their respective species and are therefore not likely to have been hybrids. In contrast the lower bird's measurements were close to those of *fortis*, and therefore the bird is suspected of being an F_1 hybrid.

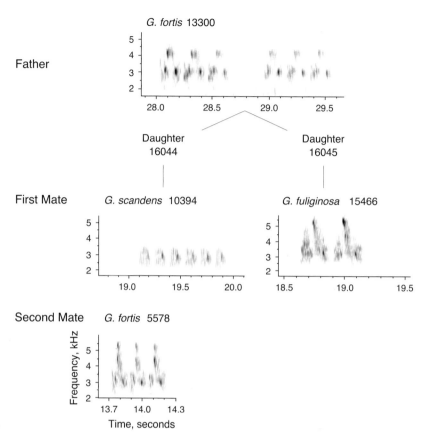

Fig. 8.8 A unique example of heterogeneity of pairing within a family. One daughter (16044) of a *fortis* male (13300) and female (13687) paired with a *scandens* male (10394), and later with a *fortis* male (5578). Another daughter (16045) paired with a *fuliginosa* male (15466). From Grant and Grant (1997a). Later genetic analysis with microsatellite DNA markers (box 10.1) confirmed the *fortis* identities of 13300 and 13687 but also revealed that 5578 was a *fortis* × *fuliginosa* F_1 and 15466 was a rare *fortis* × *fuliginosa* F_2. Although 10394 was genotyped as *scandens*, its small size suggests the possibility of it being a *fortis* × *scandens* F_1.

one hand and *fuliginosa* or *scandens* on the other, learning the song of another species can lead to hybridization, but when the size difference between species is large, the barrier to interbreeding is robust, for then even learning the song of another species does not lead to hybridization.

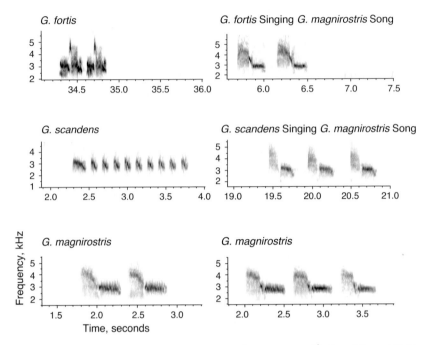

Fig. 8.9 Copying of *magnirostris* song by *scandens* 19936 in 2010 and *fortis* 17936 a decade earlier. Both bred with *fortis* females. Typical *fortis* and *scandens* songs are shown on the left, and typical *magnirostris* songs of two individuals are shown at the bottom.

Fitness Consequences of Hybridization

VIABILITY

Hybridization is important in the context of speciation because it provides an opportunity to test the degree of genetic compatibility between species. The first question to be answered is this: are the hybrids viable? In the first eight years none survived long enough to breed. *G. scandens* × *fortis* hybrids were too rare for analysis (2 pairs). *G. fuliginosa* × *fortis* hybrids were more numerous (12 pairs), and their survival was significantly weaker than age-matched *fortis* (fig. 8.10). One possible reason for this was an intrinsic physiological weakness due to their particular heterospecific combination of genes. An alternative reason was extrinsic: food conditions were not suitable for hybrids. During this time, large and hard seeds dominated the dry-season seed

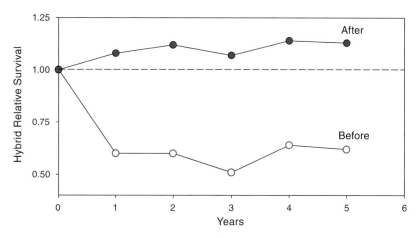

Fig. 8.10 Relative fitness of *fuliginosa* × *fortis* hybrids was lower before 1983 (open circles) than after 1983 (solid circles). Relative fitness is depicted as survival of *fuliginosa* × *fortis* F$_1$ hybrids in relation to *fortis* survival for two groups of cohorts, the 1976–81 cohorts and the 1983–87 cohorts. The broken line indicates equal fitness of *fortis* and the F$_1$ hybrids. The fitness of the 1976–81 hybrids is significantly lower than the fitness of *fortis* over the first age interval, and remained low thereafter. There are no significant differences in fitness for the 1983–87 cohorts. Sample sizes are 957 *fortis* and 32 hybrids hatched in 1976–81, and 2,376 *fortis* and 68 hybrids hatched in 1983–87. These constitute an estimated 80% (1976–81) and 97% (1983–87) of the total *fortis* and hybrid fledglings produced on the island in those years. From Grant and Grant 1993.

supply. *G. fortis* × *fuliginosa* hybrids are unable to crack large and hard seeds with beaks that are intermediate in size between the beaks of the parental species, and the small seeds they might have exploited were scarce. *G. fortis* × *scandens* hybrids with similarly intermediate beak sizes are inefficient at cracking *Opuntia* seeds, a *scandens* food, and unable to crack *Tribulus* fruits, a *fortis* food (fig. 8.11; Grant and Grant 1996b).

Subsequent observations showed the extrinsic reason to be correct. From 1983 onward, when small and soft seeds dominated the food supply, hybrids survived well, in agreement with the extrinsic hypothesis. Hybrids survived surprisingly well relative to the parental species, in fact as well as if not slightly better than them (fig. 8.12). This clearly demonstrates they are not physiologically weak, and there is no evidence here of genetic incompatibility between the hybridizing species (Grant and Grant 2008b).

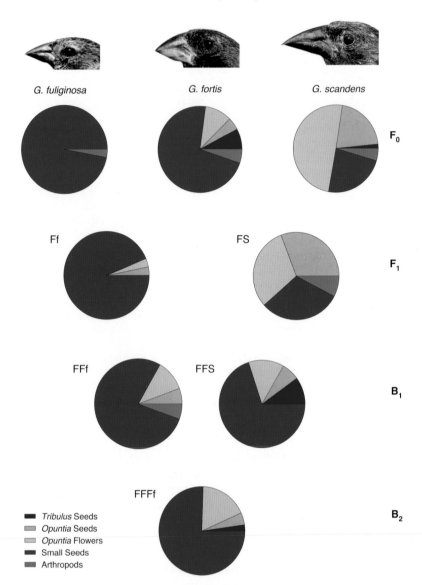

Fig. 8.11 Diets of hybrids and backcrosses. From Grant and Grant 1996b.

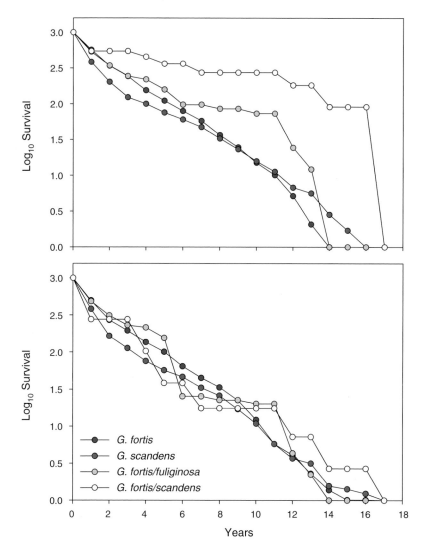

Fig. 8.12 Composite survival curves of seven cohorts of finches, hatched in the years 1978, 1981, 1983, 1984, 1987, 1991, and 1998. **Upper**: Numbers were summed across years, standardized to 1,000 at fledging, and converted to logs. **Lower**: Numbers in each cohort were first standardized to 1,000 at fledging and converted to logarithms, then summed across years and averaged. The hybrid groups comprise F_1 hybrids and first generation backcrosses. From Grant and Grant 2010b. Higher survival of the *fortis* × *scandens* hybrids in the upper panel is a consistent feature of the three cohorts with largest sample sizes: 1983, 1987, and 1991 (Grant and Grant 2008b).

Fig. 8.13 Three species, hybrids, and backcrosses. **Top row**: Representatives of the three hybridizing species, *fuliginosa* (left), *fortis* (center), and *scandens* (right). **Middle row**: F₁ hybrids, *fuliginosa* × *fortis* (left) and *fortis* × *scandens* (center and right). **Bottom row**: First-generation backcrosses, *fuliginosa* × *fortis* backcrossed to *fortis* (left) and *fortis* × *scandens* backcrossed to *fortis* (center) and to *scandens* (right). Modified from Grant and Grant 2008a.

FERTILITY

The second question is, are the hybrids fertile? The answer is yes. They bred in six of the nine years in the period 1983–91, when all breeding was monitored. They had no difficulty in acquiring mates and produced as many eggs, nestlings, fledglings, and recruits as the parental species. Hybrids and backcrosses (fig. 8.13) with *scandens* genes had apparently lower hatching success than those with *fuliginosa* genes; nevertheless in terms of fledgling success they all did about as well as the parental species and sometimes distinctly better. In short we could find no systematic difference in any of these reproductive categories between hybrids and parental species (Grant and Grant 1992b).

OVERALL FITNESS

Fitnesses of hybrids and nonhybrids produced in 1987 are compared in figure 8.14. The 1987 cohorts are chosen because they are the largest

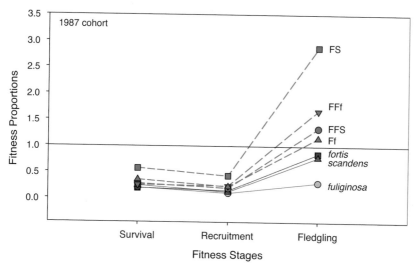

Fig. 8.14 Fitnesses of the 1987 cohorts over the first four years, expressed as survival to adulthood at stage 1, the product of survival and recruitment to the breeding population at stage 2, and production of survival, recruitment and fledglings produced in the breeding seasons of 1990 and 1991 at stage 3. A value of 1.0 for a class of finches indicates numerical replacement; the starting number of fledglings hatched in 1987 has been replaced by an equal number of fledglings hatched in 1990 and 1991. The fitness values for hybrids and backcrosses exceed 1.0 (replacement), whereas the three species all have fitness values less than 1.0. *G. fortis* and *scandens* reached replacement levels in 1992, whereas *fuliginosa* failed to replace themselves. Symbols: FS (*fortis* × *scandens*), Ff (*fortis* × *fuliginosa*), FFf (*fortis* × Ff), and FFS (*fortis* × FS). Adapted from Grant and Grant 1992b.

for making such comparisons. They were living at a time when small and soft seeds were abundant. An integrated measure of fitness is shown over the first four years as the product of survival of fledglings, recruitment to the breeding population, and production of fledglings in the breeding seasons of 1990 and 1991. Hybrids replaced themselves during this period, whereas the three parental hybridizing species did not. *G. fortis* and *scandens* reached replacement levels one year later, in 1992, whereas *fuliginosa* failed altogether to replace themselves. The fitness differences for this cohort arose largely from higher breeding success of hybrids and backcrosses.

Genetic factors, over and above those affecting beak or body size, may also contribute because the hybridizing parental populations show some deleterious effects of inbreeding (Gibbs and Grant 1989, Keller et al. 2002, Grant et al. 2003, Markert et al. 2004). Hybridiza-

Fig. 8.15 Relative fitness of inbred birds and hybrid backcrosses of the 1991 cohort. **Upper**: Theoretical expectation of fitness of the two groups relative to the fitness of noninbred birds and nonhybrids. As genetic incompatibilities between species increase in time, the relative fitness of hybrids is expected to decline, as shown by the broken line with arrow through alternatives a, b, and c. **Middle**: The expected pattern is realized in *scandens* (middle), with relative fitness measured as the difference in longevity; the data conform most closely to alternative a above. The same pattern is realized in *fortis* (**lower**), though weaker and conforming to a or b but not c. Hybrid backcross (*scandens* × *fortis*) birds were matched with nonhybrid birds of the same sex as controls in time and space; they bred at the same time, usually on adjacent territories. The same was done for inbred and noninbred pairs. The horizontal axis is expected heterozygosity for 14 microsatellite loci. Arrowheads mark the extremes of close inbreeding ($f = 0.25$) and F_1 hybrids. Mean expected heterozygosities of outbred birds ($f = 0$) are indicated by broken vertical lines. The *scandens* slope is significantly different from 0, whereas the *fortis* slope is not, and the two

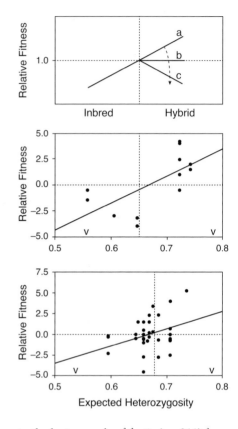

slopes are statistically indistinguishable. However, in the large sample of *fortis* ($n = 211$) from this cohort mean heterozygosity increased and inbreeding coefficient decreased with age in the first two years owing to differential mortality (Markert et al. 2004). From Grant et al. 2003.

tion can enhance fitness to different degrees by counteracting the effects of inbreeding depression through the masking of exposed deleterious alleles in heterozygotes. Alternatively deleterious alleles at one locus in one parental genome might be masked or compensated for by alleles at a different locus in the other parent's (complementary epistasis). A small sample of hybrids and inbred birds, matched with nonhybrids and noninbreds as controls, shows the expected relationships (fig. 8.15). Relative fitness increased with mean expected heterozygosity across the total spectrum of *scandens* from inbred to outbred. The trends for *fortis* are the same but quantitatively and statistically weaker, although at the same rate as in *scandens*. Purging of deleterious alleles in *scandens* may be retarded by recurrent introgression of alleles from *fortis*, not all of which will be advantageous in a *scandens* background (Keller et al. 2002).

The Mating Pattern of Hybrids

Hybrid (F_1) sons have intermediate morphology between their heterospecific parents, and sing the same songs as their fathers. Therefore they transmit ambiguous signals about their identity, like heterospecifically imprinted parental species but to a lesser degree. The importance of song in mate choice is shown by the fact that the mating rule which applies to species—females pair with males that sing the same species song as their own fathers (box 8.2)—applies to F_1 hybrids as well, despite some morphological ambiguity (Grant and Grant 1997b). Hybrids are generally too rare to breed with each other and produce an F_2 generation (see fig. 8.8 for one of only two known cases involving *fuliginosa* × *fortis* F_1s); instead they backcross to one of the parental species. Hybrids formed by the interbreeding of *fortis* and *scandens* backcrossed to the two parental species equally, the direction depending on paternal song, whereas those formed by the interbreeding of *fortis* and *fuliginosa* backcrossed entirely to *fortis*, due to the rarity of *fuliginosa*. Moreover male hybrids sang a *fortis* song, as did the fathers of the female hybrids.

Consistent with the pattern of breeding according to father's song type, all backcrosses bred with the parental species to which they were most related or with other hybrids and not with the parental species to which they were least related. For example, all 8 backcrosses to *scandens* (four males and four females) paired with *scandens*, and all 10 backcrosses to *fortis* (eight males and two females) paired with either *fortis*, backcrosses with a predominantly *fortis* constitution, or, in one case, a *fortis* × *scandens* F_1 female.

Pair formation along lines of paternal song continued in the next generation. We noticed an apparent tendency for backcrosses to pair at a relatively high frequency with other hybrid classes, that is, F_1 hybrids and backcrosses. Four of 16 first-generation *fuliginosa* × *fortis* backcrosses (0.25) paired with another hybrid class derived from *fuliginosa* and *fortis*, and 3 out of 10 *scandens* × *fortis* backcrosses (0.30) paired with hybrid classes derived from the same species. These are much higher than mean expectations for those years (1987–93) based on random sampling, which are 0.07 and 0.16 respectively. We interpret this pattern to be the result of a broad range of parental stimuli experienced by hybrids and backcrosses early in life that is translated into a broad range of acceptable mates later in life.

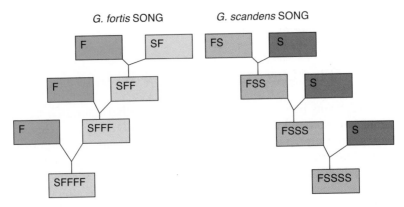

Fig. 8.16 The mating pattern of interbreeding *fortis* (F) and *scandens* (S), and their F$_1$ hybrids and backcrosses. Both sons and daughters imprint on their father's song and mate according to species song type. Hybrids do not sing intermediate or combined songs. As a result of introgressive hybridization genes flow from one species to another, but the two populations are kept apart by song differences and mating assortment based on song. FS refers to a F$_1$ hybrid, the product of a *fortis* father imprinted on a *scandens* song mated to a *scandens* female. FSS refers to a first-generation backcross, FSSS refers to a second-generation backcross, and the same system is used for *scandens* backcrosses to *fortis*. From Grant and Grant 2008a.

The chief implication of these mating patterns is that with rare exceptions the direction of gene transfer in the first generation of interbreeding is not reversed in the next two to four generations of backcrossing (fig. 8.16).

Conclusions

The poor survival and lack of breeding of hybrids prior to 1983 stand in marked contrast to the superior fitness of hybrids afterward, especially the 1987 cohort. The contrast illustrates the dependence of hybrid fitness on the environment. In a fluctuating environment fitnesses are not constant but fluctuate in response to changing ecological (food) conditions as well as social conditions (fig. 4.1). This gives rise to opposing tendencies of population fusion caused by introgressive hybridization and fission caused by directional selection (Grant and Grant 1983, 2008b). As discussed in the next chapter fusion refuels the genetic variation, enhances morphological variation, and helps to explain why genetic variation is not eroded but is maintained at the observed high level (chapter 3) or even increased.

Summary

This chapter and the next one address the question of why some populations vary in morphological traits much more than others. The present chapter focuses on the causes and fitness consequences of hybridization, and leads into the topic of morphological consequences of introgressive hybridization in the next chapter. The two main findings are, first, the normal barrier to gene exchange is behavioral and not one of genetic incompatibility, and second, mate choice is determined by learning; there is no indication of genetically based mating preferences.

G. fortis hybridizes with fuliginosa and scandens, rarely but repeatedly. A relative scarcity of mates can explain why a rare species (fuliginosa) hybridizes with a common species (fortis). Hybridization of scandens, even though it is a common species, can be explained similarly because males outnumbered females at all times and many failed to obtain mates. Hybridization of fortis is understood in some cases. Mates are chosen according to species-specific song and morphology, which are learned early in life through imprinting on parents and occasionally others. Perturbation of the normal learning process can result in imprinting on another species, and this can lead to hybridization but does not do so always. With regard to consequences of hybridization, survival of hybrids depends on dry-season feeding conditions. Survival was poor before the 1982–83 El Niño year of substantial food production. In later, relatively wet, years it was as good as, if not slightly better than, survival of the parental species. Not only did the hybrids survive well; they bred successfully and backcrossed to each of the parental species, the direction in each case depending on paternal song. As a result of episodic introgression the fortis and scandens populations oscillated between fusion caused by introgressive hybridization and fission caused by directional selection.

Variation and Introgression

Introgression may be said to have taken over the role of
excessive mutation in providing multitudes of new genetic
combinations involving total blocks of cooperating genes.

(Svärdson 1970, p. 57)

To evaluate the evolutionary significance of hybridization
requires a full accounting of the selective forces that act on
the production of hybrids, and on their reproduction and
survival. This includes consideration of the consequences
of gene flow both within and between species.

(Schemske and Morgan 1990, p. 2151)

Introduction

MANY BIRD SPECIES, approximately one-tenth of roughly ten
thousand species, are known to hybridize (Panov 1989, Grant
and Grant 1992b, McCarthy 2006). Hybridization and subsequent introgression through backcrossing can be a powerful source of
significant genetic variation because it can generate new combinations
of genes in either or both parental populations, allowing a rapid evolutionary response to new or changing environments. In the Galápagos
Archipelago closely related species occur in sympatry in various combinations (Grant 1986). This provides many opportunities for occasional hybridization with introgression to contribute to the high varia-

tion observed in morphological traits. As we mentioned in the previous chapter, our study of breeding yielded evidence of hybridization right at the beginning, in 1976 (Grant and Price 1981, Boag and Grant 1984b). This made us realize we could not afford to ignore it as a possible source of genetic input into a population and one of several factors affecting the observed levels of morphological variation in *fortis* and *scandens*.

When we began, studies of genetic variation in laboratory populations and in animal breeding considered standing variation at any one time to be the product of agents that increased or decreased variation: mutation and recombination elevate genetic variation, and stabilizing selection and random drift deplete it. The presence of abundant polygenic variation was nevertheless seen theoretically as a problem to be explained in the face of strong stabilizing selection that should rapidly eliminate it (Barton 1990). The primary evidence for strong stabilizing selection was the low fitness of extreme individuals in some studies (Endler 1986, Barton 1990). The problem disappears if mutation rates are high, for then losses due to selection are balanced at equilibrium by gains from mutation (Lande 1976).

This four-factor scheme with mutation-selection balance at its center is a useful theoretical framework for empirical studies and is sufficient for closed populations. In nature more is needed because populations there are open to exchange of individuals, environments are not constant in time or space, and selection regimes other than stabilizing may operate (Grant and Grant 1989). Variation at any one time is set by a resolution of all these factors. The point of resolution for Darwin's finches is likely to vary from one island to another as a result of variation in the relative strengths of the factors, and to vary from time to time because environments are not constant (Grant and Grant 1989, 2000a; fig. 9.1). For these various reasons the question of why populations display different levels of variation is as much an ecological genetics problem, dependent on the environment, as it is a population genetics problem, dependent on demography (Grant and Price 1981). Both genes and environment play essential roles (fig. 4.1), and the critical and debated question is how they do (e.g., Bürger 1999, Zhang 2012).

The goal of this chapter is to consider how introgressive hybridization might affect the maintenance of variation in individual traits, as well as covariation in correlated traits. Our study is not comprehensive but narrowly focused on introgression. To understand variation in its absence would require knowledge of how it arises from genetic mechanisms—mutation, gene duplication and conversion, recombination, pleiotropy, and linkage (Stern 2000, Barton 2001, Hill 2010, Hill

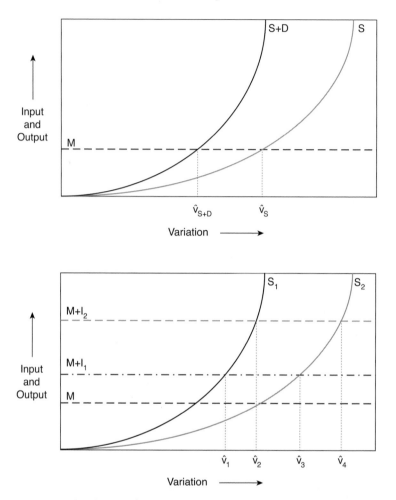

Fig. 9.1 A graphical model for the maintenance of variation in a continuously varying trait. Input from mutation (M) and introgressive hybridization (I) is balanced by
output from selection (S) and random genetic drift (D). Selection is either stabilizing or directional. **Upper**: An equilibrial level of variation (V) is determined by a
constant rate of input of genetic novelty through mutation, counterbalanced by loss
from selection alone or selection combined with drift. At constant or occasionally
low population sizes, where drift is likely to be most effective, variation is relatively
low ($V_{S+D} < V_S$). **Lower**: Two levels of constant introgression (I_1, I_2) and two levels of
selection intensity (S_1, S_2) are added to the scheme above to show their joint determination of equilibrial levels of variation. Drift, not shown, could be added equally
or unequally to the selection functions. Conspecific gene flow, not shown, could be
added to the I + M levels of variation, or could be considered as a component of
introgression more broadly defined. From Grant and Grant 2000a.

and Kirkpatrick 2010)—that are beyond the scope of this field study. Their relative roles in the maintenance of variation are currently uncertain (Hill 2010).

Morphology of Hybrids

Hybrids are generally morphologically distinct (fig. 8.13), especially in the F_1 generation but also in the first two generations of backcrosses (Grant and Grant 1994). They are almost always intermediate, on average, between the means of the parental groups that produce them, as expected from the polygenic nature of trait inheritance in the parental species. The *scandens* × *fortis* hybrids are the only F_1 group departing significantly from expectation. Those identified from the pedigrees were not midway between the means of the parental species, on average, but relatively small (fig. 8.13), perhaps because of a negative interaction between the two sets of parental genes in the growth of the hybrid offspring, or because the parents happen to be below average in size, or because they do not function well as a pair and the feeding of the offspring suffers in consequence. High fitness of hybrids (fig. 8.12) makes the last explanation the least likely.

Heritable variation in the hybrids is substantial. To estimate it, we had to combine F_1 hybrids with backcrosses because individual sample sizes were small. We found that variation in all six measured traits was heritable, with averages of 0.802 for *fuliginosa* × *fortis* heritabilities and 0.865 for *scandens* × *fortis* (Grant and Grant 1994). These values are comparable to the heritabilities of the parental species, *fortis* and *scandens* (chapter 3).

Effects of Hybridization on Variation

Hybridization results in backcrossing in diverse and complex ways (fig. 9.2). For example, even though *fuliginosa* and *scandens* have never been known to interbreed, each has bred with *fortis*. *G. fortis* acts as a conduit (Grant 1993) or bridge (Clarke et al. 1998, Broyles 2002) for the passage of genes from one species to another. A similar situation has been described for three species of *Nothonotus* fish in a North American river, with the difference being that in this linear environment only two coexist at any one place (Keck and Near 2010).

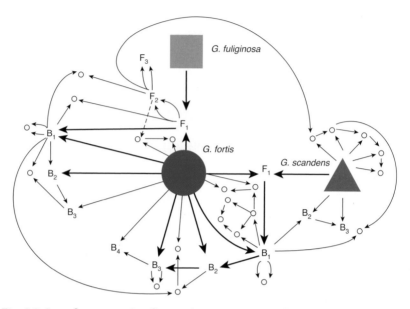

Fig. 9.2 Gene flow network reflecting known and quantified introgressive hybridization. All three species are connected by exchanging genes, but the relatively rare *fuliginosa* has not hybridized with *scandens*, and *fuliginosa* × *fortis* hybrids have not backcrossed to *fuliginosa*. Thick lines indicate the primary pathways of genes from one population to another. From Grant and Grant 2010b.

Hybridization without loss of fitness should enhance variation (fig. 9.1). A sample of the parental species that contains a few hybrids should be more variable than one without them. The degree to which the expected enhancement is expressed and statistically detectable depends on the heritabilities of the traits, which are generally high, the mean differences between the species, which vary among the traits, and the frequencies and composition of the hybrids. We examined the expectations by first calculating variances for the samples of parental species that were known from pedigree data to lack hybrids, adding the hybrids and backcrosses to the samples, and then recalculating the variances. Additions were made to match the proportions in the population at large (Grant 1993). This entailed adding *fuliginosa* × *fortis* and *scandens* × *fortis* hybrids to *fortis*, but only *scandens* × *fortis* hybrids to *scandens*.

As expected, introgressive hybridization increases phenotypic and genetic variation in all traits (table 9.1). Hybridization affects traits differently according to how much the hybridizing species differ from each other. Thus increases are small in beak depth and width in *scan-*

TABLE 9.1

Proportional effects of including F_1 hybrids and backcrosses on estimates of heritabilities (h^2) and coefficients of phenotypic (CV_P), additive genetic (CV_A), and residual (CV_R) variation

	h^2	CV_P	CV_A	CV_R
G. scandens				
Mass	+33.9	+16.4	+32.2	+4.1
Wing length	+3.1	+15.0	+16.7	+43.0
Tarsus length	+3.3	+10.3	+10.2	+8.5
Beak length	+11.2	+22.4	+29.2	+2.7
Beak depth	−1.4	+3.6	+2.7	+4.8
Beak width	−9.0	+6.2	+1.0	+10.7
G. fortis				
Mass	+19.2	+37.9	+50.6	+18.9
Wing length	+40.0	+30.1	+53.9	+5.9
Tarsus length	+29.3	+15.4	+31.0	+4.8
Beak length	+7.8	+22.1	+26.8	+2.5
Beak depth	+11.0	+26.9	+33.8	+11.4
Beak width	+10.1	+28.4	+34.7	+12.0

Note: Two *fortis* × *scandens* F_1 families, five first generation backcross families and two second-generation backcross families, were added to the total of 149 *scandens* families lacking known hybrids and backcrosses; hybrids and backcrosses constituted about 5% of the total, which approximates their incidence in the population. The sample of *fortis* without hybrids was 167 families for which all four grandparents were known. To this were added 12 F_1 families, 13 families involving first-generation backcrosses to *fortis*, and another 15 second-generation backcross and other families that involved descendants of interbreeding (see fig. 9.2). Among the "other" category were two families in which one of the grandparents was a *fortis* × *scandens* F_1 hybrid and the other was a *fortis* × *fuliginosa* F_1 hybrid. Hybrid and backcross individuals constituted about 20% of the total, which approximated their frequency in the population in 1992. From Grant and Grant (2000a).

dens because the average size of these traits is almost identical in *fortis* and *scandens*. The increase in *scandens* beak length variation is much larger, and this corresponds to the large difference in averages between the species. Introgressive hybridization affects *fortis* variation more uniformly than *scandens* because it occurs mostly with *fuliginosa*, and *fuliginosa* and *fortis* differ to approximately the same extent in all six traits.

COMPARISON OF *G. FORTIS* AND *G. SCANDENS*

Additive genetic variances of the two species were approximately at equilibrium from 1976 to 1991 (fig. 3.4), with at most minor perturbations that could be attributed to intermittent selection. The level of variation is consistently higher in *fortis* than in *scandens*, and the same is true for alleles at microsatellite loci (fig. 9.3).

There are three reasons why *fortis* varies more than *scandens*. First, introgressive hybridization has a larger influence on variation in *fortis* than in *scandens*, partly because the frequency of hybridization was higher in *fortis* in the early part of the study (Grant 1993) and partly because *fortis* receives genes from two different species, whereas *scandens* receives genes from only one. Second, *scandens* breeding populations are consistently smaller than *fortis* populations, their genetically effective population sizes are smaller (<100 ignoring gene flow: Grant and Grant 1992a), and as a result they are likely to lose variation through drift at a higher rate than *fortis* (Houle 1989, Caballero and Keightley 1994).

The third reason is ecological. Even though there is little spatial heterogeneity on the island, the food supply is diverse and allows a partitioning of the niche of the generalist granivore species, *fortis*, by phenotypes that are partially specialized on seeds of different sizes (chapter 2). This individual specialization of a generalist species is depicted as model 1b in figure 8.1. The between-phenotype component of the total niche width of the population has been estimated to be 11% (Price 1987). In contrast to *fortis*, *scandens* feeds in a uniform manner on seeds and cactus products (figs. 2.17–2.20), and all members of the population are specialized in similar ways: model 2 in figure 8.1.

Differences in variation between the species are not fully accounted for by our estimates of introgression. For example, the coefficient of additive genetic variation of *fortis* beak depth in the absence of hybrids and backcrosses (6.39) is almost twice as large as the comparable coefficient of *scandens* even with their hybrids included (3.43). Thus hybridization has a short-term effect of increasing variation, but differential introgression in the short term is not large enough to explain the actual differences in variation between the species. The most likely explanation for the discrepancy is a greater cumulative (storage) effect of past hybridization in *fortis* than in *scandens*. *G. fortis* hybridized with *fuliginosa* more frequently than with *scandens* before the El Niño event of 1982–83 (chapter 8), and perhaps for a long time before the study began. Other possibilities, such as higher mutation rates in *fortis* than *scandens*, seem less likely.

Returning to the model in figure 9.1, *fortis* receives more genetic input from introgression than *scandens*; therefore its variation is at position V_2 or V_4, and *scandens* is at V_1 or V_3. Which of the alternatives is correct depends upon the long-term selection functions (S). However, *fortis* variation remains much higher than *scandens* when

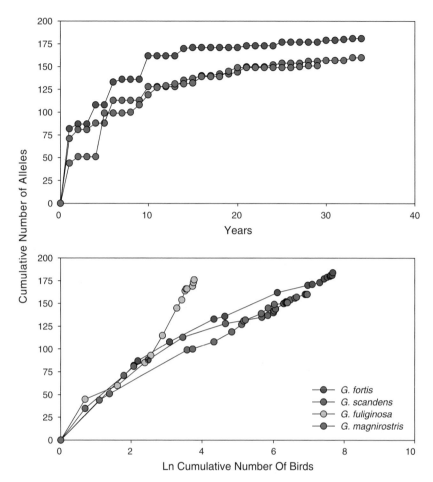

Fig. 9.3 Increase in allelic diversity at microsatellite loci with an increase in number of sampled individuals. **Upper**: Different asymptotes are approached by the different species. **Lower**: Addition of *fuliginosa* shows the effect of continuing immigration from a diverse source or sources. A sample of 24 *fuliginosa* from Santa Cruz Island alone has 168 alleles in total, almost the same as in the Daphne sample of 44 *fuliginosa* (176). Samples of the other three species are 2,169 *fortis*, 1,054 *scandens*, and 601 *magnirostris*.

hybrids are removed from the *fortis* sample but retained in the *scandens* sample. Therefore *fortis* cannot be at V_2 and must be at V_4. By a similar argument *scandens* must be at V_1. The equilibrial variation in figure 3.4 implies that *fortis* is subjected to stronger forces of stabilizing or directional selection than *scandens*, balancing the stronger genetic input from introgression. Evidence in support of this expectation is given in chapter 11.

Conspecific Gene Flow

Gene flow from conspecific populations on other islands does not alter the above comparisons. It is rare and contributes less to standing variation than does introgressive hybridization, but it may be sufficient to counteract losses from random genetic drift (Grant and Grant 2010b, Farrington and Petren 2011). Estimating it is not easy, because there are no diagnostic phenotypic traits possessed by *fortis* and *scandens* immigrants (fig. 9.4), nor do we have island-specific genetic markers. The best we can do is to use assignments tests with microsatellite DNA markers to identify probable immigrants among the *fortis* and *scandens* adults that lacked bands when first captured in the years 1976–98 (Grant and Grant 2010b; fig. 9.5). By this means we found that *fortis* and *scandens* receive genes more frequently by hybridizing than by breeding with conspecific immigrants. The net effect of hybridization upon morphological variation must be larger than implied by the difference in frequencies because species differ genetically from each other more than do populations of the same species.

Hybridization versus Mutation

Mutation and introgression both contribute to the maintenance of variation (fig. 9.1), but how do they compare quantitatively? To address this question, we obtained an average value for mutational variance from several studies of laboratory organisms reported in the literature (Grant and Grant 1994): it is $10^{-3} V_e$, that is, expressed as a fraction of the environmental variance of a trait. Genetic variance resulting from introgressive hybridization is of approximately the same order of magnitude as the environmental variance (10^{-0}), and is three or more orders of magnitude greater than variation introduced by new mutations per generation (10^{-3}). The reason for the great difference is that introgressive hybridization affects many more loci than mutation. This means that the effects of mutation on standing variation are dwarfed by effects of hybridization (fig. 9.1). Therefore it is reasonable to consider standing variation to be in equilibrium primarily because selection balances introgression.

Fig. 9.4 Immigrants and residents. **Upper**: Immigrant *fortis*, characterized by curvature of the beak, size, and genotype. **Middle**: Resident *fortis*. **Lower**: A pair of resident *magnirostris* derived from immigrants.

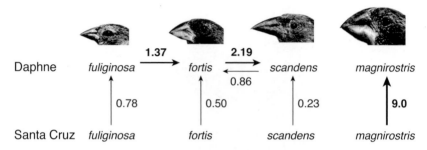

Fig. 9.5 Summary of gene flow through immigration and introgression on Daphne. Numbers refer to immigrants or hybridizing individuals per generation: those greater than 1.0 have been highlighted in boldface. Genes flow from *fortis* to *scandens* when the father of an F₁ hybrid sings the *scandens* song, and vice versa for gene flow from *scandens* to *fortis*. The species hybridize on Santa Cruz Island to an unknown extent (Grant et al. 2005). G. *magnirostris* hybridizes with *fortis* on Santa Cruz (Huber et al. 2007) but not on Daphne. A rare *fortis* immigrant is illustrated in figure 9.4. From Grant and Grant 2010b.

Correlations

Treating genetic variances of each trait alone as an indication of the potential to evolve ignores the fact that traits are correlated, so evolutionary change in one trait is affected by the correlation with other traits. Chapter 4 discussed this point in connection with natural selection. The six traits are intercorrelated in a similar way in all three species: strongly and positively. However, *scandens* differs from the other two species in beak proportions, and in the fact that beak length is relatively weakly correlated with beak depth and width. Correlations are important: when hybrids and backcrosses are formed, the correlations between pairs of traits depend upon the allometric relations of the hybridizing species (fig. 9.6).

Phenotypically, beak depth and width are strongly and positively correlated in all species and hybrid groups (Grant and Grant 1994). Introgressive hybridization has no effect on the beak correlations of *fortis* but does weaken the correlation between length and depth in *scandens*, by 11.7%. Genetically, all traits are positively correlated, strongly, in all species. Introgression strengthens the genetic correlations, more so in *fortis* than in *scandens*, with one difference between them being particularly striking. Addition of F₁ hybrids and backcrosses increased the genetic correlation between depth and width by 50.7% in *fortis* but only by 4.5% in *scandens* (Grant and Grant 2000a).

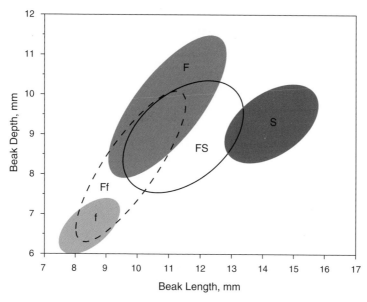

Fig. 9.6 Beak proportions of *fortis* (F), *scandens* (S), and *fuliginosa* (f), together with *fortis* × *scandens* hybrids (FS, ellipse with thick line) and *fortis* × *fuliginosa* hybrids (Ff, ellipse with broken line). Hybrids were identified from pedigrees as F_1, but some were probably backcrosses. Density ellipses approximate 99% of individuals. Based on Grant and Grant (1994).

Evolutionary Potential

Enhanced variation facilitates directional evolutionary change, subject to constraints arising from genetic correlations between characters (chapter 4). These constraints become stronger when species with similar proportions hybridize, thereby rendering evolution in a new direction less, not more, likely. Only with species of transposed allometries, such as beak dimensions of *fortis* and *scandens*, are genetic correlations likely to be weakened or eliminated by hybridization. With enhanced genetic variation but reduced genetic constraints, a population has the potential to evolve in new directions more easily than in the absence of introgression. Thus the creation of individuals in novel genetic and morphological space could provide the starting point of a new evolutionary trajectory that is not easily reached by mutation alone (Miller 1956, Svärdson 1970, Lewontin and Birch 1966, Van Valen 1976, McDade 1990, Chiba 1993) or by directional selection acting on constrained genetic variation (Grant and Grant 1994, 2000a).

Discussion

In the previous chapter we introduced the topic of morphological variation with an adaptive hypothesis to explain why some populations vary
more than others in continuously varying traits such as beak or body
size. The hypothesis proposes that high variation is maintained by diversifying natural selection in a heterogeneous environment. It is analogous to adaptive radiation theory. Both assume genetic variation. Both
invoke natural selection from diverse ecological pressures in a heterogeneous environment. In both cases different units have different ecological niches as a result of those pressures; the units are species in an
adaptive radiation and morphologically extreme individuals in adaptive
population variation.

However, variation is not adaptive in the model of figure 9.1. Variation that is elevated by introgressive hybridization is not maintained by
diversifying selection; on the contrary increases in variation are opposed by stabilizing or directional selection. In reality, too, selection is
not diversifying but directional (appendix 9.1). Either small or large
birds may be at a selective advantage but not at the same time. In fact
the selective advantage of one extreme size class arises from the selective disadvantage of the other.

A second reason for not accepting the adaptive variation hypothesis
is that extreme individuals of either species do not have unique diets.
There is no component of the dry-season diets of either *fortis* or *scandens* that is exploited only by extreme individuals (fig. 9.7), and therefore nothing comparable to the unique ecological niches of species in
an adaptive radiation.

Nonetheless the increase in population variation caused by introgressive hybridization is adaptive in a more restricted sense. The *fortis*
population exploits the environment more fully as a result of being
more variable. The evidence for this is the greater feeding efficiency of
fuliginosa × *fortis* hybrids and backcrosses (small phenotypes) when
feeding on small and soft seeds compared with nonhybrid *fortis* (Grant
and Grant 1996b): hybrids consumed seeds faster. The difference between phenotypes supports model 1 in preference to model 2 in figure
9.7 as a representation of feeding efficiency in relation to seed size and
hardness. Moreover *scandens* × *fortis* hybrids were more efficient than
fortis when feeding on *Opuntia* seeds, although less efficient than
scandens. The hybrids are generally at the small end of the body and
beak size spectrum (fig. 9.6). At the opposite end of the size spectrum
the largest individuals are the most efficient in extracting seeds from
Tribulus fruits (figs. 2.15 and 4.7), although we are unable to say if they
are large because of hybridization with *scandens*.

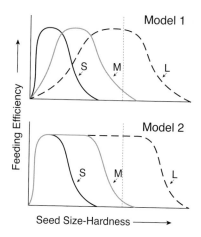

Fig. 9.7 Different ways in which feeding efficiency varies among small (S), medium (M), and large (L) phenotypes in relation to the size and hardness of seeds in the diet. The key difference between the models is in how small and soft seeds are exploited by the different phenotypes. *G. fortis* phenotypes and seeds on Daphne span the range to the left of the broken line. On other islands seeds (Abbott et al. 1977) and beak sizes (fig. 2.1) are larger. For example, on Santa Cruz only the largest *fortis* are able to crack the seeds of *Cordia lutea* ($DH^{1/2} > 13.0$).

Consideration of diets leads to the question of whether temporal variation in environmental conditions is a factor favoring a high level of morphological variation, depending on the exact nature of the variation, as has been suggested sometimes for other systems (e.g., Sasakai and Ellner 1997, Zhang 2012). Annual variation in food supply affects morphological variation in *fortis* more than *scandens*. Extreme excursions from long-term average morphology in *fortis*, caused by selection in either direction (figs. 7.3 and 7.6), are rare, and this gives the impression of long-term relaxation of selection or environmental tolerance of extreme individuals. However, this does nothing to promote, let alone generate, morphological variation, because new food items are not added to the diet when selection does or does not occur. The same range of resources is exploited at all times.

If not temporal variation, then how about spatial variation? Strong spatial variation in habitat features can promote diversifying selection, as has been shown with freshwater fish (e.g., Robinson and Schluter 2000, Nosil and Reimchen 2005), but such selection is not possible on Daphne, because habitat heterogeneity there is far from comparable to the vertical stratification of habitats in lakes in degree or extent.

In conclusion, introgressive hybridization increases the level of morphological variation in the *fortis* population, and enhanced feeding efficiency of the morphologically extreme individuals maintains it.

Summary

To understand population variation in nature, two questions must be addressed: what is the origin of the variation, and what maintains it?

The first is a question of genetics, and the second is a question of ecology. In this chapter we concentrate on the role of one genetic factor, introgressive hybridization, and one ecological factor, food. *G. fortis* hybridizes with two species, immigrant *fuliginosa* and resident *scandens*, whereas *scandens* hybridizes only with *fortis*. We explore the morphological consequences of hybridization by comparing the features of a combined sample of parental species and hybrids with those of the parental species alone. For this we use observations of breeding, pedigrees, morphological measurements, and microsatellite DNA measures of genetic effects. Hybrids and backcrosses by themselves were found to be intermediate in mean morphology between the respective parental species, and their variation is highly heritable. Introgressive hybridization results in an increase in phenotypic and genetic variances, and a moderate increase in heritabilities. Further increases in variation are opposed by selection. *G. fortis* varies more than *scandens* in all body and beak traits, largely as a result of receiving a more diverse genetic input from hybridization, and varies more ecologically. Enhanced variation facilitates directional evolutionary change, subject to constraints arising from genetic correlations between characters. These constraints become stronger when species with similar proportions hybridize, such as *fortis* and *fuliginosa*, thereby rendering evolution in a new direction less likely. *G. fortis* and *scandens* have different beak allometries—they are transposed and have different slopes—and interbreeding of these two species tends to weaken the genetic correlations, thereby increasing their evolutionary scope.

PART 3

Hybridization and Speciation

Long-Term Trends in Hybridization

Hybridization and backcrossing to one or both of the
parent types can result in incorporation of alleles from
one taxon into the gene pool of the other.

(Harrison 1993, p. 11)

Introgressive hybridization may, then, be a passing stage
in the process of species formation. On the other hand,
the adaptive value of hybrids may be as high as that of
their parent; introgressive hybridization may lead to
obliteration of the differences between the incipient
species and their fusion into a single variable one, thus
undoing the result of the previous divergent development.

(Dobzhansky 1941, p. 350)

Introduction

THE MAIN IMPLICATION of high hybrid fitness and continuing hybridization described in the previous two chapters is that interbreeding species should converge toward an intermediate morphology, and, if unchecked, this convergence would result in complete fusion of the previously reproductively isolated populations (Dobzhansky 1941, Clarke et al. 1998, Grant et al. 2004, Taylor et al. 2006, Grant and Grant 2008b, 2010b, Seehausen et al. 2008, Behm et al. 2010, Webb et al. 2011). It would then represent speciation in reverse, or despeciation of sister species such as *fuliginosa* and *fortis* (Grant et al.

2004). A better metaphor for the situation on Daphne is reproductive (Alerstam et al. 1978, Grant 1986) or genetic (Carney et al. 2000) absorption: *fuliginosa* are absorbed into the *fortis* population by hybridizing and backcrossing solely to *fortis*, with rare recurrent immigration replenishing *fuliginosa* numbers. *G. fortis* and *scandens* are not sister species; nevertheless, regardless of whether this could be regarded as a process of despeciation or not, they serve as a model for understanding the dynamics of recently formed species with incomplete barriers to gene exchange. A similar situation exists on the island of Genovesa with the interbreeding of a different set of three *Geospiza* species, *difficilis*, *conirostris*, and *magnirostris* (Grant and Grant 1989). In both cases the dynamics are entirely natural, in contrast to the several cases of speciation in reverse that have been attributed to human disturbance of the environment such as eutrophication, in fish (Seehausen et al. 2008, Behm et al. 2010, Webb et al. 2011, Vonlanthen et al. 2012) and planktonic *Daphnia* (Keller et al. 2008, Brede et al. 2009).

In this chapter we explore the implications of convergence through hybridization. We ask to what extent did the predicted convergence occur, was it sustained, and what effect did it have on morphological traits and their relationships? The chapter differs from the previous two in encompassing the full 40 years of study, which requires a meshing of heterogeneous data. Up to 1998 we were able to use pedigree information to assign individuals to breeding populations (species), whereas afterward we had to rely on song, morphology, and genetics. We begin by discussing the problem of individual assignments, and our solution, and then consider the effects of hybridization on phenotypic means, variances, and the differences between species over the long term. In the following chapter we discuss changes that take place as a result of selection over 40 years. These occur in dry years of no breeding, and for that reason can be separated from effects of hybridization: the first is associated with mortality, the second with reproduction. The remaining chapters use information on breeding biology, feeding ecology, and morphological evolution to consider how speciation occurs. The present chapter therefore represents an important transition from detail-rich ecology to a broad view of long-term trends.

A Question of Identity

In 1973–76 *fortis* and *scandens* could be readily distinguished by their beak measurements. Distributions of their measurements on two beak axes were discretely separated (fig. 10.1). However, the continuing in-

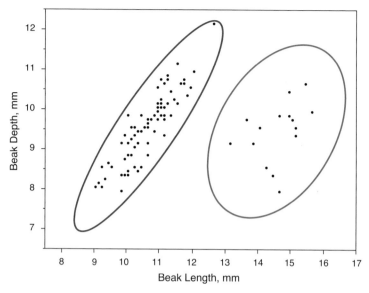

Fig. 10.1 Distributions of *fortis* (blue: $n = 73$) and *scandens* (red: $n = 16$) beak measurements in early 1973, drawn with 99% density ellipses.

trogressive hybridization in the first 25 years brought the two species closer together in morphology and genetics, blurring the distinction between them. In the absence of pedigree information, convergence of the species gave rise to a problem in assigning a few individuals to species, especially in the last half dozen years. Our solution was pragmatic (box 10.1).

BLURRING OF GENETIC DISTINCTIONS

Hybridization blurs the genetic distinction between species because F_1 hybrids are neither one species nor another but genetic mixtures of the two. They belong to the population of one parental species or the other according to which one they choose to breed with (box 8.1). In changing the frequencies of alleles in the receiving population they reduce the difference between the populations. Successive backcrossing to one of the parental species and random mortality dilutes the contribution of introgressed genes and results in loss by genetic drift (and selection). In species like the finches that may have been exchanging genes for many generations loss due to drift tends to be counterbalanced by replenishment through introgression (and selection). New combinations of genes will be formed. For example an individual may possess a few

Box 10.1 Assigning Individuals to Species

Morphological criteria were used in the field to classify individuals to species (box 8.1). These criteria became increasingly difficult to apply with complete confidence because the species gradually converged as a result of hybridization. When genetic (microsatellite) data became available after 1987, we used them to identify species and hybrids in three steps. First, an Admixture model in STRUCTURE (Pritchard et al. 2000, 2007) was run to estimate the probability of each individual belonging to *fortis*, *scandens*, or *fuliginosa*. All captured birds were genotyped and therefore included in the analysis. Twenty-four individuals of *fuliginosa* from Santa Cruz were included as a standard for comparison with Daphne birds. Forty-seven Daphne birds (and all Santa Cruz *fuliginosa*) were assigned to one cluster at probabilities of 0.900 or higher. Not all agreed with morphological criteria (box. 8.1), and three were transferred from *fuliginosa* to *fortis* because they were larger than all *fuliginosa* on Santa Cruz Island.

In a second STRUCTURE analysis without *fuliginosa* all birds were predefined as *fortis* and *scandens*, and the program assigned each individual to one species or the other by the majority rule ($p \geq$ 0.500). Third, to identify hybrids the Admixture model was run with all individuals of all three species ($n = 3{,}125$). Each individual was entered as one of the three predefined species, and one- or two-generations-back options were chosen to allow hybrid identification. Results of these runs were lists of birds with probabilities of assignment to *fortis*, *scandens*, *fuliginosa*, and categories with mixed ancestry in parental or grandparental generations. Generally these last two are F_1 and B_1, that is, backcross generation 1. Individuals were considered to have mixed ancestry when support for the species designation fell below $p = 0.9$. Hybrids among the mixed ancestry categories were operationally defined by the majority rule. The totals were 236 *fortis* with *fuliginosa* ancestry in the parental generation, 44 *fortis* with *scandens* ancestry, and 83 *scandens* with *fortis* ancestry. Three immigrants and the unusual birds described in chapter 13 were excluded from these analyses.

alleles of another species inherited from a great grandparent that hybridized. By breeding with another individual of similar genetic constitution it may, through chance Mendelian segregation of alleles, produce an offspring with strong genetic features of that other species, and be assigned to that species by assignment tests.

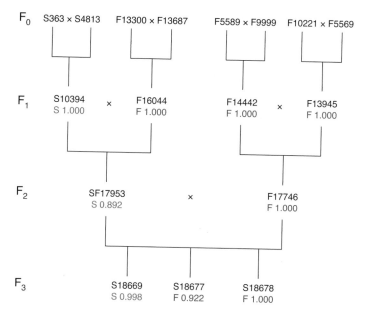

Fig. 10.2 Genetic results of hybridization: heterogeneity among sibs of the F_3 generation. Parentage of all F_2 and F_3 generation hybrids was confirmed genetically. Genetic assignments were made by the program STRUCTURE, with an Admixture model. Only *fortis* and *scandens* (not *fuliginosa*) were entered (box 10.1). Probabilities of belonging to *fortis* (F) or *scandens* (S) populations are shown in color. Figure 8.8 shows the songs of 10394 and the father of his mate.

Two examples from the pedigrees illustrate this effect of latent alleles that come together in chance combinations. In the first example, genetically assigned hybrid offspring (box 10.1) are formed from the genomes of parents that are both assigned genetically to one species with a probability of greater than 0.98. There are five *fortis* × *fuliginosa* hybrids in the pedigree with such *fortis* parents (paternity confirmed genetically). The second example illustrates the possibility of two species being produced from the breeding of two hybrids. A *fortis* × *fuliginosa* backcross to *fortis* (B_1), female 15779, bred with a male on a territory adjacent to her natal territory. The male was cuckolded by her F_1 hybrid father (10376). Assignment probabilities to *fortis* of daughter and father were 0.873 and 0.878 respectively, yet the assignment probabilities of the two offspring, 17637 and 17638, were 0.924 and 0.020, leading to the conclusion that the first is a *fortis* and the second is a *fuliginosa* (see also fig. 10.2)! The example illustrated in figure 10.2 shows that such genetic anomalies are produced also by *fortis* breeding with *scandens*.

These findings have two implications, one methodological, the other biological. The first is that genetic characterization of individuals does not always reflect the population- or species-based pedigree. A few errors of assignment may be made whether one uses genetic criteria (box 10.2) or morphological criteria (box 10.1). The biological consequence is the potential fusing of two populations into one and the loss of their separate identity.

BLURRING OF MORPHOLOGICAL DISTINCTIONS

The size and proportions of the beaks of *fortis* and *scandens* at the start of the study provide a point of reference for the changes that took place afterward. In 1973 it was possible to draw a line of demarcation down the middle of the gap between the distributions of the two species. This is illustrated with 1975 data in figure 10.3 because the samples sizes are larger than in 1973 and the frequency distributions re-

Box 10.2 Genetic Hybrids

Darwin's finches have 38 pairs of chromosomes (Jo 1983). An individual receives a random sample of one of each of the 38 chromosomes from each parent. By chance an offspring may receive more alleles indicative of the species of one parent than of the other. Our 16 microsatellite markers—14 autosomal and 2 sex-linked—are on 15 at most of the 38. Thus the chances of a moderate parent-of-origin bias in the genetic constitution of some individuals are quite high against a background of small differences between populations in the frequency of most alleles and a scarcity of fixed differences. The average difference in allele frequencies between *fortis* and *fuliginosa* is only a third of the difference between *fortis* and *scandens* frequencies.

Concordance between genetic and pedigree assignments is 66% for *fortis* with *scandens* ancestry ($n = 53$ offspring) and 69% for *fortis* with *fuliginosa* ancestry ($n = 52$ offspring). The low values are to be expected with closely related species (Vähä and Primmer 2006). Broader sampling of markers on more chromosomes would reduce the discrepancies, but given the available genetic markers, we must recognize that genetically identified hybrids are not necessarily pedigree hybrids (F_1 and B_1). For this reason we refer to them as genetic hybrids.

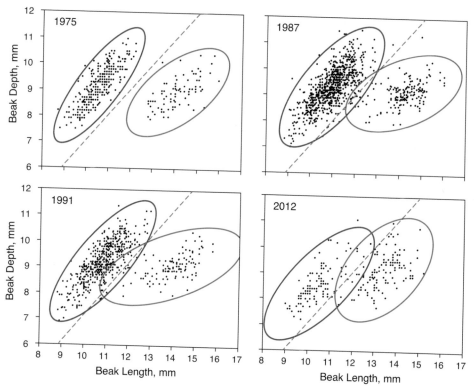

Fig. 10.3 A line of demarcation is drawn by eye midway between the beak distributions of *fortis* (left: *n* = 316) and *scandens* (right: *n* = 87) in 1975. Ellipses are 99% densities as in figure 10.1. The first sample in 1973 (illustrated in fig. 10.1) shows the same pattern, but sample sizes are much smaller. The other three plots show the progress of the species toward overlapping the line of demarcation as well as each other. Note the change in orientation of the *scandens* ellipses.

mained the same over the three years (fig. 10.1). The ellipse enclosing the predicted 99% of the *fortis* measurements crossed the line of demarcation for the first time in 1978, and by 1987 the ellipses of both species had crossed the line and overlapped each other. Overlap then increased, with the result that by the end of the study the original line of demarcation ran right through the middle of the overlap zone (fig. 10.3): each species had invaded the morphological space of the other. *G. fortis* and *fuliginosa* were not clearly separable right from the beginning (box 8.1), and they also penetrated further into each other's morphological space through introgressive hybridization (fig. 9.6).

Morphological Convergence

G. fortis and *scandens* became more similar to each other in all dimensions. Convergence was strongest and most regular in beak shape (fig. 10.4). Both species changed in beak dimensions, to approximately the same degree but in different ways. The average beak depth of *fortis* was 9% smaller at the end of the study (2012) than at the beginning (1973), whereas beak length had decreased by only 2% on average. The opposite was true of *scandens*; average beak length was 8% smaller at the end, but beak depth was only 1% smaller. Episodic selection and random sampling effects contribute to some of the year-to-year changes but do not produce the systematic trend illustrated in figure 10.4. Instead, the explanation for the regular increase in similarity—an hypothesis based on observed hybridization (chapters 8 and 9)—is that it was caused by the gradually accumulating effects of introgressive hybridization.

Differences between the species are most pronounced on a beak-shape axis (PC2-beak: length/depth or width). By the end of the study

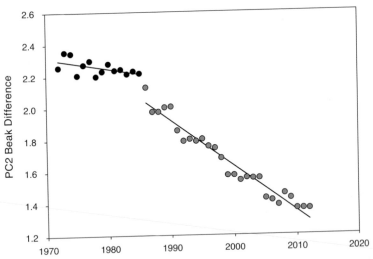

Fig. 10.4 Convergence of *fortis* and *scandens* in beak shape (PC2-beak) from 1986 onward. This is attributable to hybridization in 1983 and 1984 with subsequent backcrossing. To emphasize the contrast before (black dots) and after (purple dots) hybridization, two ordinary least-squares regression lines are fitted. This is done for heuristic and not statistical purposes since sequential points are not independent. By simple extrapolation, the species will be identical morphologically in the year 2057.

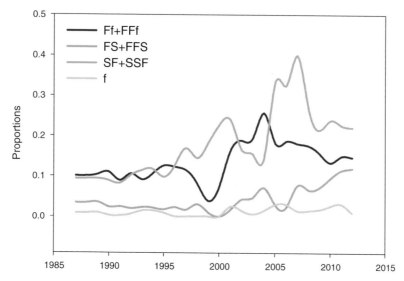

Fig. 10.5 The frequency of *fuliginosa* (f) and three categories of hybrids (F₁ and backcrosses combined). Frequencies of *fuliginosa* (f), *fortis* × *fuliginosa* (Ff + FFf), and *fortis* × *scandens* (FS + FFS) hybrids are expressed as proportions of the population of *fortis* in which they bred. Frequencies of *scandens* × *fortis* (SF + SSF) hybrids are expressed as proportions of the population of *scandens* in which they bred. Cubic splines are fitted to the data. Sample sizes are in table 10.1.

the difference between the species was only 57% of the difference at the beginning. If the documented convergence continues at the same rate, the two species will become identical in 43 more years, in the year 2057. This extrapolation assumes all else is equal, including an unchanging environment, which seems highly unlikely. The dangers of such extrapolation are well known (appendix 10.1).

Frequencies of Hybrids

The hypothesis that convergence was caused by introgressive hybridization assumes that hybrids were present in most if not all years of breeding. This is indeed so (fig. 10.5). Genetically identified hybrids were present among newly captured birds in every year from 1987 onward, and in some of the preceding years back to 1978. Average frequencies of hybrids (F₁ and backcrosses combined) over a period of 26 years are 0.111 for Ff (*fortis* × *fuliginosa*) and 0.038 for FS (*for-*

tis × scandens) in the fortis breeding population and 0.154 for SF (scandens ×fortis) in the scandens breeding population. Overall 14.9% of fortis and 15.4% of scandens had mixed ancestry as detected by our assays.

Three features stand out from the temporal pattern of hybrid frequencies. First, hybrids increased in frequency from about the year 2000 onward. Second, F_1 hybrids formed by fortis breeding with scandens backcrossed to scandens more frequently than to fortis in every year, to judge from the relative frequencies. This stands in contrast to the observed higher frequency in the opposite direction at the beginning of the study (chapters 8 and 9). Third, the frequency of fuliginosa ×fortis hybrids was consistently intermediate between the frequencies of the two classes of fortis × scandens hybrids. All three features match what we know from pedigrees from 1987 to 1998.

New, genetically identified fuliginosa appeared in the samples in most years. They were either immigrants or hybrids with a strong complement of fuliginosa genes (box 10.2). The maximum was 10, in 1993, but most of them never bred. They were always rare, and never exceeded 3.1% of the fortis population.

Morphological Variation

The hypothesis of introgressive hybridization without selective penalty predicts an increase in morphological variation with continuing hybridization (chapter 9). In the post-pedigree period of the study an increase in frequency of hybrids in the samples (fig. 10.5), and the inferred increase in frequency of hybridization in the last decade, lead us to expect an increase in morphological variation. We use coefficients of variation (100 × s.d./mean) to assess this because average size of several traits changed across years, especially in the latter half of the study (chapter 7).

In agreement with the prediction, variation in the composite measure of beak shape (PC2-beak) increased systematically in both fortis and scandens (fig. 10.6), and at about the same rate. G. fortis started and finished about twice as variable as scandens. The difference between species in magnitude of the variation (different intercepts) is explicable in terms of two sources of introgression into the fortis population, from scandens (larger) and fuliginosa (smaller), but only one into the scandens population, from fortis (chapters 8 and 9). Another likely factor is greater introgression into the fortis population (from fuliginosa) than into the scandens population early in the study (chapter 8).

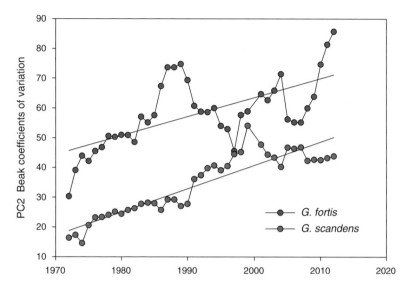

Fig. 10.6 Increases in variation in beak shape (PC2-beak) in *fortis* (blue) and *scandens* (red) across years. Ordinary least-squares regression lines of best fit are shown for heuristic and not statistical purposes.

Similarity in the rate of increase of variation in the two species (equal slopes) is expected from the approximately equal frequencies of birds with mixed ancestry during the study (p. 192), combined with minor genetic contributions from conspecific immigrants (fig. 9.5). Annual variation in individual beak dimensions fluctuated, in both species, but combined to produce the trend in shape. Noteworthy in figure 10.6 is a strong increasing trend in *fortis* variation after 2007 to the point where variation in the last three years exceeded all preceding values. The terminal increase corresponds with an increase in genetic input from *scandens* (fig. 10.5).

Annual values of variation fluctuate around the trend lines of best fit, more strongly for *fortis* than for *scandens*. Stochasticity contributes to the fluctuations because the numbers of hybridizing birds are always low. In addition hybridization varies in time. According to the pedigrees introgression occurred earlier in *fortis* than in *scandens*. Figure 10.6 shows two periods of substantial increase in *fortis* variation in the early part of the study, the first in the 1970s before pedigree data had been obtained, and the next in the 1980s when backcrossing to *fortis* was observed to be higher than in *scandens*. Two small increases in *scandens* variation occurred at times of known hybridization, 1991 and 1998.

With hybridization followed by backcrossing and introgression caus-ing an increase in variation, there should be a positive association be-tween the coefficient of variation and the number of hybrids and back-crosses in the samples from 1987 to 2012. The two are strongly correlated in *scandens* ($r = 0.643$, $p = 0.0004$, $n = 26$), whereas the *for-tis* correlation is on the borderline of significance ($r = 0.385$, $p = 0.0520$, $n = 26$).

A third test employs the technique of hybrid removal. In chapter 9 we compared samples with and without hybrids and known back-crosses to quantify the effects of hybrids on genetic and phenotypic variation. This was done in the early part of the study with pedigree data, when *fortis* hybridized more with *fuliginosa* than with *scandens*. By adding hybrids to samples that lacked them, we found evidence for an increase in both phenotypic and genetic variation in proportion to the magnitude of the difference between the traits of interbreeding spe-cies. Here we repeat the procedure with phenotypic data from 1987 onward, when hybrids could be identified genetically, but this time by removing hybrids from the total samples.

The manipulation of removing hybrids from the analyses had three notable effects on the trends in coefficients of variation of beak shape (fig. 10.7). First, *scandens* variation was reduced across the 26-year period. This agrees with expectation. Second, despite the large effect, variation still increased over the period, albeit at a lower level. Third, *fortis* variation was scarcely affected. The second and third conse-quences do not contradict the hybridization hypothesis, but they do show that the trends are not produced solely by recent, identifiable hybridization. After 1993 genotyped individuals rarely amounted to more than 50% of each breeding population, so we suspect that the samples of genotyped birds, after the known hybrids had been re-moved, contained alleles from unidentified hybrids.

ALLOMETRY

As described in chapter 9, beak dimensions of adults are correlated positively with each other in all three species, but they vary in how they are correlated as well as in the strength of the correlations: their allometric relations differ. In a sample of measurements of adult *fortis*, beak depth and length increase at roughly the same rate as birds get larger, and the same applies to *fuliginosa*. The dimensions are approxi-mately isometric. In contrast, *scandens* beak length increases at a faster rate than beak depth; the two dimensions scale with size allo-metrically. When two species with different allometries hybridize, their

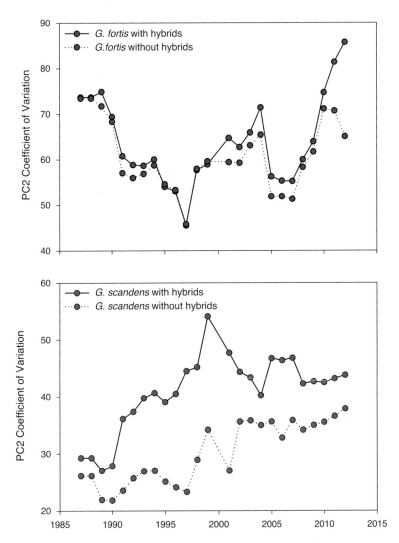

Fig. 10.7 The effects of hybrids and backcrosses on coefficients of variation of beak shape are much stronger in *scandens* (below) than *fortis* (above). This is shown by the difference between samples with hybrids (solid line) and without hybrids (broken line). Adapted from figure 10.6 for the period when hybrid frequencies (including backcrosses) were quantified (fig. 10.5). Introgressive hybridization also creates a small degree of skewness in the *scandens* frequency distributions as shown by the same removal procedure. Grant and Grant 2002a.

allometries change. Interbreeding of *fortis* and *fuliginosa* should have little effect on the relationship of beak length and depth, because the two species lie approximately on the same line of allometry. Interbreeding of *fortis* and *scandens*, on the other hand, leads to an allometric change because the slopes of their lines of allometry are different and because one is displaced (transposed) relative to the other on the beak-length axis. This is a strong prediction, but the direction of change is not easy to predict, because it depends on the particular individuals that interbreed.

The predicted change is observed in figure 10.3. While the orientation of the *fortis* distribution of measurements on two beak size axes remained approximately constant, the *scandens* distribution changed in two ways: initially the slope became shallower; then it became steeper. The first change is interpretable in terms of pedigree data in 1991, when the first extensive backcrossing occurred from *scandens* × *fortis* hybrids to *scandens*. The second shift occurred when we had no pedigree data, and therefore we do not know how it occurred. It resulted in the generation of individuals in previously unoccupied morphological space. Notice in figure 10.3, for example, that in 1991 there were no *scandens* with beak lengths of 13–14 mm that had beak depths greater than 9.5 mm, but in 2012 there were plenty. The most likely explanation is a change in mating patterns of hybridizing birds, from a disproportionate pairing of small birds (chapter 9) to a pairing of large birds.

Allometric slopes changed unidirectionally in *fortis* and bidirectionally in *scandens* (fig. 10.8). The *fortis* change is consistent with a small genetic influence of *scandens*. However, removal of the F_1 hybrids and backcrosses from both *fuliginosa* and *scandens* (not shown in the figure) had almost no effect upon the trend. The biphasic shift in *scandens* allometry is consistent with an altered pattern of hybridization explained in the previous paragraph. Removal of hybrids from the *scandens* sample has the same effect as in figure 10.7, dampening but not eliminating the trend. Remarkably the allometric slopes of the two species, so different in the 1970s, were identical in 2012.

Genetic Convergence

Considering the whole period for which microsatellite data were available, both species underwent genetic changes. The microsatellite loci are presumed to be selectively neutral, and therefore serve as markers for random drift. Gradual, long-term, and unidirectional increases in allele frequencies could be caused by genetic drift or close linkage with

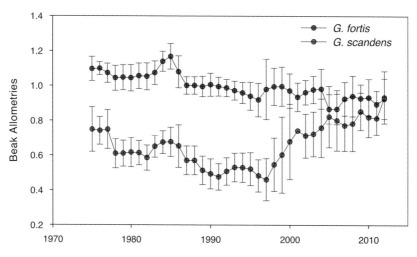

Fig. 10.8 Reduced major-axis slope coefficients for beak depth regressed on beak length in each year from 1975 onward. Samples of *scandens* (red) in 1973 and 1974 were too small for reliable analysis. *G. fortis* estimates are in blue. Coefficients are shown with 95% confidence limits.

selected genes, but also by continuing introgression. The best candidates for introgressed alleles start out as common in one population and initially rare or absent in the other population, and gradually rise in frequency in the latter. Temporal dynamics of two are illustrated in figure 10.9.

Four long-term trends are evident. First, introgressive hybridization between *fortis* and *scandens* is episodic on a scale of decades. This is indicated by initial scarcity of *fortis* alleles in the *scandens* population (fig. 10.9), and lack of morphological convergence prior to the mid-1980s (fig. 10.4). As suggested on page 192, *fortis* and *scandens* may not have been exchanging genes much if at all before the study began and in the first few years. Consistent with this suggestion, *scandens* experienced inbreeding depression that was counteracted by interbreeding with *fortis* (fig. 8.15), whereas no such effect was detected in *fortis*, probably because *fortis* and *fuliginosa* had been hybridizing.

Second, *fortis* and *scandens* species became increasingly different in genetic constitution from their respective populations in 1982 (fig. 10.10 upper). At the same time each became more similar to the starting (1982) genetic constitution of the other (fig. 10.10 lower). These trends reflect bidirectional gene flow, and they are stronger in *scandens* than in *fortis* because change in *fortis* with respect to *scandens* was dampened by genetic input from *fuliginosa*.

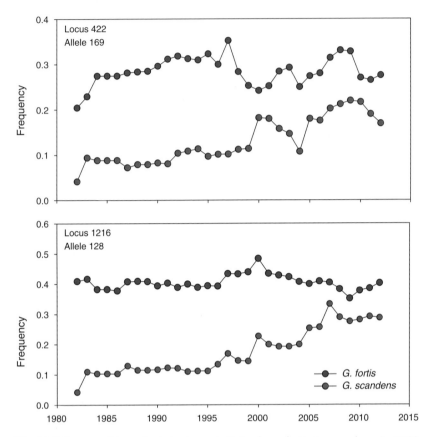

Fig. 10.9 Two candidates for introgressed alleles from *fortis* to *scandens*. In 1982 they were present in *fortis* but almost absent in *scandens* (single individuals only).

Third, the species converged genetically (fig. 10.11). This happened most rapidly between 1998 and 2007. The trend was slightly reversed in the last five years, but we are more impressed by the overall congruence between genetic and morphological convergence over the 30 years (fig. 10.11 lower).

Fourth, average multilocus heterozygosity remained stable in both *fortis* and *scandens* (fig. 10.12), even in the face of gene exchange, though with a possible slight increase after 2004 (fig.10.5). Stability of the hybridizing species contrasts with a decline in heterozygosity in the nonhybridizing *magnirostris* at a time when immigration rate declined and inbreeding may have increased (chapter 11; also Markert et al. 2004). Hybrids are more heterozygous than their parental species

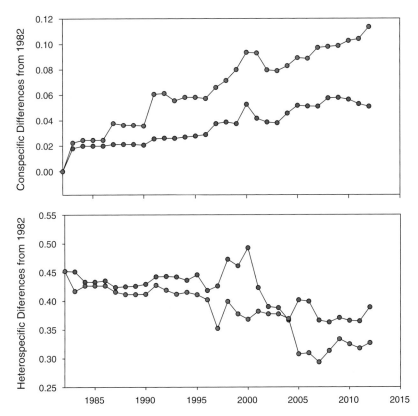

Fig. 10.10 **Upper**: *G. fortis* and *scandens* became increasingly different in genetic constitution from their own respective populations in 1982, as measured by Nei's genetic distance (*D*) with 14 microsatellite loci based on the frequency of shared alleles. The two species changed at different rates. **Lower**: The two species' genetic constitutions became increasingly similar to each other's in 1982.

(table 10.1). Being relatively uncommon, removing them from the total samples results in only a slight reduction in average heterozygosity (fig. 10.12).

Two Species or One?

Are *fortis* and *scandens* now one species? The answer to when two species become one has to be arbitrary under the biological species concept (de Queiroz 1998, Harrison 1998, Helbig et al. 2002) because

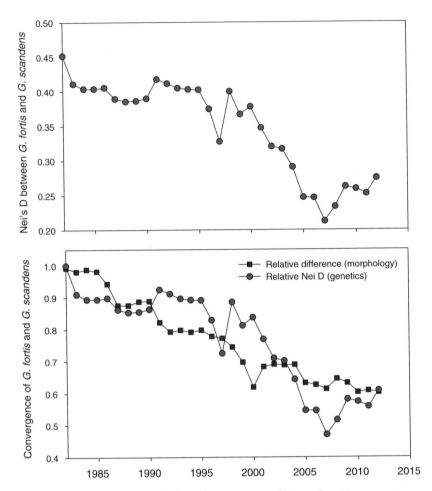

Fig. 10.11 Genetic and morphological convergence. **Upper:** Genetic convergence. Genetic distances (*D*) between *fortis* and *scandens* are calculated for each pair of annual samples of allele frequencies. **Lower:** Morphological convergence from figure 10.4 has been superimposed on the genetic convergence in the upper figure after standardizing both by giving a value of 1.0 to the difference between the species in each trait in 1982. Extrapolations to the x-axis give the expected year when the species will be identical, assuming unchanging rates of convergence. This is 2057 for morphological convergence and 2049 for genetic convergence.

despeciation, like speciation, is a gradual and continuous process and is not a categorical switch from one state to another. We consider them still to be two species because they constitute two populations separated almost completely by song. But clearly they are heading in the direction of one species, on Daphne.

TABLE 10.1

Differences in mean heterozygosities at 14 autosomal microsatellite loci between the three species and three classes of hybrids (F_1 and backcrosses combined)

Group	Sample size (n)	Mean ± SD
fortis (F)	1,891	0.661 ± 0.127
fuliginosa (f)	47	0.696 ± 0.152
scandens (s)	821	0.658 ± 0.127
Ff + FFf	255	0.745 ± 0.107
FS + FFS	86	0.737 ± 0.128
SF + SSF	149	0.717 ± 0.121

Note: In each case the breeding population to which each group of hybrids belongs is indicated by the first initial. SD = standard deviation.

Plumage and Behavior

In this and the preceding chapters we have focused on morphological effects of introgressive hybridization. There are other possible effects that deserve a brief mention. The first, behavior, is conjectural, whereas the second, plumage, is quantitative and more substantive.

In the first two decades only *fortis* approached us while we occupied caves for meals or for banding. This changed, gradually, and by the end of the study *scandens* were just as likely as *fortis* to approach or even hop onto our boots or shoulders. The change in *scandens* behavior could be due to copying *fortis*, to inherited factors through introgressive hybridization from about 1991 onward, or both. G. *magnirostris* never showed the same behavior on Daphne, which is curious because on Genovesa they were regular camp birds (Grant and Grant 1989). The inheritance of tameness is an unexplored topic with these birds.

Male ground finches acquire increasing amounts of black in their plumage with successive molts (Grant 1990). G. *scandens* acquire fully black plumage faster than *fortis*. The difference may be due to inherited factors that display dominance, because backcrosses to either *fortis* or to *scandens* acquire fully black plumage as fast as *scandens* and faster than *fortis* (appendix 10.2).

Discussion

Some changes take place slowly. Two decades of measurement were needed for us to be confident that population variation in beak shape

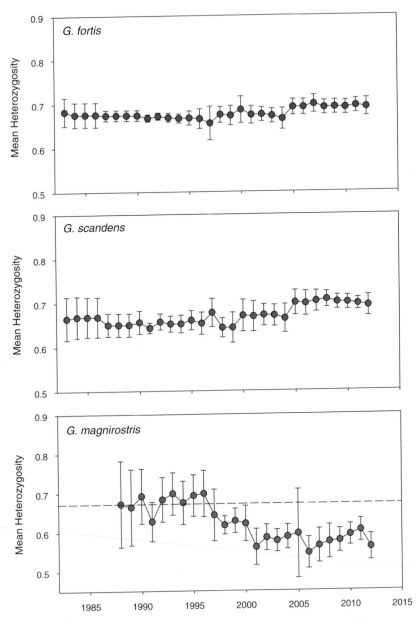

Fig. 10.12 Average observed multilocus heterozygosity (with 95% confidence intervals) at 14 autosomal microsatellite loci in three species. Stability of *fortis* and *scandens* values, with possible increases in the last few years, is emphasized here by the contrast with a strong decrease in heterozygosities in the nonhybridizing population of *magnirostris* from 1996 to 2001. The long-term average for *fortis* and *scandens* is shown on the lower figure as a broken horizontal line. Removal of

was increasing (fig. 10.6). However, changes over the last 15 years were not simply an unaltered extension of changes in the first 25 years. Some trends changed direction (figs. 10.7 and 10.8), while others became more pronounced (fig. 10.6). Frequencies of hybrids increased (fig. 10.5), indicating an increased frequency of hybridization and relatively high survival of hybrids (chapter 8), and new hybrid morphologies were produced (fig. 10.3), suggesting changed mating patterns. We conclude that introgression caused striking convergence of *fortis* and *scandens* over the 40-year period.

Despite support for the hypothesis, differences in variation between the species were not fully accounted for by our estimates of introgression. The *fortis* trends were affected to a minor extent by removal of the F_1 hybrids and backcrosses from samples, and although the *scandens* trends were reduced, they were nevertheless seen after removal of hybrids from the analyses. Thus short-term effects of known hybridization are not sufficient to explain the trends of increasing variation. We encountered the same in analyses restricted to data from the pedigrees (chapter 9). There and here we suggest the most likely explanation is a cumulative (storage) effect of past hybridization, particularly in *fortis*. Consistent with this suggestion, *fortis* × *fuliginosa* hybrid offspring determined genetically are sometimes formed from the genomes of parents that are both assigned genetically to *fortis* with a probability of greater than 0.98 (p. 187). This surprising outcome is probably attributable to the random segregation of alleles and the combination by chance of heterospecific alleles in high frequency.

Simple diffusion models applied to beak dimensions or allele frequencies show that a pair of hybridizing species may fuse into one panmictic population at a rate that depends upon the hybridization frequency (Boag and Grant 1984a, Clarke et al. 1998). At constant rates of hybridization the approach to identity, which is inherently frequency-dependent, is rapid at first and declines with increasing similarity. We found an approximately linear trend with time. This may simply reflect the rapid initial phase of progress toward identity. Alternatively, introgression rates may not be constant but increase with increasing similarity. There is evidence for this in the higher frequency of hybrids in samples of measurements after 2000 (fig. 10.5), although the genetic convergence trend was reversed in the last few years (fig. 10.11).

In the first half of the study heritable variation underlying beak and body size traits of both species appeared to be roughly stable (chapters

Fig. 10.12 (*continued*) hybrids from *fortis* and *scandens* samples reduces average heterozygosities by only about 0.01, and therefore they are not shown. Sample sizes are 2,133 *fortis*, 1,024 *scandens*, and 588 *magnirostris*.

3 and 9), in agreement with the introgression-selection balance model in figure 9.1. This appears to have changed in the second half of the study, in the light of a doubling of phenotypic variation in both species from beginning to the end (fig. 10.6) and an unmeasured increase in additive genetic variance from introgression. The chief implication is that selection, which opposes increases in variation (fig. 9.1), must be relatively weak, and differences between species in variation must be more a function of differences in genetic input than output. Consistent with this reasoning, we could find no hybrid disadvantage in the last 30 years (chapter 8; also next chapter).

Summary

In this chapter we explore the implications of convergence through hybridization. It differs from the previous two chapters in encompassing the full 40 years of study, and represents a transition from chapters that are rich in ecological detail to ones with a broad view of long-term trends. Hybridization led to morphological convergence through introgression and the blurring of both morphological and genetic distinctions between *fortis* and *scandens*. The difference between species in beak shape gradually decreased to a value of 57% of the starting difference, due principally to decreases in beak depth of *fortis* and beak length of *scandens*. Hybridization increased in the last decade, as inferred from an increase in the frequency of genetically identified F_1 hybrids and backcrosses in samples. Beak shape variation increased from start to finish, at the same rate in the two species but more smoothly and evenly in *scandens* than in *fortis*. The increases resulted in an approximate doubling of variation in 40 years in both species. Removal of the known genetic hybrids from the samples had a strong effect on *scandens* samples, reducing the level of variation in all years, but not eliminating the trend. Removal of the known hybrids from the *fortis* sample had scarcely any effect on the trend. Beak shape allometries changed, due principally to bidirectional change in *scandens*: initially toward lower slopes of beak depth in relation to beak length, followed by a strong shift in the opposite direction. Most but not all of these observations are consistent with an hypothesis of immediate, short-term effects of introgressive hybridization on morphological and genetic variation. In addition there may be a cumulative (storage) effect of past introgression, particularly in *fortis*.

Long-Term Trends
in Natural Selection

Unlike physics, every generalization about biology
is a slice in time.

(Bronowski 1973, p. 309)

Thus, the nature, causes, and effects of variability in
ecological systems can be understood only through
extended study over far longer time periods than those
typical of . . . most other biological sciences.

(Cody 1996, p. 4)

Introduction

Long-term field studies help to bridge the gap between two
very different temporal scales of evolution, the micro and the
macro. Fossils reveal temporal patterns in size, shape, and complexity of organisms as diverse as trilobites, oysters, and dinosaurs over
millions of years. The patterns include long-term linear trends with or
without reversal (Darwin 1859, Simpson 1953), long-term stasis punctuated by rare, abrupt, and substantial change (Eldredge and Gould
1972, Eldredge et al. 2005) and bursts of diversification following major
tectonic activity (Williams and Duda 2008), concentrated periods of

extinction (Bapst et al. 2012), or entry into new environments (Hou et al. 2011). How are these macroevolutionary patterns produced? Micro-evolutionary studies of biological mechanisms and natural processes in contemporary time provide answers (e.g., Bell and Foster 1994), and are especially helpful when they are coupled with studies of fossils of the same organisms (Bell et al. 2006).

The purpose of this chapter is to take a longer view of morphological evolution in contemporary time than is usual, drawing upon information on what happened year by year for 40 years. In doing so we revisit the evolutionary trajectories in figures 1.6 and 1.7 and the questions they raise, but expand the scope to consider not just beak depth but all measured morphological traits. We ask whether there were net trends in evolutionary change as a result of selection, introgression, and drift, whether species changed in tandem or out of step, and whether by the end of the study they were morphologically the same as at the beginning. We interpret changes brought about by selection in terms of feeding ecology. In the next chapter we make the link between changes over 40 years and the early stages of adaptive radiation over a million years or more.

Annual changes in morphological means are caused by selective losses, as a result of mortality and emigration, and selective gains, as a result of natality and immigration. All other things being equal, the means of these traits are expected to remain constant over the long term except for minor deviations caused by our sampling (Grant and Grant 2002a). All other things are decidedly not equal in this fluctuating and unpredictable environment. In chapter 5 we discussed morphological changes occurring as a result of differential breeding success, found little evidence for an association between the two, and concluded that reproductive success was only weakly affected by morphology. In contrast we showed in chapter 4 that morphology makes a strong contribution to differential survival of *fortis* in the short term when the environment changes, and in chapter 7 we described the combined effects of food shortage and interspecific competition on survival causing evolutionary change in *fortis* morphology.

Here we provide the long-term perspective by showing how *fortis* and *scandens* are affected by natural selection repeatedly and differently on a scale of decades. *G. magnirostris* is treated separately later in the chapter because recurring immigration complicates the interpretation of morphological trends. We conclude by describing allometric changes. All three species underwent changes in proportions, each taking a different route.

Selection

G. FORTIS

Climatic fluctuations and the changes they cause to both the vegetation and finch foods imply episodic selection that fluctuates in strength and direction (chapter 4). Viewed over a period of four decades, this expectation is observed strongly in *fortis*, and in all three traits; body size, beak size, and beak shape (fig. 11.1). The main pattern is this: in most years there is little or no directional selection, and interspersed are a few years of very strong selection.

The strongest selection events, in 1976–77 and 2003–5, were associated with droughts and high mortality due to starvation (chapters 4 and 7), but these were not the only ones (fig. 11.1). Body size was a selected trait in six years. In four of those years beak size was also a selected trait, and in the same direction, resulting from the strong and positive correlation between the two traits. Beak shape is independent of both and shows a contrasting pattern of selection. At the two times of strongest selection on body size and beak size (1976–77 and 2003–5) there was no selection on beak shape. In only one year, 1985, were all three traits selected; survivors were small and had pointed beaks. At the next time when birds with pointed beaks had a selective advantage, in 2009, there was no significant selection on the size traits. Thus different combinations of traits were selected in different years.

G. SCANDENS

Selection regimes on *scandens* were different from selection on *fortis* in two respects (compare figs. 11.1 and 11.2). The most important difference is the absence of fluctuating selection. Instead selection on both body size and beak size was unidirectional; small birds with small beaks were repeatedly at a selective disadvantage. Therefore in some years the two species experienced different selective mortality. For example, in 1985 the birds at a disadvantage were large *fortis* and small *scandens*. The second difference between species is the almost complete absence of selection on beak shape in *scandens*.

ECOLOGICAL CAUSES

Differences between the species in patterns of directional selection derive from their different diets and territorial systems. The causes of selective mortality in *fortis* were discussed in chapters 4 and 7. In

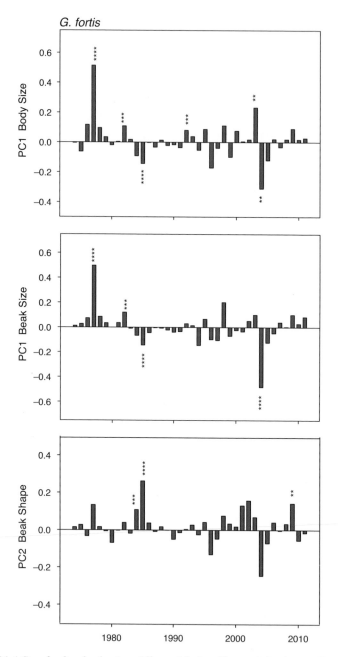

Fig. 11.1 Standardized selection differentials (coefficients) for *fortis*. They are calculated for each sample of adults surviving from one year to the next by dividing
the difference in their means by the standard deviation of the first sample. Positive

drought years many finches die of starvation. Survivors have beak sizes and shapes appropriate for handling the particular foods that are available as the supply dwindles. Body size may play an additional role in affecting survival through social dominance at contested food sources. In contrast to times of scarcity, when seeds are abundant at the beginning of a drought, survival is higher and there is no selection. For example, *fortis* mortality was relatively low in both the first (32%) and second (27%) of two consecutive years of drought, 1988 and 1989 (fig. 4.12), when only 5 mm of rain fell. The droughts followed a highly productive year of El Niño (fig. 4.12), and there was no association with morphology (figs. 7.6 and A.9.1).

Diets of *scandens* are not the same as *fortis* and nor are the causes of selective mortality. In this species social dominance is of overriding importance because it determines access to *Opuntia* fruits and flowers. These are fixed, spatially localized, on or below dispersed cactus bushes, and are defended year-round by males (Millington and Grant 1983). *G. fortis* are territorial only in the breeding season (Boag and Grant 1984b), and at other times they search for seeds on the ground that are widely distributed across the island. Thus their respective food supplies affect the two species differently.

When food is limiting, seed-handling ability is the major factor determining survival of *fortis*, whereas access to cactus foods governs survival of *scandens*. *G. scandens* are cactus specialists, able to perform two tasks with long and pointed beaks. They are proficient at opening buds and fruits by tip biting (figs. 2.18 and 2.19), and are able to probe deeply into flowers to reach the basal nectar. Opening a bud gives them first access to a rich supply of pollen, and opening a fruit gives them first access to more than a hundred seeds surrounded by fleshy arils (Grant 1996). *G. fortis*, with blunt beaks, do not have the skill of *scandens* in opening buds or fruits, for mechanical reasons

---

Fig. 11.1 (*continued*) values indicate selection for large size or pointed beaks on the PC2-beak axis. Males and females were combined with adults of unknown sex because separate selection analyses of males and females give similar results (chapter 4). The magnitude of the coefficients is indicated by the size of the histogram bars, and their direction is shown by their placement above (positive) or below (negative) the baseline of zero. The difference between the means of survivors and nonsurvivors was tested with two-tailed t-tests that do not require equal variances, and the significance of each coefficient is indicated with from two to four asterisks: ** ($p < 0.01$), *** ($p < 0.005$), **** ($p < 0.0001$). Six differences significant at $p < 0.05$ are not shown, because the observed six are expected by chance in the 111 tests performed. Sample sizes range from 21 to 735 (median = 200) for the years 1974 to 2011 (table 11.1 in appendix 11.1). For further details see Grant and Grant 2002.

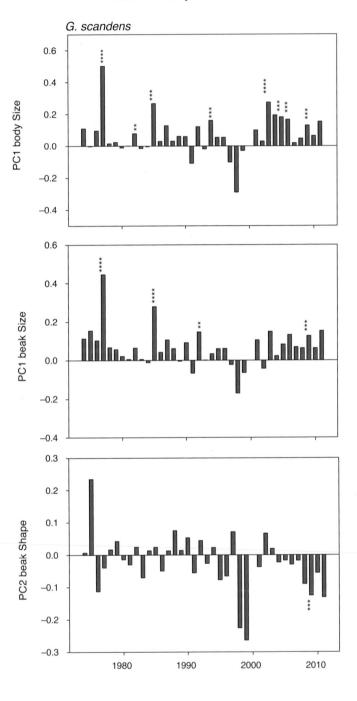

(Bowman 1961), and can only probe flowers when they have begun to open widely or have been opened by *scandens* (Grant and Grant 1981). The two species compete for food at times (chapter 8), but here is an example of the opposite, that is, nonmutual facilitation. *G. magnirostris* likewise benefits from *scandens* opening *Opuntia* fruits.

STABILIZING SELECTION

Because it is a resource specialist, *scandens* might be subject to stabilizing selection on morphology more frequently than *fortis*. Assessed by variance tests, however, stabilizing selection is rare in both, and the species do not differ in the frequency of either stabilizing or diversifying (disruptive) selection (appendix 9.1). Polynomial regression analysis (Fairbairn and Preziosi 1996) gives a similar result. In those years of significant directional selection (figs. 11.1 and 11.2) there are no significant quadratic terms in the selection equations, with one marginal exception that is opposite to the expected: in 1977 *scandens* experienced disruptive selection on beak shape (PC2-beak) at $p = 0.0317$.

Morphological Trends

G. FORTIS

The expected consequence of selective mortality is morphological change. Specifically, fluctuating selection, together with high heritability of morphological traits, implies fluctuating morphological trajectories through time and not morphological constancy. Morphological trends of *fortis* over four decades meet this expectation (fig. 11.3). Trends are characterized by (1) long periods of stability, (2) short periods of abrupt change, and (3) a net change from beginning to end. If body size, beak size, and beak shape remained constant over periods as

---

Fig. 11.2 (*facing page*) Standardized selection differentials (coefficients) for *scandens*. Positive values indicate selection for large size or pointed beaks on the PC2 beak axis. Calculations and symbols as in figure 11.1. Significance of the coefficients is indicated by two to four asterisks: ** ($p < 0.01$), *** ($p < 0.005$), **** ($p < 0.0001$). Six differences significant at $p < 0.05$ are not shown, because the observed six are expected by chance in the 108 tests performed. Data were inadequate for analyses in 1973 and 2000. Sample sizes range from 31 to 300 (median = 93; table A.11.1).

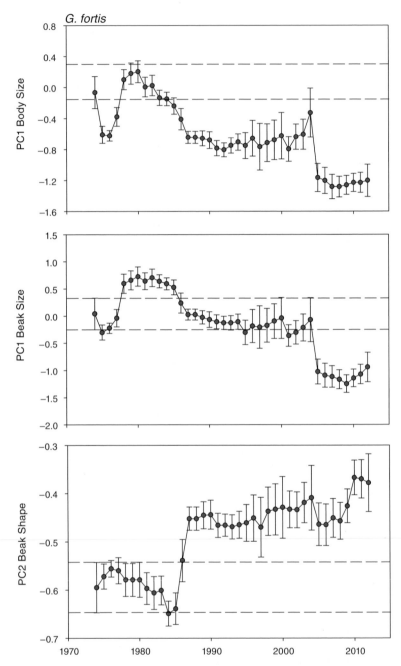

Fig. 11.3 General trends in *fortis* morphology based on principal components scores. Beak shape becomes increasingly pointed from bottom to top. Means and 95% confidence limits are shown for all birds alive in each year: in total 1,489

long as decades, we would see random fluctuations in our estimates of the means largely confined to the initial confidence limits on mean estimates. The initial confidence limits are highlighted in figure 11.3 by a pair of horizontal lines. In fact trajectories cross the boundaries, sometimes abruptly, and then remain outside.

G. *fortis* changed in average body size and beak size in the mid-1970s and then again about 30 years later. We have no evidence that change occurred as a result of selection for or against hybrids (chapters 8–10). For example, when small birds were at a strong selective advantage in 2004–5, genetically identified hybrids ($n = 20$) survived no differently from nonhybrids ($n = 52$; $\chi_1^2 = 0.02$, $p \approx 1.0$). The strongest change in beak shape occurred at a different time, in the mid-1980s (fig. 11.3). G. *fortis* became smaller over the whole period, and beaks became more pointed on average. Figure 11.4 shows how this was brought about. In the mid-1980s beak depth and width decreased without a concomitant change in beak length. Twenty years later the trend was reversed, abruptly, when all three dimensions changed at a time of selection against birds with large beaks. The pattern of changes reflected episodic directional selection much more strongly than the gradual changes caused by introgressive hybridization (chapters 9 and 10). Figure 7.6 shows the evolutionary responses to selection.

G. SCANDENS

Small-bodied birds were repeatedly at a selective disadvantage (fig. 11.2), yet average body size declined over the 40-year period (fig. 11.5)! How could this be? Where this kind of contradiction has been found elsewhere, it has been explained as the result of a deteriorating environment (Larsson et al. 1998, Merilä, Kruuk, and Sheldon 2001, Merilä, Sheldon, and Kruuk 2001, Pemberton 2010) or unmeasured selection opposing the known selection (Connor 2012). In the present case there is a simple alternative: introgressive hybridization, with genes transferred predominantly from *fortis* to *scandens* (chapter 9).

Selection and hybridization have opposite effects on some morphological features of *scandens*, and at different times, selection being

---

Fig. 11.3 (*continued*) males, 968 females, and 1,593 birds of unknown sex. Parallel horizontal lines mark the upper and lower 95% confidence limits on the first estimate of a mean based on a large sample size (in 1973). Annual samples range from 30 to 932 (median = 270; table A.11.2). Statistical tests of annual heterogeneity use measurements of each bird only once, and they were grouped in years of known or inferred hatching: for body size $F_{23, 3652} = 19.1497$, $p < 0.0001$; for beak size $F_{23, 3794} = 19.9620$, $p < 0.0001$; and for beak shape $F_{23, 3794} = 12.1604$, $p < 0.0001$. Number of years = 24; in other years there was little or no breeding.

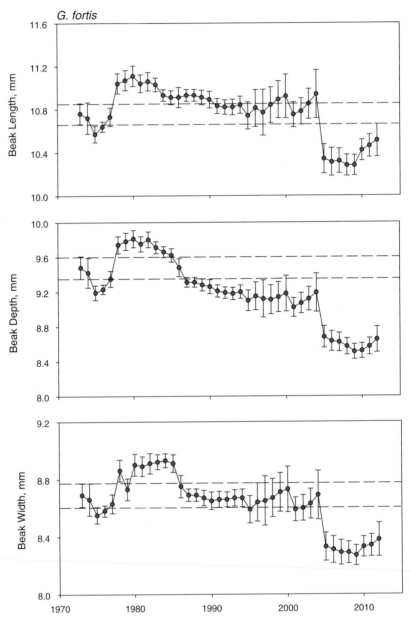

Fig. 11.4 Trends in *fortis* beak traits. Means and 95% confidence limits are shown for all birds alive in each year. Parallel horizontal lines mark the upper and lower 95% confidence limits on the first estimate of a mean based on a large sample size (in 1973). Sample sizes are in table A.11.2. Statistical tests of annual heterogeneity use measurements of each bird only once, and they were grouped in years of known or inferred hatching: for beak length $F_{23, 3843} = 12.6158$, $p < 0.0001$; for beak depth $F_{23, 3796} = 30.0389$, $p < 0.0001$; and for beak width $F_{23, 3843} = 16.3765$, $p < 0.0001$. Number of years = 24; in other years there was little or no breeding.

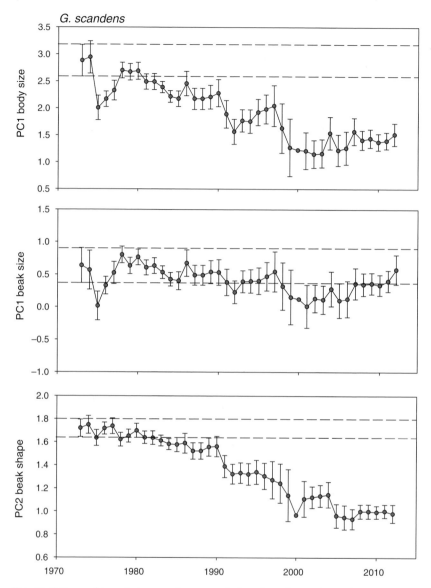

Fig. 11.5 General trends in *scandens* morphology based on principal components scores. Beak shape becomes increasingly pointed from bottom to top. Means and 95% confidence limits are shown for all birds alive in each year: in total 573 males, 390 females, and 458 birds of unknown sex. Parallel horizontal lines mark the upper and lower 95% confidence limits on the first estimate of a mean based on a large sample size (in 1973). Annual samples range from 15 to 346 (median = 104; table A.11.2). Statistical tests of annual heterogeneity use measurements of each bird only once, and they were grouped in years of known or inferred hatching: for body size $F_{24, 1283} = 13.4603$, $p < 0.0001$; for beak size $F_{24, 1383} = 5.1526$, $p < 0.0001$; and for beak shape $F_{24, 1383} = 23.3570$, $p < 0.0001$. Number of years = 25; in other years there was little or no breeding.

measured as differential mortality in dry seasons and introgressive hybridization being production of offspring in wet seasons. Selection is episodic, introgression is continuous, and effects of introgression predominated, unlike in *fortis*, and determined the downward trend. Not only did mean body size decline, beak shape became increasingly robust (fig. 11.6), like that of *fortis* (fig. 11.3), but in this case there was no selection on beak shape, either opposing or congruent except in 2009 (fig. 11.2; fig. 7.5 shows the evolutionary response). Up to 1991 size and shape traits of *scandens* had remained approximately constant for 15 years after an initial decrease in size in the first four years. In 1991 the first breeding of *fortis* × *scandens* F_1 hybrids occurred. They backcrossed to the *scandens* population (chapter 9), and after that all traits changed as introgression continued (fig. 8.13). Body size decreased, and stayed small, while beaks became progressively blunter (fig. 11.6). The temporal pattern of variation in individual beak traits (fig. 11.7) reveals that shape change was almost entirely due to a shortening of the beak. Beak length is the trait by which the species differ the most. It is the more important beak dimension of tip-biting *scandens* for opening cactus fruits and buds and probing flowers for nectar (Grant and Grant 1981).

Backcrosses to *scandens* must have some advantage that offsets the social and feeding disadvantage at cactus flowers of being small and having a relatively short beak. At least part of the answer may be that females do not suffer the disadvantage as much as males do, and may have some cryptic advantage in avoiding conflict with males. Male *scandens* generally survive better than females, especially in the first year, and become recruits at a higher rate, but in hybrids (F_1 and backcrosses combined) the relative success of males and females is reversed: 10 males and 16 females that were identified from pedigrees were recruited as F_1 and B_{1-3} generations.

G. MAGNIROSTRIS

The population of *magnirostris* underwent morphological change, but the interpretation of the changes is not as straightforward as it is with *fortis* and *scandens*. *G. fortis* and *scandens* are residents, and there is negligible immigration of these species from other islands (chapter 9). Presumably there is little emigration too (chapter 4), and so selective losses to their populations on Daphne can be confidently attributed to local mortality. The situation with *magnirostris* is different because the population is open to immigration to an unknown extent (chapter 6), possibly subject to losses from emigration (return to natal island), and therefore interpretation of morphological trends is less clear-cut.

Fig. 11.6 Changes in beak size and shape of Cactus Finches. **Upper:** Representative of beak shape in the first part of the study. **Middle:** Representative of a later, more blunt, beak. **Lower:** Even blunter beak at the end of the study.

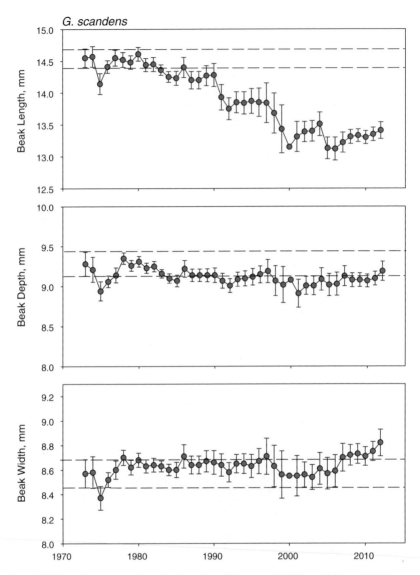

Fig. 11.7 Trends in *scandens* beak traits. Means and 95% confidence limits are shown for all birds alive in each year. Parallel horizontal lines mark the upper and lower 95% confidence limits on the first estimate of a mean based on a large sample size (in 1973). Sample sizes are in table A.11.2. Statistical tests of annual heterogeneity use measurements of each bird only once, and they were grouped in years of known or inferred hatching: for beak length $F_{24, 1394} = 21.7832$, $p < 0.0001$; for beak depth $F_{24, 1383} = 5.1526$, $p < 0.0001$; and for beak width $F_{24, 1394} = 3.9776$, $p < 0.0001$. Number of years = 25; in other years there was little or no breeding.

Three periods can be recognized in the history of *magnirostris* during the study: a prebreeding phase up to 1982 with recurrent immigration (I), a low-density breeding phase up to 1997 with occasional immigration (II), and a generally high-density phase (fig. 6.1) thereafter (III) with little immigration or addition of new alleles (fig. 9.3). Toward the end of the high-density breeding phase we became aware of an apparent change in beak size and found, when measurements were analyzed, that indeed the birds were not the same as they had been at the beginning of the study (fig. 6.5). Average beak size had increased despite fluctuations from year to year and continued to do so (fig. 11.8). Beak shape had become slightly more robust (fig. 11.8), as a result of greater increases in depth and width than in beak length (fig. 11.9). In contrast to these trends average body size had remained unchanged (fig. 11.8), which shows that the beak-size trend cannot simply be attributed to correlated effects of a change in overall body size as in *fortis*.

Potentially, the beak-size trend could be caused by variation in growth if adult beaks increase in size with age and if old birds dominated the later samples. This possibility can be ruled out (appendix 3.2). Another factor, introgressive hybridization, can be ignored because *magnirostris* never hybridized on Daphne during the study.

THE POSSIBLE ROLE OF SELECTION ON *G. MAGNIROSTRIS*

Trends in beak size and shape were more the result of gradual accumulated change over several years than the result of occasionally strong change from year to year, as happened in *fortis*. We draw this conclusion from analyses of year-to-year selection, recognizing that sample sizes of adults are generally not large, and so the absence of demonstrable selection may simply mean we have failed to demonstrate it. For example, the 7 individuals that survived the drought of 2004 and bred in 2005 were large in beak and body size, and had robust beaks, compared with the 71 that died, but the selection coefficients were not significant (fig. 11.10). In two years the evidence for selection on beak size is strong. Birds with large beaks survived better than those with small beaks in 1993 and again in 1994. In three other years, 2006–8, the evidence is statistically weaker but points in the same direction. In two different years, 1992 and 2011, the evidence for selection on beak shape is strong (fig. 11.10). Generally small fluctuations that are random in direction predominate in other years.

The samples in 1993 and 1994 were large because they comprised several new immigrants together with the products of local breeding (next section). Thus the selective disadvantage of individuals with rela-

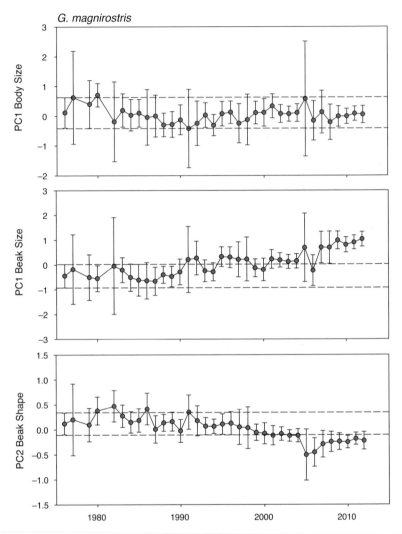

Fig. 11.8 General trends in *magnirostris* morphology based on principal components scores. Beak shape becomes increasingly pointed from bottom to top. Means and 95% confidence limits are shown for all birds alive in each year: in total 204 males, 141 females, and 291 birds of unknown sex. Annual sample sizes are in table A.11.2. Considering birds only in their first year of measurement, there is statistical heterogeneity in beak size ($F_{29, 602} = 3.0369$, $p < 0.0001$) and beak shape ($F_{29, 602} = 2.3211$, $p < 0.0001$) but not body size ($F_{29, 602} = 1.2153$, $p = 0.2042$). Linear regression analyses of annual means give the same results. There are no first-capture data for 1978, 1986, 1990, 1997, 2000, and 2005. Note that PC1 body scores are correlated with PC1 beak scores ($n = 632$, $r = 0.6267$, $p < 0.0001$) but not with PC2 beak scores ($n = 632$, $r = 0.0671$, $p > 0.1$).

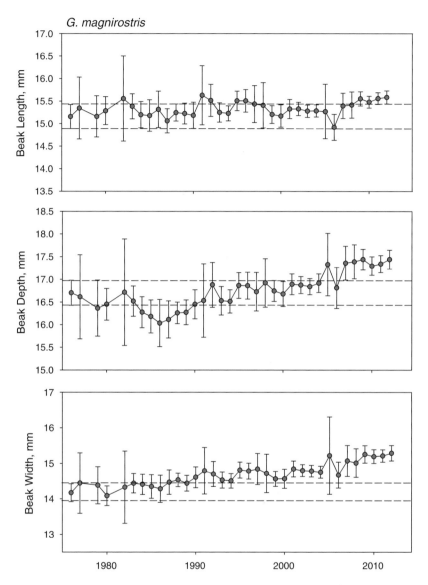

Fig. 11.9 Trends in *magnirostris* beak traits. Means and 95% confidence limits are shown for all birds alive in each year as in figure 11.8. Sample sizes are in table A.11.2. There is statistical heterogeneity in beak depth ($F_{29, 602} = 3.7106$, $p < 0.0001$) and beak width ($F_{29, 602} = 3.9776$, $p < 0.0001$) but not beak length ($F_{29, 602} = 1.3359$, $p = 0.1141$). Linear regression analyses of annual means give the same results. Slopes are almost identical for beak depth (0.0279 ± 0.0061 s.e.) and width (0.0281 ± 0.0053). Statistical significance is not altered in any analysis after excluding means from four years with sample sizes of fewer than five individuals.

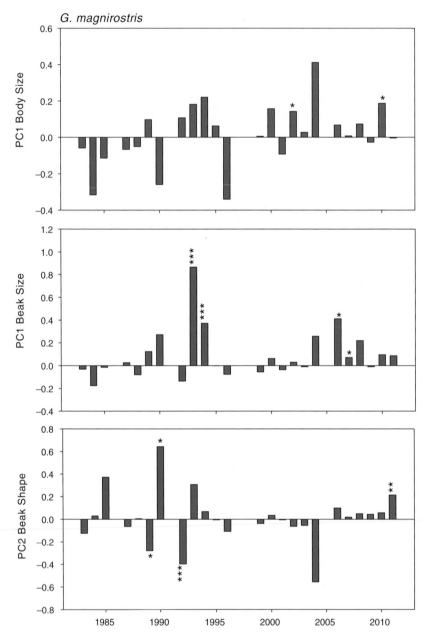

Fig. 11.10 Standardized selection differentials for *magnirostris*, calculated for each sample surviving from year x to year $x + 1$. Positive values indicate selection for large size or pointed beaks. Males and females were combined with adults of unknown sex because separate selection analyses of males and females give similar

tively small beaks should be understood to mean disappearance through death or emigration. In contrast, we believe there were relatively few immigrants on the island in 2006–8, when weaker beak-size selection occurred, and therefore birds that disappeared probably died on Daphne.

IMMIGRATION

Immigrants may have contributed to the gradual changes in mean beak morphology. Unambiguous identification of immigrants is possible only in the years up to 1982–83, when the birds first bred on the island. We can be almost certain of their identity at two times after this. The first was in 1988–89, when many birds appeared on Daphne, in numbers far in excess of the one or two individuals hatched in 1987 that we might have missed when banding offspring. The second time was in 1993–94, when again there was a large influx of birds too numerous to have hatched on the island in 1992. Immigration after 2000 was infrequent, as indicated by the small number of new alleles and low average heterozygosity (figs. 6.6 and 10.12).

At the first of these two times the presumed immigrants ($n = 41$) did not differ in mean body size, beak size, or beak shape from the known offspring produced on Daphne that year ($n = 5$). At the second time, however, the presumed immigrants ($n = 80$) were smaller in average beak size than the known resident offspring ($n = 14$; $t = 4.116$, $p < 0.0001$), and therefore were not responsible for the increase in mean beak size. They did not differ in body size or beak shape from the residents either ($p > 0.1$).

However, the immigrants were also more variable in beak size than the local offspring (Levene's test, $F_{1,92} = 11.424$, $p = 0.0011$), which is consistent with a heterogeneous origin from different islands. Large variation in beak size of the immigrants provides ample scope for nonrandom settlement and establishment on the island. Settlement may depend in part on beak size if, for example, those with large beaks have a feeding advantage and are in better energetic condition than those with small beaks. Indeed, immigrants that persisted on Daphne for at least

---

Fig. 11.10 (*continued*) results. Sample sizes in 20 years range from 20 to 103 (median = 42; table A.11.1). There are no selection coefficients for 1986, 1991, 1997, 1998, and 2005, because sample sizes were smaller than 20. Significance levels are shown without correction for multiple testing or lack of independence: * $p < 0.05$, ** $p < 0.01$, *** $p < 0.005$. Sixty tests were performed, so 3 are expected to yield differences that are significant at $p < 0.05$. Since more than twice this number were observed (8), they are shown in the figure.

one year after being banded ($n = 31$) had larger beaks than those that did not ($n = 49$; $t = 4.800$, $p < 0.0001$). Local persistence was therefore not random (see also fig. 11.10). To investigate this further, we repeated the test of residents versus immigrants with just those that survived locally for a year. As expected, this eliminated the difference in beak size ($t = 0.171$, $p = 0.8652$), while the difference in variances remained (Levene's test, $F_{1,43} = 10.629$, $p = 0.0022$). The relatively small birds that disappeared either died (selection) or dispersed (emigration).

THE CAUSE OF THE BEAK-SIZE TREND

G. magnirostris with large and robust beaks may gain an energetic and fitness (survival) advantage from their ability to rapidly crack open the large and hard fruits of Tribulus cistoides and seeds of Opuntia echios. The finches that came to Daphne, mainly from Santa Cruz Island, and remained to breed may have been adapted to a different and diverse set of foods on the source island, and less adapted to the foods on Daphne. They and their descendants were then subject to selection. G. magnirostris are relatively rare on Santa Cruz island, and the feeding data to test these ideas have not been obtained. Individuals of either sex with large beaks may also have gained an additional mating advantage over those with smaller beaks.

Allometry of Means

In single samples from a population all traits of all species increase in size as body size increases, whether body size is measured by mass, by a proxy dimension such as wing length or tarsus length, or by the synthetic variable PC1. Nevertheless traits do not scale with body size in an identical way (Boag 1984). The previous chapter provided examples of allometric relations between variables in samples at one time. Here we use averages of two beak traits to illustrate the changes in proportions that occur over many years.

Samples of adults in different years differ in their (static) allometries of means, and all three species are unique (fig. 11.11). Beak length and beak depth scale with body size similarly but not equally in *fortis* (fig. 11.11 legend). Beak length increases with body size in *scandens* much faster than beak depth, which scarcely increases at all, whereas the opposite is true for *magnirostris*. The contrast between species is interesting because length is the more important beak dimension of tip-biting *scandens* for opening cactus buds and probing flowers for nectar, whereas depth is the more important beak dimension of seed-cracking

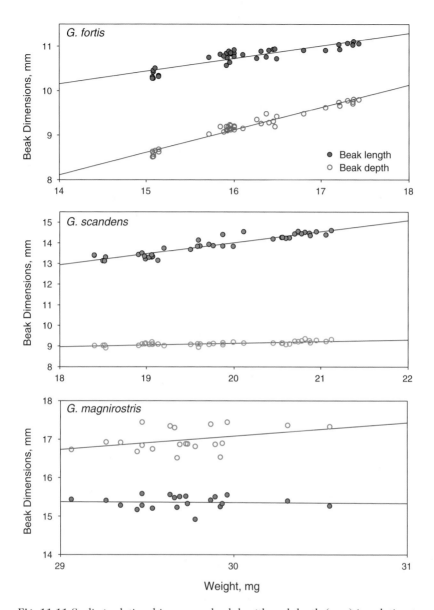

Fig. 11.11 Scaling relationships: mean beak length and depth (mm) in relation to mean mass (weight, g) in each year from 1973 to 2012. Lines are least-squares best fits from linear regressions. Allometric slopes are estimated with logarithmically transformed data. Beak depth (0.903 ± 0.031 s.e.) increases faster with weight than does beak length (0.432 ± 0.034) in *fortis*, whereas the opposite occurs in *scandens*: beak length (0.754 ± 0.046) increases faster than beak depth (0.183 ± 0.026) with weight. All four slopes differ from zero at *p* < 0.0001, and from each other. The slopes for *magnirostris* (1992–2012 only) are not significant (*p* > 0.1). Reduced major-axis regressions have different slope coefficients, but the contrasts are the same as in logarithmic analyses.

magnirostris. The figure brings out the additional point that depth is more pronounced in relation to length in *magnirostris* when compared with the other two species.

Annual changes in *fortis* took place along a single line of allometry of means as a simple consequence of scaling until 1985, when an abrupt shift took place (fig. 11.12). After 1985 annual changes occurred along an altered but parallel line of allometry, even when both beak dimensions underwent a major reduction in average size in 2004–5. Thus changes in beak shape occurred in three ways, along two different but parallel lines of allometry and across an abrupt transition between them. Annual changes in *scandens* and *magnirostris* show similar abrupt transitions, but later and at different times. These are examples on a microevolutionary scale of a macroevolutionary pattern referred to in the introduction to this chapter: long-term stasis punctuated by rare, abrupt, and substantial change.

Natural and Artificial Selection

Unlike some regimes of artificial selection, natural selection trims but does not truncate one end of a size-frequency distribution: some individuals of all sizes survive (fig. 7.6). The next generation has a frequency distribution around a larger or smaller mean. Recombination gives rise to some individuals beyond the previous range of morphologies and genotypes. With continued artificial selection in the same direction the mean continues to increase in that direction, generation after generation, until the variation runs out or is replenished by new mutations. This may be a good model for evolution when the environment changes substantially, and for sustained directional coevolution between a consumer and the consumed (Benkman et al. 2013) or predator and prey (Thompson 2013). However, episodic and fluctuating natural selection, as observed here with *fortis*, lasts for less than one generation and results in little net divergence. Continued directional introgression, on the other hand, produces continued directional change (fig. 10.11).

Conclusion

The most significant finding of our study has been the dynamic nature of morphological variation in all three species. Morphologically the finch species are not static, not fixed at some long-term average of body

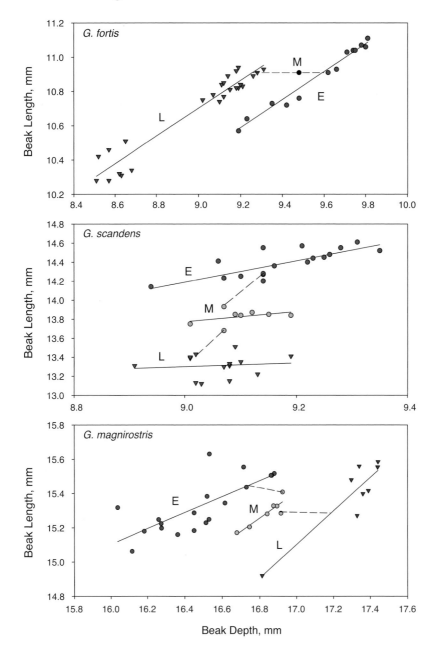

Fig. 11.12 Changes in the relationship between average beak length and average beak depth (mm) in the three species. E, M, and L refer to early, middle, and late years in the 40-year study. Solid lines are least-squares best fits from linear regressions, and broken lines connect early, middle, and late phase morphologies.

size or beak size that could be interpreted as an optimum, but dynamic and changing in response to a changing environment. All terrestrial environments undergo change (Bell 2010), some more strongly than others, some more erratically than others. Therefore similar microevolutionary change is to be expected in other organisms, although the magnitude of those changes may be much smaller than the changes we have documented in Darwin's finches over several decades.

Summary

This chapter provides a long-term perspective to the evolutionary changes produced by natural selection described in previous chapters. *G. fortis*, *scandens*, and *magnirostris* each changed morphologically over four decades, but in different ways. Strong changes in body size, beak size, and beak shape occurred in *fortis*, rarely, and in opposite directions, primarily as a result of fluctuating directional selection associated with droughts and only to a minor extent from introgressive hybridization. Selection differentials were large. The relative importance of selection and hybridization was reversed in *scandens*. Despite large size being selectively favored repeatedly in this population, average body size declined, because effects of introgressive hybridization with *fortis* and high relative fitness of hybrids outweighed effects of selection. *G. magnirostris* changed gradually toward large beak size and robust beak shape, but not in body size, partly as a result of selection in small steps and possibly also by size-differential immigration. The three species display striking differences in beak shape when analyzed as a function of time. Both *fortis* and *magnirostris* underwent abrupt transitions in the allometric relationship between beak length and beak depth: average beak depth changed, but beak length did not. Change in *scandens* allometry was just as strong but occurred primarily as a result of change in beak length and not depth. Morphological differences between species were associated with different feeding modes and diets. *G. fortis* and *magnirostris* are primarily seedeaters. Key to the change in beak dimensions of *fortis*, and possibly *magnirostris*, was their abilities at handling seeds of different sizes and hardnesses. *G. scandens* is a specialist on cactus products, and in this species access to foods in addition to food-handling skills is likely to have determined survival. Large size is advantageous because it confers social dominance at these restricted food sources.

Speciation

Looked at in the broadest possible way the species
problem thus becomes the problem of the establishment
of group discontinuities in the evolutionary continuum.

(Huxley 1938, p. 254)

If one accepts some version of the biological species
concept, then the central problem of speciation is
understanding the origin of those isolating barriers that
actually or potentially prevent gene flow in sympatry.

(Coyne and Orr 2004, p. 57)

Introduction

CHAPTER 1 INTRODUCED THE MAIN PROBLEM that has motivated research on Daphne: the problem of explaining how one
species of Darwin's finch became many by repeated speciation
through morphological and ecological diversification. Speciation occurs when one species splits into two populations or sets of populations
(fig 1.2) that do not interbreed or interbreed rarely with little genetic
consequence (Grant and Grant 2008a). We observe the products, species, and the challenge we face is to reconstruct the process by which
they were formed. One purpose of this chapter is to use information
from the study on Daphne to reconstruct and interpret the morphological transformation of species to answer the question of how speciation
occurs.

A second purpose of the chapter is to discuss the interactions that take place when previously separated and divergent populations come into contact. According to the allopatric model of speciation, splitting of a species into two is initiated allopatrically, for example, when individuals disperse from one island to another and establish a new population (fig 1.1). They evolve by natural selection and genetic drift, and become adapted to each ecologically distinctive environment. A sympatric phase may then follow, as a result of dispersal of individuals from one island to another occupied by a related population. The outcome depends on the degree of prior divergence: reproductive isolation if divergence was large or interbreeding with residents if it was small. Theory is clear on the essential ingredients—natural selection, genetic variation, ecological divergence, and reproductive isolation (Coyne and Orr 2004)—but not on what tips the balance toward fission or fusion. This is where a study on a single island such as Daphne can help, by throwing light on interactions at the secondary contact stage and later.

Morphological Transformation in Speciation

The ground finches differ in size and shape. The three, coexisting, granivore species on Daphne differ by at least 15% in one beak dimension, which is equivalent to approximately two standard deviations either side of each mean value (Grant 1986). Beak dimensions change across decades as a result of fluctuating selection (chapter 11), but how much selection is required to transform one species into another by changing a mean by 15%? This question can be approached at two different but complementary life history stages: adults or embryos. The adult approach uses genetic variances and covariances for the species on Daphne and the phenotypic (morphological) differences between them (Price et al. 1984a) to estimate the minimum cumulative or net forces of selection required to effect the transformation of adults (Lande 1979, Price et al. 1984a, Schluter 1984). The other approach examines the underlying genetic developmental program leading to different beak sizes and shapes. We will discuss the two approaches in turn, and consider how together they reveal the way changes in size and shape can occur.

Size transformation occurs readily because there is a large amount of genetic variance for size. Genetic covariances and correlations between *fortis* traits are uniformly strong and positive (chapter 3). This implies (Grant 1981c, 1983), and analysis shows (Price et al. 1984a, Schluter 2000), that transformations involving primarily overall size require less selection than those involving primarily shape; from the *fortis* starting point the *magnirostris* and *fuliginosa* morphologies can be reached more easily than can the *scandens* morphology. Simultaneous changes in shape tend to be opposed by the strong, positive genetic correlations between traits, and in this sense they encounter genetic resistance (Schluter 1984, 1996, 2000); changes in shape are constrained but not prevented (Connor 2012).

SPECIES THAT DIFFER IN SIZE

These points are illustrated with calculations of residual shape change that remains after size changes have been accomplished in the transformation of one species to another (box 12.1). A low value of residual shape change in standard deviation units implies that the transition from one species to another has been largely accomplished by selection on size alone. A high value implies that selection on shape itself is needed. In this exercise we are not attempting to reproduce the actual transitions in both size and shape, nor are we using a known, statistically well-supported phylogeny, which is still not available (Petren et al. 2005). Instead, we use the calculations to illustrate the fact that some morphological transformations are much more easily accomplished than others, perhaps more likely to have occurred than others (Schluter 1996), and possibly to have occurred in less time. In doing so we assume that the genetic covariance matrix changes relatively little during the transformations, which may not always be correct (Grant and Grant 2000a, Björklund et al. 2013). We also assume a continued and sufficient supply of genetic variation, which is supported by experience of long-term responses to selection in animal and plant breeding (Hill and Kirkpatrick 2010). Strong directional selection, continued for up to 50 or 100 generations, has produced continued evolutionary responses in flies (Weber 1996), mice (Keightley 1998), and corn (Walsh 2004). This has been interpreted as the result of new mutations arising continuously and supplementing standing genetic variation (Hill and Kirkpatrick 2010).

Selection on a size factor alone can easily transform species differing in size but scarcely in shape (fig. 12.1). The transformation from *fortis* to *fuliginosa* morphology is a clear example of this. The little variation in shape that is left after selection on size is actually less than the observed change in beak shape from selection in 1985 and 1986 (fig. 11.12). The *fortis* to *magnirostris* transition differs in that the correlated effect on shape of selection on size has been too strong. Genetic facilitation rather than genetic resistance has caused an overshoot of the shape target.

Box 12.1. Species Transitions in Morphology

The core species with genetic and phenotypic data are *fortis* and *scandens* from Daphne (chapter 3). We lack quantitative genetic data from *fuliginosa* and *magnirostris*, and therefore used *fortis*, the most similar species, as the starting point for reconstructing the transitions to *fuliginosa* and *magnirostris* morphologies. We used equal numbers of families for a correlation-based principal components analysis to characterize structural size variation among pairs of these species as the first component. Between 63.0% and 84.9% of the total variation is accounted for in each analysis by PC1, and in all analyses the six morphological traits have approximately equal loadings, which justifies using it as a measure of overall size. The second principal component accounts for most of the remaining variation, 7.8%–25.7%. In each analysis, loadings on PC2 for beak depth and beak width are large and positive, whereas the loading for beak length is negative, and loadings for the other traits are generally small. Therefore, PC2 is mainly a beak-shape factor, varying from pointed to blunt (also Boag 1983, Grant and Grant 1989; chapter 4).

Next we calculated heritabilities for PC1 and PC2 scores for *fortis* and *scandens* and genetic correlations between them. Even though the two axes of variation are uncorrelated in the combined data, the individual species do sometimes display positive correlations. Heritable variation for the size and shape factors and the additive genetic correlation between them vary strongly among the species (table 1.10 in Grant and Grant 2000a). Third, we calculated the net forces of selection on the size factor (alone) involved in each of the four possible (reciprocal) species transformations using these genetic parameters and the phenotypic distances in standard deviation units of principal components scores. Shape changes arising from selection solely on size were calculated from these measures of selection and genetic variation and covariation; they are the product of the net forces of selection on size, the square root of the heritability of size, and the genetic correlation between size and shape (Lande 1979, Price et al. 1984b, Schluter 1984). What remains is residual shape differences between species.

SPECIES THAT DIFFER IN SHAPE

Among the four species on Daphne those that differ most in shape, *fortis* and *scandens*, are the least interconvertible by selection. Asterisks in the lower part of figure 12.1 show that selection on size of either

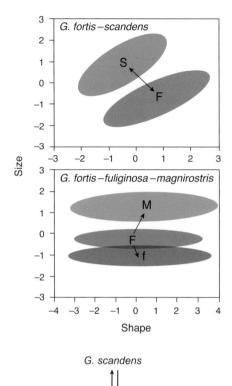

Fig. 12.1 Evolutionary transitions along body size (PC1) and beak shape (PC2) axes (upper), and residual shape changes in standard deviation units after the size changes alone have been effected (lower). A negative sign indicates that change in shape associated with a change in size has gone too far, and residual shape change must be in the reverse direction. An asterisk shows that the difference in shape between two species has increased as a result of a change in size. From Grant and Grant 2000a.

species results in a correlated shape change in the opposite direction, so that the difference between species in shape becomes magnified as selection decreases the difference in size, and vice versa. We reason from this conflict that the ancestor was possibly smaller than both of the extant species and possessed a pointed beak, for then each transformation would have proceeded with little genetic resistance. This reasoning is consistent with the molecular phylogeny (Petren et al. 1999b, 2005, Sato et al. 1999) that shows pointed-beaked species, and specifically *Geospiza difficilis* (Sharp-beaked Ground Finch), to be basal to cactus and granivorous ground finches.

Despite strong positive correlations between the primary traits of beak length, depth, and width, changes in shape do occur, as shown by both *fortis* and *scandens* but in different ways (fig. 11.12). Changes in beak proportions of *fortis* were produced by natural selection (chapter 4), which shows that heritable beak proportions and not just individual beak dimensions are sometimes the target of selection. Changes in beak proportions of *scandens* were produced by introgressive hybridization with *fortis* in the absence of selection. Therefore genetic constraints on changes in shape are not absolute in these species; they can be circumvented. Furthermore the genetic constraints themselves change when the genes governing development of different beak dimensions undergo change.

Genetic Transformation in Speciation

In the absence of genetic incompatibilities, the genetics of speciation is the genetics of beak divergence. What are the genetic changes involved in the morphological changes taking place during speciation? The place to look for them is in development (Price and Grant 1985, Björklund 1993, Grant et al. 2006). The four species differ in beak proportions at the time of hatching, which implies differences in embryonic growth, and they also differ in relative growth rates of different dimensions after hatching (Grant 1981c). Hence the genetic divergence that accompanied species formation involved evolution of developmental differences. Recent discoveries in developmental genetics have identified genes affecting embryonic beak development in the finches, and have characterized differences in expression patterns among closely related species (fig. 12.2). These are parallel to discoveries in plants and animals in other systems, in both plants and animals, of key regulatory genes that have undergone change by mutation (Shapiro et al. 2004), introgression (Kim et al. 2008), or polyploidization (Barrier et al. 2001), and been subject to selection.

G. *magnirostris*, *fortis*, and *fuliginosa* differ in the size and robustness of their beaks. This variation in adult beak morphology is paralleled by variation in the expression of a gene responsible for producing a signaling molecule, *Bmp4*, in the prenasal cartilage (*pnc*) in all three species at days 5 to 6 during 12 days of embryonic development—earlier and more intensely in the development of *magnirostris* than in the other species (Abzhanov et al. 2004; also Wu et al. 2004, Campàs et al. 2010). The association is more than just a correlation: experiments

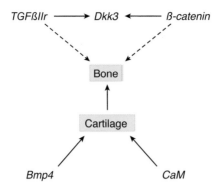

Fig. 12.2 Growth of beak dimensions is affected by the expression of different genes at different times during development. Bone morphogenetic protein 4 (*Bmp4*) affects beak depth and width, and Calmodulin (*CaM*) affects beak length in the developing prenasal cartilage. At the later premaxillary bone stage a pathway involving different signaling molecules affects development of beak length and depth only. Based on Mallarino et al. 2011.

have demonstrated a causal role of the molecule (Abzhanov et al. 2004). Thus agents that govern the timing and intensity of expression of the gene, such as other gene products and feedback loops, modulate the development of these beaks differently in the three species. Later in embryonic development the premaxillary bone (*pmx*) is laid down, and another interacting set of genes—*TGFβIIr*, *β-catenin* and *Dickkopf-3*—have similar differential effects on beak development of the three species (Mallarino et al. 2011). Relatively simple changes in gene regulation and the timing of gene expression must have occurred along the *fuliginosa-fortis-magnirostris* axis of the Darwin's finch radiation.

Development of *scandens* is different in two respects: the nature of the molecule and where it is expressed. At the prenasal cartilage stage another signaling molecule involved in a different pathway, calmodulin (*CaM*), is expressed in the growing tip of the embryonic *scandens* beak, but weakly if at all in the other three species (Abzhanov et al. 2006). The longer beak of adult *scandens* is the product of this influence, together with the later-acting set of three molecules at the premaxillary bone stage (Mallarino et al. 2011). The *pmx* trio of genes is expressed dorsally in the granivores and distally in *scandens*. A significant consequence of the combined differences in gene expression is that adult beak length in *scandens* is less strongly correlated genetically with depth and width than are depth and width with each other (chapter 10), and has more freedom to evolve independently (Grant and Grant 1994).

From the static comparison of species differences a picture is beginning to emerge of the evolutionary changes in developmental genetics that accompany speciation. Beak depth, width, and length can be altered one at a time by differential expression of the five identified molecules in three different biochemical pathways, either singly or in combination (Mallarino et al. 2011): length alone by calmodulin (*CaM*), depth and width (but not length) by bone morphogenetic protein 4 (*Bmp4*), and, a little later in development, depth and length (but not width) by *Dickkopf-3*, *TGFβIIr*, or *β-catenin*. Undoubtedly there are many more important molecules yet to be discovered: some candidates, in addition to *Bmp4*, have been discovered in the North American House Finch *Carpodacus mexicanus* (Badyaev et al. 2008, Badyaev 2010, 2011) and Caribbean tanagers related to Darwin's finches (Mallarino et al. 2012).

We would like to know how genes such as *Bmp4* are regulated because changes in regulatory networks are the causes of evolution in the size and shapes of beaks: the *Bmp4* gene itself varies very little among Darwin's finch species (A. Abzhanov pers. comm.). For example, a change in the timing of activation of *Bmp4* in *fortis*, resulting in an earlier onset of tissue-specific expression, potentially results in a deeper and wider adult beak, which is a shift in the direction of *magnirostris*. A question for future research is whether such change requires a new mutation, or whether a shuffling of existing variation is all that is necessary to produce just the right combination of genes in the formation of a new species. The next advance in our understanding will come with the identification of genetic differences in development among individuals within a population. Then it will be possible to discover the genetic consequences of selection on adults, because directional selection on adults is indirect selection on developmental programs (Price and Grant 1985). Equally valuable will be determining the genetic and developmental consequences of combining the genomes of species as different as *fortis* and *scandens* through introgressive hybridization.

Growth after Hatching

Nothing is known about gene function after hatching, but there are hints of interspecific differences in expression patterns from differences in growth trajectories, first as nestlings and then as fledglings (Grant 1981c, Boag 1984, Price and Grant 1985, Grant et al. 2006). As adults the three granivore species differ conspicuously in size and to a lesser extent in proportions (chapter 11), so we might expect that they

Fig. 12.3 Deformed beaks: disturbance to the normal process of growth cessation. **Left**: Mild hypertrophy of a side of the lower mandible. **Right**: Extreme hypertrophy. This bird picked up small seeds at the base of its beak on the left side. It was able to live for two to three months at the end of the wet season in the presence of plentiful food but failed to survive the following dry season of declining food supply. Birds elsewhere develop similar abnormalities. These include, for example, cardueline finches (Badyaev 2010) and titmice (parids; Handel et al. 2010) in North America.

grow along the same growth trajectory but achieve different adult proportions by stopping at different points along their trajectory (e.g., Badyaev et al. 2001). This turns out to be wrong; each species has a unique trajectory. A significantly faster proportional elongation of the beak occurs in *fuliginosa* than in the other two species. Growth of beak length relative to depth is slowest in *fortis*. Compared with *magnirostris*, the *fortis* trajectory is displaced in the direction of a greater beak length for a given beak depth. Perhaps this reflects a past history of introgression of genes from *scandens*. On Santa Cruz island, where hybridization appears to be much rarer, there is no such displacement (Boag 1984).

G. *magnirostris* (on Genovesa, and presumably Daphne) is unique in that beak depth initially increases at a faster rate than the other beak dimensions, and then there is a switch, and beak length increases at the faster rate during terminal growth (Grant 1981c). Two associated features of *magnirostris* imply a profound reorganization of timing and amount of resource allocation to the development of these dimensions: reversal of the relative sizes of adult beak lengths and depths, and the unique pattern of static allometry in which beak depth, but not length, varies in proportion to body size among years (fig. 11.11). The lack of scaling, and the same for *scandens* beak depth, is unexpected. It will remain a puzzle until more is known about gene expression in late development, the possible influence of fledgling diets (Genbrugge et al. 2011), and the factors that normally bring growth to an end (fig. 12.3).

Thus expression of developmental genes is likely to vary throughout the growth period in both scale-dependent and scale-independent ways, and not just in early embryonic growth. Although we know that species differ in gene expression and regulation, we are not yet in a position to quantify the differences that accompany speciation.

Rapid Tempo of Speciation

The four finch species on Daphne originally formed and diverged in less than a million years (Petren et al. 2005, Grant and Grant 2008a). Hypothetically the morphological transition from one species to another differing mainly in size, e.g., *fuliginosa* to *fortis*, could take place very rapidly, perhaps in a few hundred years, according to some simple arithmetic. For species of similar beak shape the difference between them in one dimension is about 2 standard deviations of each species' distribution, or 4 standard deviations in total. An evolutionary response of approximately 0.5 standard deviations followed selection events in *fortis* in 1977 and 2004 (table 7.2, fig. 7.5); therefore eight such events could effect a transition from one species to another.

The same conclusion of potentially rapid speciation has been reached in a study of three species of crossbills (*Loxia*) in Europe. They differ in body size and beak size in a manner parallel to the *fuliginosa-fortis-magnirostris* sequence, and differ ecologically too (Marquiss and Rae 2002), yet no difference in mitochondrial and microsatellite DNA has been found, a failure that cannot be attributed to introgressive hybridization (Piertney et al. 2001). Reproductively and ecologically they are three species, whereas genetically, at the studied loci, they are one (Piertney et al. 2001). These examples share with each other and with several other closely related species of birds (e.g., Lack 1944, Schoener 1984) the characteristic of feeding on foods of different size. As more examples of rapid speciation come to light (Milá et al. 2007, Moyle et al. 2009, Kirchman 2012), more will be sought and more will be found.

The two favorable conditions for the Darwin's finch radiation were ecological opportunity and genetic responsiveness. Like archipelagos elsewhere (Hawaii, Caribbean), Galápagos are unusually suitable for rapid diversification because they comprise many islands, well isolated from numerous competitors and predators on continents, and supporting populations in diverse habitats (Ricklefs and Bermingham 2007). Moreover the Galápagos environment changed over the last two million years, with a trend toward aridification in the last million years and presumably colonization of new plant species adapted to those conditions (Grant and Grant 2008a), which permitted exploitation of new food resources in new ways, especially by the granivores. Being phylogenetically young under favorable environmental conditions for diversification, finch populations evolved rapidly owing partly to elevated levels of genetic variation from introgressive hybridization. Rapid evolution of Darwin's finches (fig. 12.4) stands in contrast to slower patterns of diversification elsewhere on continents (Price 2010, Weir and Price 2011). Slower diversification has been characterized by Uyeda et al. (2011) as evolutionary fluctuations apparently bounded within a stable adaptive zone for one million years, and interpreted as local variations in niche optima due to restricted environmental variation. Causes and characteristics of the contrasting finch radiation are discussed further in chapter 15.

Interactions in Sympatry

A critical stage in the evolution of beak differences between species is the encounter of two populations derived from a common ancestor. There must have been countless dispersal and colonization events throughout the archipelago in the course of the Darwin's finch radiation. We witnessed one in 40 years. It gave us insights into both ecological and reproductive interactions between colonists and residents at this critical stage of the speciation cycle when divergence can be enhanced or reversed. Such interactions, occurring frequently, may have contributed to the rapid diversification of the ground finches.

Colonization by *magnirostris* led to ecological character displacement of *fortis* (chapter 7). This demonstrates an evolutionary adjustment of competitors to a limited resource. The two species did not interbreed either at initial contact or any time in the following 30 years, and therefore they are not strictly an example of secondary contact of differentiated populations as in figure 1.1. A hypothetical example of this process would be colonization of Daphne by unusually large *fortis* from another island such as San Cristóbal. They are 20% larger in average beak depth on this island than on Daphne, and with such large beaks they would be, like *magnirostris*, superior exploiters of *Tribulus* fruits.

Lack of interbreeding between *fortis* and *magnirostris* on Daphne shows that they were sufficiently different in beak morphology and song. Both cues are important in mate recognition and choice, but one piece of evidence suggests morphology can be the more important. A minimum of nine *fortis* males sang *magnirostris* song, presumably

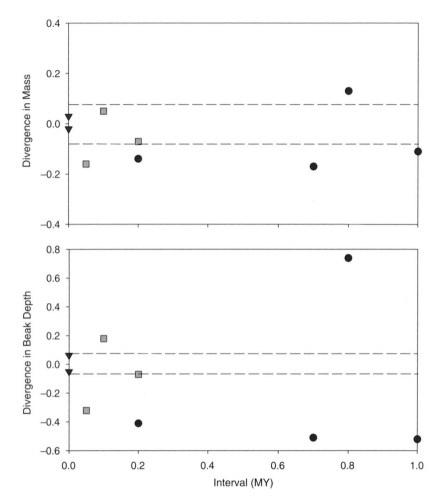

Fig. 12.4 Finches in the genus *Geospiza* have diverged relatively rapidly in beak depth, less so in body size. Divergence is measured as the difference between the means of log-transformed size in pairs of species, and, in the case of mass, weight is divided by 3 to correct for dimensionality. The broken lines show the maximum divergence up to one million years in a large data set of birds in Uyeda et al. 2011: after one million years of independent evolution divergence increases in this set. Squares identify pairs of tree finch species (*Camarhynchus* species), and circles identify pairs of *Geospiza* species. Morphological data are from Grant et al. 1985, and phylogenetic data are from Petren et al. (2005). The two strongest observed selection events in *fortis* are shown by solid triangles: before and after selection means are contrasted.

through learning the song during their sensitive period (figs. 8.6 and 8.9), yet none of them bred with *magnirostris* females. *G. magnirostris* males recognized their songs and responded by chasing the *fortis*. We interpret these observations to signify that the beak and body size differences between the species were large enough to prevent interbreeding, regardless of their song.

Nevertheless interactions involving song differences are also important at this stage of secondary contact. The songs of both *fortis* and *scandens* underwent a change after *magnirostris* numbers had increased, starting in the 1990s (Grant and Grant 2010c). Songs of both species became faster in rate of note repetition or trill rate, and in doing so they diverged from the songs of *magnirostris* but not from each other (fig. 12.5). Divergence began well before the character displacement shift in *fortis*, and therefore cannot be simply accounted for as a correlated effect of a change in beak morphology that affects song characteristics (Podos 1997, 2001, Huber and Podos 2006). It happened without a change in the structural habitat more than one meter above ground where birds sing, therefore it cannot be explained as a change in the sound transmission properties of the environment (Wiley 1991, Luther and Wiley 2009). Since songs of two species, *fortis* and *scandens*, changed in the same direction and at the same time, divergence is unlikely to have been caused by random changes (cultural drift). The most likely explanation is acoustic interference between species (Nelson and Marler 1990, Luther 2009). *G. magnirostris* are large and aggressive, frequently interfering with and attempting to take over the nests of *fortis* and *scandens*. Their songs are louder and are in the same frequency bandwidth as the songs of *fortis* and *scandens*. According to this explanation, divergence reduces acoustic and physical interference (Grether et al. 2009), as well as improving communication through learning (Peters et al. 2012, Ríos-Chelén et al. 2012) and transmission (Luther and Wiley 2009, Nemeth et al. 2013) in a noisy environment.

A MECHANISM PRODUCING SONG DIVERGENCE

Divergence from *magnirostris* songs appears to be an example of the phenomenon known as a peak shift (ten Cate et al. 2006, ten Cate and Rowe 2007). During the process of discriminatory learning and production, an individual learns to respond to an exaggerated form of the desired signal or part of the signal in a direction away from a negative stimulus. This happens if there is a penalty to interactions with a heterospecific individual, such as the larger, socially dominant *magnirostris*. For example, a son, surrounded by singing heterospecifics during

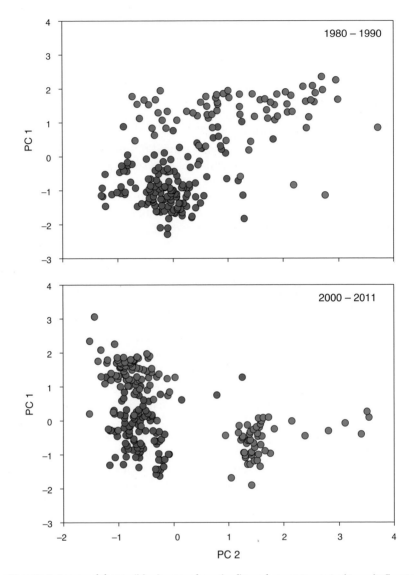

Fig. 12.5 Songs of *fortis* (blue), *scandens* (red), and *magnirostris* (green). Songs were recorded from 1980 to 1990, when there were few or no singing *magnirostris* present, and again from 2000 to 2011, when *magnirostris* were common. All individuals were recorded once only. Comparison of the two time periods shows the extent of divergence of *fortis* and *scandens* songs away from *magnirostris* songs. Variables used in the analysis were trill rate, song duration, and frequency bandwidth. Both *fortis* and *scandens* songs became shorter and trill rate (notes/sec) became faster (PC2), whereas frequency bandwidth (PC1) did not change. The songs of two *scandens* recorded in 2011, 19527 and 19518, were outliers and were removed from the analysis. Both had very fast songs of 35 and 36 notes per second respectively, and were furthest removed from *magnirostris*. Adapted from Grant and Grant 2010c.

his sensitive period of learning or production of song, would be expected to learn and produce elements of conspecific song that were clearly distinct and not blocked by sympatric heterospecific song.

We investigated the mechanism of change and found the origin was in the song learning process (Grant and Grant 2010c). Although sons copy songs of their father quite faithfully, we discovered they produced slightly faster songs than their fathers during this period of change. For those individuals of each species tested in adulthood twice, trill rate was faster at the second time in the 1990s than at the first time in the 1980s. This trend occurred in all of 10 *fortis* individuals and in 8 of 9 *scandens* individuals. Other characteristics of the song, such as song length and number of notes, did not change.

We found no evidence of a female mating preference for males that sang an exaggerated form of conspecific song that is clearly distinct from *magnirostris* song, even though one might be expected. Thus the song shift was restricted to males, a form of agonistic reproductive character displacement (Grether et al. 2009).

Summary

Speciation occurs when one species splits into two populations or sets of populations that do not interbreed or interbreed rarely with little genetic exchange. We observe the products, species, and attempt to reconstruct the process by which they were formed. In this chapter we use information from adult morphology of Daphne finches and genes expressed in the development of beaks to reconstruct and interpret the transformation of species. The three granivore species, *fuliginosa*, *fortis*, and *magnirostris*, constitute an allometric series of large differences in size accompanied by small changes in beak shape. Transformation of these morphologies in either direction can be effected mostly by selection on highly heritable size variation alone; beaks change in shape as a correlated effect. *G. scandens* differs from the granivorous species in beak proportions, and correspondingly the change from one adult morphology to another is less easily effected by selection on size variation alone. A different transformation is suggested by the genetic factors in beak development of *scandens*: they differ in kind and in location of expression from those governing beak shape development of the granivores. The three granivore species differ primarily in timing and intensity of expression of a few developmental genes.

Other species are part of the selective environment of each species. When previously separate populations encounter each other, they may diverge as a result of selection, which promotes speciation, or converge through introgressive hybridization, causing speciation to collapse. Separate populations encountered each other when Daphne was colonized by *magnirostris*. In addition to causing ecological character displacement in *fortis*, *magnirostris* affected the songs of *fortis* and *scandens*. Their songs diverged from the songs of *magnirostris*, which we interpret as a learning response that resulted in a reduction of both acoustic and physical interference from a socially dominant species.

Speciation by Introgressive Hybridization

But the chief cause of our unwillingness to admit
that one species has given birth to other and distinct
species, is that we are always slow in admitting
any great change of which we do not see
the intermediates.

(Darwin 1859, p. 481)

All our knowledge has its origins in our perceptions.

(Leonardo Da Vinci; Richter 1985, p. 4)

Introduction

OVER THE PAST 20 YEARS hybridization has been seen increasingly as an important factor in speciation (reviewed in Schwenk et al. 2008, Abbott et al. 2013). An exchange of genes between related species or even genera was once thought to be a phenomenon largely restricted to plants, but is now known to be widespread and to occur from bacteria (Jain et al. 1999, Ochman et al. 2000) and corals (Veron 1995, Willis et al. 2007) to arthropods (Mallet 2005), birds (Grant and Grant 1992b, McCarthy 2006), and primates (Arnold and Meyer 2006, Schwenk et al. 2008). Even in our own evolutionary lineage there is evidence of gene transmission from Neanderthals to our direct ancestors (Reich et al. 2011). The important question is what effect does gene exchange have on the receiving populations, and in particular what is the relevance to speciation?

Hybridization can give rise to a new species through the production of fertile offspring possessing a different number of chromosomes from either parent (Coyne and Orr 2004, Mallet 2007). Polyploid species thus formed are well known in plants and in some arthropods (Bullini 1994) and cold-blooded vertebrates (Dawley and Bogert 1989), but not in birds. Hybridization of birds is more likely to cause a merging of two species as a result of introgression (chapters 9 and 10), or a transformation of them through enhanced genetic variation, than the evolution of a new species, unless the hybrids occupy a different habitat or location as happens in some plants (Huxley 1942, Rieseberg, Raymond, et al. 2003) and in a few groups of mammals (Larsen et al. 2010), salamanders (Pereira and Wake 2009), fish (Nolte and Tautz 2009, Hudson et al. 2011), flies (Schwarz et al. 2005), and butterflies (Gompertz et al. 2006, Mavárez et al. 2006, Kunte et al. 2011, Brower 2013). Two examples of bird species where this is suspected have been reported recently, and both involve a large degree of geographical isolation of the putative hybrid population. The Italian Sparrow (*Passer italiae*) is morphologically intermediate in plumage, and genetically intermediate, between the House Sparrow (*Passer domesticus*) and the Spanish Sparrow (*Passer hispaniolensis*), largely isolated reproductively and geographically from both of them, being the sole occupant of most of the Italian peninsula (Hermansen et al. 2011). In North America Audubon's Warbler (*Setophaga auduboni*) on the western side of the Rocky Mountains is genetically and phenotypically intermediate between Myrtle's Warbler (*S. coronata*) farther east and the Black-fronted Warbler (*S. nigrifrons*) farther south (Brelsford et al. 2011, Milá et al. 2011; reviewed in Rheindt and Edwards 2011).

In the case of Darwin's finches hybridization could lead to the formation of a new species if hybrids from one island fly to another, ecologically different island, establish a breeding population, and undergo evolutionary divergence that culminates in reproductive isolation from the parental species (Grant et al. 2005). The colonists need not be solely introgressed hybrids for speciation to occur; they could be a mixture of F_1 hybrids, backcrosses, and members of the parental species. Hybrids and parental types might interbreed, and the increased genetic variation would enhance the potential for change in a new direction (chapter 9). Alternatively the parental types might have poor post-colonization success and contribute less and less than the hybrids to future generations.

We now describe an example of incipient hybrid speciation on Daphne (Grant and Grant 2009a). Like the establishment of a breeding population of *magnirostris* (chapter 6) it began with immigration to the island, but unlike that event there was only a single immigrant. Its descendants formed a population that became reproductively isolated from the other finches on the island. We refer to this as introgressive speciation because backcrossing was an important part of the initial separation of the lineage from the parental species. We describe how this unlikely outcome transpired and interpret why it happened.

A Hybrid Arrives on Daphne

In 1981 Trevor Price captured a strange bird (number 5110) while doing routine netting of offspring of known parents for heritability analysis. It was young, in brown plumage, yet unusually large. In weight (29.7 g), beak length (14.1 mm), and other dimensions it was clearly beyond the range of measurements of all known resident *fortis* and *scandens*. It was more similar to *fortis* in beak proportions than to *scandens*. At the time of its arrival we, together with our assistants, had measured more than 90% of all resident *fortis* and *scandens*, and observed the breeding of the remainder, so we were confident that it had not hatched on the island. We considered it to be an immigrant *fortis*, and suspected it had hatched earlier in the year on neighboring Santa Cruz Island (Grant and Grant 2008c).

In 1990 we recaptured 5110, known by now to be a male and in shiny black plumage (fig. 13.1), and took a blood sample for microsatellite DNA analysis. Many years later we had assembled blood samples from several neighboring islands and were able to assign 5110 to a population of origin by using the newly developed program STRUCTURE (chapter 10; Pritchard et al. 2002, 2007). First, we confirmed it did not originate in either the *fortis* or *scandens* populations on Daphne. Second, it was assigned with more than 90% probability to the population of *fortis* on Santa Cruz Island. Third, in a separate analysis with the addition of Santa Cruz *scandens*, STRUCTURE assigned a significant fraction of the genome to *scandens* (0.341), the remainder being assigned to *fortis* (0.659). We concluded the immigrant was a hybrid, possibly a backcross to *fortis* of a *fortis* × *scandens* F_1 hybrid (Grant and Grant 2009a). We designate this FFS as explained in figure 8.16.

Fig. 13.1 Two lines of descent from an immigrant. **Upper left**: The original immigrant hybrid *fortis*, 5110. **Upper right**: A son, 15830, produced on the *A* line of descent. **Lower left**: 19256, a member of the Big Bird lineage on the *B* line of descent (generation f_4). **Lower right**: 17618, a second-generation member of the *A* line of descent and daughter of 15830's brother.

Descendants

PHASE I: THE START OF A NEW LINEAGE

Bird number 5110 was not only unusual morphologically; he was unusually fit. He lived 13 years, paired with six different females, and fledged at least 18 offspring of which 5 became breeding recruits (fig. 13.2). He bred for the first time in 1983, and while many birds experienced high reproductive success in this extraordinary El Niño year (chapter 5), he only managed to acquire a mate toward the end of the prolonged breeding season. In the next El Niño year (1987) his success was much higher.

Initial choice of mates appears to have been highly selective in that the first three mates were, like him, all FFS backcrosses, but produced on Daphne; two of them were sisters. Backcross individuals were very

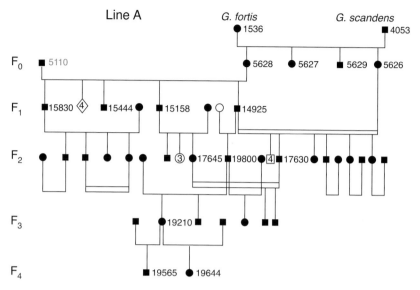

Fig. 13.2 The A line of descent from the immigrant male 5110 when paired with hybrid (backcross) females. Bird number 5110 initially bred in 1983 and 1984 with 4734, a female with a similar *fortis* × *fortis-scandens* (FFS) genetic constitution. This is omitted from the figure because only four offspring fledged from one of five nests with eggs, and they all died within a year. As shown in the figure, he then bred with the genetically similar 5628 (FFS) five times in 1987 and fledged eight young. He bred unsuccessfully with the sister 5626 in 1990. From generation 1 onward the members of the pedigree bred with each other (close inbreeding, indicated by parallel horizontal lines) or with members of the *fortis* population. Note: generations 3 and 4 are underestimated because the study of breeding from 1992 onward was not as comprehensive as beforehand. Male 19800 was the only known member of the pedigree to survive the 2003–4 drought. One member (19210) was alive in 2012. Symbols: squares, males; circles, females; diamonds, birds of unknown sex; filled, known by observation and/or genotype; open, unknown but inferred (see text). Multiples of nonbreeders of the same or unknown sex are shown by numbers within symbols. Birds 5110 and 15830 are illustrated in figure 13.1, and *scandens* 4053 is illustrated in figure 8.7.

rare in the population, as these and another sister and a brother were the only ones known to us in 1987 from pedigree analysis. The second three mates of 5110 (1991–93) were all *fortis* (fig. 13.3).

Thus there were two lines of descent from the two groups of females, an A line from the backcrosses and a B line from *fortis* (fig. 13.1). They were connected at only one point. Paternity analysis with microsatellite data revealed that 5110 was cuckolded when breeding with 5821

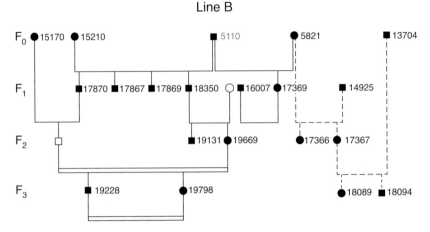

Fig. 13.3 The *B* line of descent up to generation 3 from the immigrant male 5110 when paired with three *fortis* females. In 1991 he bred with 5821. Two of three fledged young were sired by 14925, 5110's son from the *A* line (fig. 13.2), and this pathway is indicated by broken lines. In 1992 he bred with 15210 and fledged at least five young. In 1993 he bred with 16594. This is not shown, because hatching and fledging success of their two eggs is unknown. Symbols as in figure 13.2. Birds 5110 and 5821 are illustrated in figure 13.10.

(line *B*) by a son (14925) from line *A*. There are several instances in the pedigree of close inbreeding but no other known instances of cuckoldry in figure 13.1.

THE PHENOTYPIC UNIQUENESS OF **5110**. Bird number 5110 was unusual in the two features identified in chapter 6 as being important in the choice of mates: morphology and song. In beak shape he resembled *fortis* more strongly than *scandens*, as expected given his genetic constitution. Large size and beak shape are likely to have been factors in the choice of 5110 by FFS females in the initial pairings, possibly also factors in the choice by *fortis* females. The role of song in choice of mates is not so easily assessed. His song combined a rapid down-sweep of the first note with repetitions of four to six notes (fig. 13.4). This resembles the songs of *scandens* to some extent but also one type of song sung by *fortis*, distinctly slower than *scandens* songs, which we refer to in the field as "chey-yey-yeh" (fig. 13.4) (Millington and Price 1985, Gibbs 1990, Grant and Grant 1996d). The origin of 5110's song is unknown. We have never heard or recorded the song on other islands, nor did Robert Bowman (1983) in his extensive survey on many islands, and nor have Jeff Podos and Sarah Huber, who have spent much time recording ground finch songs on Santa Cruz Island (Huber

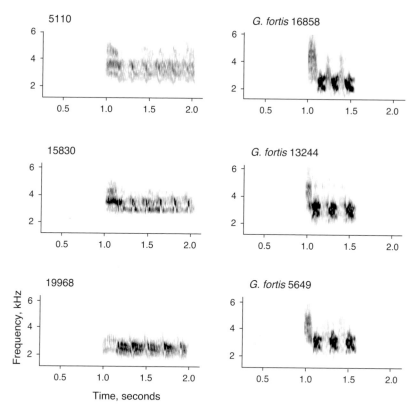

Fig. 13.4 Songs of the hybrid lineage and the most similar type of song sung by *fortis*, type III ("chey-yeh-yeh"; Gibbs 1990). A few songs of *scandens* resemble the hybrid lineage songs, perhaps as a result of copying, as they were not present on the island until 1993. From Grant and Grant 2009a.

and Podos 2006 and pers. comm.). The most likely explanation is that 5110 learned songs from his father and other birds of both species in his natal area, and produced a modified version of paternal song on Daphne, influenced to some degree by what he heard on Daphne during the crystallization phase of song production in his first breeding season in 1983. G. *fortis* males typically sing a perfect copy of their father's song, but some learn and copy another type of song sung by other members of the population (fig. 8.3) (Grant and Grant 1996d).

PHASE II: GENERATIONS 1 AND 2

Not all members of the next two generations (F_1 and F_2) were genotyped; nevertheless it appears from our reconstruction that only the *B* line contributed to the lineage of large birds (fig. 13.3). This is surprising because 5110 was genetically more similar to members of the *A* line, which initially were a mixture of *fortis* and *scandens* genes, than to members of the *B* line. Our conclusions regarding lines of descent in these generations and later ones are based on observations of breeding pairs and genetic inferences (box 13.1 and appendix 13.1). The genetic inferences make use of the fortunate fact that 5110 was homozygous for a rare allele (*183*) at microsatellite locus *Gf.11*. The rarity of the genotype can be appreciated by comparison with other species: only 2 of 2,302 *fortis*, 10 of 956 *scandens*, and no *fuliginosa* (*n* = 43) or *magnirostris* (*n* = 590) were homozygous for the *183* allele at this locus.

In the *B* line of descent two of four sons (F_1 generation) produced by 5110 × 15210 survived long enough to breed and produce the F_2 genera-

Box 13.1. Pedigree

Here we explain the procedure for identifying missing members of the pedigree. When the genotypes of both parents are known but possible offspring have not been recorded at a nest, the procedure is straightforward. A search is made among the genotypes of all unknown birds for an individual with the combination of alleles that is consistent with inheritance from the known parents. The procedure is different when one parent and an offspring are known but the other parent is not known. All the alleles contributed by the known parent to its offspring are subtracted from the offspring's genotype. The unknown parent must have contributed the remaining alleles. A search is then made among the genotypes of all known potential parents for these remaining alleles. Potential parents are excluded on the basis of mismatches of more than two base pairs at one or more loci. For example, by this subtraction-exclusion procedure we identified 19669 (generation 2) as the mother of 19228 and 19798 (generation 3; fig. 13.3). Importantly, no other bird's genotype matched both of the genotypes of the two generation 3 birds except for 19131, a male that was not known to be alive in the breeding season of 2002. The father of 19669 was identified as 18350, but the mother could not be found among the genotyped birds. All known females of the *A* lineage were excluded. As explained in the text, song and size provide additional information for identifying positions of males in the pedigree. Appendix 13.1 provides more detail on the identification of breeders. Previous reconstructions of the pedigree have been revised as more connections have been established (Grant and Grant 2008c, 2009a).

Fig. 13.5 Generation 3 and a Big Bird family. **Upper left**: Father, male 19228. **Upper right**: Mother, female 19798. **Lower left**: Male 19228 cracking *Tribulus cistoides* fruits using the same technique as *magnirostris* (fig. 6.10). **Lower right**: Male 19228 feeding fledgling son 9807.

tion. One of them (18350) bred with an unidentified female that may have been a *fortis*. The other (17870) bred with an identified *fortis* female (fig. 13.3). There was no known or suspected interbreeding with *scandens*.

PHASE III: ENDOGAMY AND REPRODUCTIVE ISOLATION

The fates of the two lines of descent differed, and the *A* line will be discussed later. All members of the *B* line died in the devastating drought of 2003–4 except for two, a male (19228) and a female (19798) (fig. 13.5). Their genotypes matched the genotype of 19669. For this reason, and because they were homozygous (*183/183*) at locus *Gf.11*, we consider them to be siblings, members of the F_3 generation.

The pedigree of generations F_3 to F_6 is shown in figure 13.6. All individuals in these generations were homozygous (*183/183*) at locus *Gf.11*. All those we were able to capture as adults and measure were large in body and beak measurements (figs. 13.7 and 13.8). All adult males sang

Big Bird Lineage

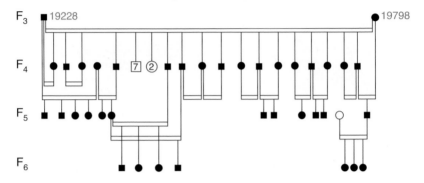

Fig. 13.6 The Big Bird lineage, an extension of the *B* line of descent from the original immigrant 5110 (fig. 13.3). Symbols as in figure 13.3. One pair has been omitted for simplicity: 19352 × 19742. They did not produce offspring. Note the original male 19228 bred with two daughters. In each case they were paired with brothers, so the offspring were extra-pair young.

the song sung by 5110 (figs. 13.4 and 13.9), and those that reached a mature black plumage were shiny black like 5110. The consistency in all four traits—genetics, morphology, song, and plumage—indicates that members of the lineage bred with each other (endogamy) and were thus reproductively isolated from *fortis*. In view of their large size in all dimensions we refer to them as the Big Bird lineage (fig. 13.10).

Direct observations of breeding and behavior verify the conclusion of reproductive isolation. All pairs observed breeding in the years 2006 to 2012 were made up of members of this lineage: we observed no mixed pairs. As early as 1993, when the *A* line of descent was at its most numerous, most males established territories adjacent to each other (fig. 13.11). After 2005 the same pattern of contiguous territories was evident among males of the Big Bird lineage (fig. 13.12), with the majority of males in acoustic contact with at least one other. Playback of tape-recorded song and observations showed them to be aggressive to other members of the lineage that entered their territories but not to members of the other three species, which they ignored unless they approached to within a few meters of an active nest or a favored song perch. In these respects their behavior resembled the intraspecific aggressive behavior of *fortis*, *scandens*, and *magnirostris*.

All evidence thus points to reproductive isolation of the Big Bird lineage. Nevertheless males may have engaged in cryptic extra-pair mating with females of the other species (exogamy). If so we would expect offspring that carried the *183* allele to match their fathers at every one of the other 15 microsatellite loci. To test this we examined

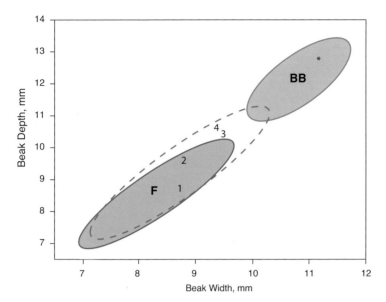

Fig. 13.7 Beak sizes of the Big Bird lineage in comparison with *fortis* present in the years 2006–12. Symbols: Big Birds in purple; *fortis* before the 2004 drought shown with a broken line (95% density ellipse) and after the drought in blue; asterisk 5110. Numbers refer to four female parents: 1 = 5628, mother of the *A* lineage (fig. 13.2); 2 = 15210, mother of the *B* lineage (fig. 13.3); 3 = 5821, another mate of 5110 whose offspring did not survive; 4 = 15170, mate of 5110's son 17870 and grandmother of the Big Bird lineage (figs. 13.3 and 13.6).

Fig. 13.8 Big Bird in comparison with a Medium Ground Finch. **Left:** Big Bird 9807, distinctly larger than *fortis* and with a large, blunt beak despite being an immature only one to two months old, as indicated by the multicolored lower mandible. Figure 13.5 shows it being fed by its father. **Right:** *fortis* 19181, of typical size and proportions.

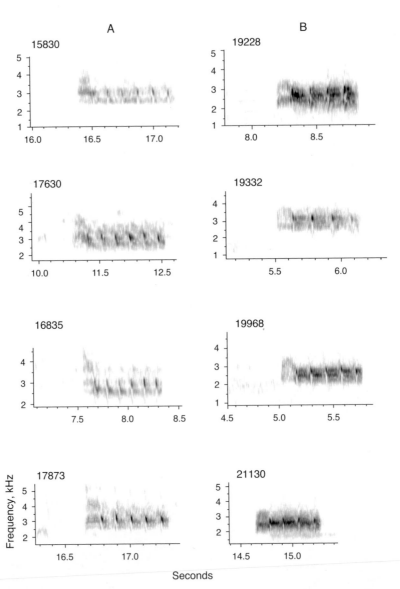

Fig. 13.9 Songs of several members of the two lines of descent (*A* and *B*) from the original immigrant male 5110, and the song of one *scandens* (21130) that sang a *B* lineage song in 2011. On the *A* line 15830 was a son of 5110, 16835 was his son, and 17630 was another grandson of 5110. Male 17873 was a grandson raised by 5110 but fathered by 5110's son 14925 (see figs. 13.2 and 13.3). On the *B* line 19228 was the male in generation 3, and 19332 and 19668 were his sons. In addition 3 of 10 recorded *A* line males sang a *fortis* type 1 song (see fig. 8.2), whereas all 17 *B* line males (some not recorded) sang the same song as 5110. Note the subtle difference in tempo and pitch between the *A* and *B* line songs. The difference probably arose in generation 3 because there were no other *B* lineage males present when 19228 first sang.

Fig. 13.10 Big Bird lineage: males on the left, females on the right. **Upper left**: 5110, the original immigrant male. **Upper right**: 5821, one of 5110's *fortis* mates, although not the mother of the lineage (photo not available)—it is included to indicate size and shape of the beak. **Upper middle left**: 19256 (generation f_4). **Upper middle right**: 19566 (generation f_4), long-term mate of 19256. **Lower middle left**: 19410 (generation f_4). **Lower middle right**: 19321 (generation f_5: offspring of 19256 × 19566 f_4). **Lower left**: 19721 (generation f_4). **Lower right**: 19364 (generation f_4).

Fig. **13.11** A local community of the hybrid lineage in 1993. The original immigrant, 5110, established a territory here in 1983 and held it until 1994. Two of his sons that hatched in 1987 established territories in adjacent locations in 1991: 15158 and 15830. A third son, 14925, established a territory next to his brother 15158. Two of 14925's sons that hatched in 1991, 16833 and 16835, established adjacent territories in 1993, and one of them, 16833, bred with a sister, 16834, from the same natal nest. Two other sons of 14925 established territories respectively ~50 m to the north (17630) and ~100 m to the northeast (17948) on the inner slope near the crater. All individuals were members of the A line of descent. One member of the B line, 17870, hatched in 1991 and established a territory next to his father 5110, while a brother (18350) hatched in 1992 and established a territory on the outer slope in sector 11 (fig. A.3.1). All males of the two lines sang the song of 5110. The pattern of aggregated and contiguous territories of relatives is highly unusual. In contrast only 4 of 251 *fortis* offspring (2 males and 2 females) that hatched in 1987 (1.6%) bred on territories adjacent to their parents' territories. For *scandens* the numbers are 2 (both daughters) out of a total of 50 (4.0%). There is only one record in the sample of 80 *fortis* and 8 *scandens* grand-offspring of grandsons (2) establishing territories adjacent to parents and grandparents. The grandfather was a *fortis* × *fuliginosa* F$_1$ hybrid. One of the grandsons bred with a granddaughter of 5110, thereby combining the genes of *fortis*, *scandens*, and *fuliginosa* in the offspring (Grant and Grant 2010b). Filled circles indicate nest locations. From Grant and Grant 2009a.

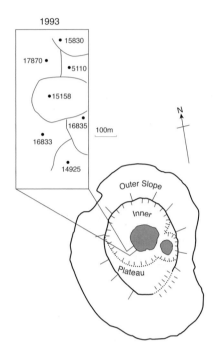

Fig. **13.12** Contiguous territories of Big Birds (line B) in 2010. Only one male of the A line (not shown) was known to be present at this time; it sang the type 1 song of *fortis* and held a territory on the outer slope in acoustic isolation from males of the Big Bird lineage. The distribution illustrates the fact that territories are held on the inner and outer slopes of the crater but not on the plateau (fig. 3.11), where the three other species breed.

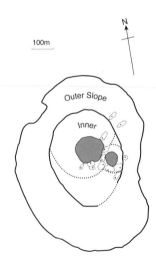

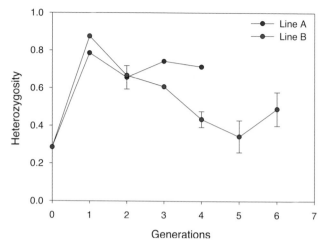

Fig. 13.13 Average multilocus heterozygosity of birds hatched in successive generations in the *A* line and the *B* line (Big Bird lineage), beginning with 5110 at generation 0. For each generation the fractions of 14 autosomal loci that are heterozygous have been averaged across individuals. A strong decrease in average heterozygosity in the *B* line followed a population bottleneck of a single breeding pair at generation 3 (figs. 13.3 and 13.6). Sample sizes for the following generations are 4, 19, 5, and 2 for the *A* line and 4, 6, 2, 27, 11, and 7 for the *B* line. Even though sample sizes are generally low, the difference between lines is very large when the largest samples are compared: generation 2 of the *A* line (*n* = 19) has twice the average heterozygosity of generation 5 of the *B* line (*n* = 27). Ninety-five-percent confidence limits are shown for samples larger than 6 individuals. For comparison, the long-term average heterozygosity of *fortis* and *scandens* is approximately 0.67 (fig. 10.12). Averages for each of the last three generations of the *B* line are significantly lower than the average at generation 2 for the *A* line.

several hundred individuals of known genotype of *fortis* (*n* = 665), *scandens* (*n* = 432), and *magnirostris* (*n* = 308) captured in 2006–12. The frequency of the *183* allele was low in all three species: 0.04 in *magnirostris*, 0.08 in *fortis*, and 0.17 in *scandens*. None of them with the *183* allele matched any member of the lineage at the remaining loci. Moreover none of the *fortis* or *scandens* were large. We conclude that members of the Big Bird lineage bred with each other and not with members of the other species. Significantly, all known and suspected extra-pair fertilizations were within lineage (fig. 13.6). If this is not simply a reflection of the tendency to mate with extra-pair neighbors (appendix 5.1), the pattern is strong evidence for within-group mate choice. Close inbreeding prevailed, and average heterozygosity declined (fig. 13.13), but without loss of any of the 28 alleles present in the combined genotypes of the brother (19228) and sister (19798) of generation F_3.

Origin of Reproductive Isolation

Bird number 5110 was distinctive in morphology and song. These characteristics were passed on to successive generations through genetic and cultural transmission respectively. Viewed from an interspecific perspective, the differences between the Big Bird lineage and other species are twin components of a barrier to interbreeding. Tracing back the lineage from generation F_5, we can ask when did the barrier to interbreeding arise? The answer cannot be the F_1 generation, because sons had no surviving sisters, as far as we know, and therefore they had to breed with females of the *fortis* population. The barrier arose at generation 2 when 17870's son bred with 18350's daughter (19669). This was the first generation to have a choice between a relative and members of the *fortis* population. Reproductive isolation of the B lineage from *fortis* continued from then on. Intragroup mate recognition and endogamy in generations F_3 to F_5 may have been facilitated by the scarcity of potential *fortis* mates of appropriate size as a result of selection in 2003–4 (chapter 7).

Reproductively isolated from the other species, the B line individuals are behaving as a separate species. Regardless of whether they should be considered a separate species or not (box 13.2), they provide a valuable example and insight into one way in which a new species forms. The lineage originated allopatrically and became reproductively isolated sympatrically (box 13.3) at the F_2 generation. Both members of the breeding pair were third-generation backcrosses, assuming correct identification of 5110 as a first-generation backcross. Subsequent generations of the lineage retained that genetic composition by breeding endogamously and forming an inbreeding swarm.

This is speciation of a special sort: introgressive speciation.

Fate of the *A* Line of Descent

Further insight into how the reproductive barrier arose is gained by comparing the two lines of descent. From generation 2 onward there was no interbreeding between members of the *A* and *B* lines; they were reproductively isolated from each other. Members of the *A* line, in contrast to the *B* line, bred with *fortis* and became integrated into the *fortis* population. Size may have played a role: members of the *A* line had a smaller (hybrid) mother than members of the *B* line (fig. 13.7), and from generation 1 onward were smaller on average than members of the *B* line (fig. 13.14) as a result of the repeated breeding with *fortis*

Box 13.2 A Question of Species

Discretely different from *fortis* in song, and almost so in beak size, the Big Bird lineage has functioned as a biological species through being reproductively isolated from *fortis* as well as the other species. Therefore the question arises as to whether it is appropriate to refer it as a new species. Several concepts of species, including the biological species concept, share the characteristic of a lineage that maintains continuity and integrity through successive generations. They have been unified and referred to as the lineage concept of species (de Queiroz 1998, 2011). The Big Bird lineage conforms to this concept of a species, and therefore it would be justified to give it a scientific name, such as *Geospiza strenuirostris*. The only problem with doing so is the short life of the lineage as a reproductively isolated entity. How many generations of exclusively within-group mating are needed before the group is recognized as a separate species? The answer has to be arbitrary. If, say, the number 10 is chosen, then after 10 generations the conclusion will be reached that the lineage has achieved the status of a separate species, and it actually did so at generation 2! Faced with this conundrum, we maintain only that the Big Bird lineage is functioning as a reproductively isolated species, a species "in the making," and has done so for five generations. Of greater interest and importance than its taxonomy and nomenclature are the insights it provides into how a new species arises and either persists or becomes extinct.

(fig. 13.1). In contrast to the morphological difference between the two lines, males of both lines sang 5110's song (fig. 13.9). Thus it appears that beak and body size were more important than song in choice of mates and hence as components of the barrier to interbreeding. A similar barrier occurs between *fortis* and the much larger *magnirostris* (chapter 8).

Initially members of the *A* line were partly isolated reproductively from *fortis* and apparently heading in the same direction of complete isolation taken by the *B* line. The tendency for the *A* line male territories to be contiguous, even clustered, in 1993 (fig. 13.11) suggests that the males recognized each other as distinct from the other species. Song may have played a major role in this recognition. Females possessed the same recognition as the males, as suggested by four pairings of close relatives. These two features of territory contiguity and close inbreeding also characterized the *B* line. However, isolation of the *A*

Box 13.3 A Question of Speciation

The reproductive and ecological isolation of the Big Bird lineage on Daphne has allopatric and sympatric elements. It does not fall neatly and discretely into categories of allopatric or sympatric species, because the hybrid lineage initiated by the interbreeding of *fortis* and *scandens* on one island became reproductively isolated from the two parental species on another island. The isolation owes its origin to introgressive hybridization on Santa Cruz Island. The hybrid immigrant possessed the traits, passed on to subsequent generations with little alteration, that conferred success on the lineage. This conforms to the classical allopatric model of speciation, in which evolution in allopatry, specifically adaptive divergence, facilitates coexistence in sympatry at the secondary contact stage of a speciation cycle (e.g., Sobel et al. 2010, Harrison 2012), with or without continuing gene flow or further divergence in sympatry (chapter 1). In other respects speciation was essentially sympatric, because the assortative mating and the barrier to interbreeding arose only in sympatry (e.g., see Dieckmann and Doebeli 1999, Ryan et al. 2007, Huber et al. 2007, Ito and Dieckmann 2007, Grant and Grant 2009b, Hendry et al. 2009, van Doorn et al. 2009). The traits that constitute the barrier, distinctive morphology and song, were possessed by the immigrant. The barrier was not sufficient to prevent interbreeding at generation 0, for had it been so the immigrant would have died without reproducing. Instead the barrier became effective only after a potential mate (daughter) had been produced at generation 1 by outbreeding. In other words the reproductive barrier was latent. There was no evolution of morphology or song from generation 1 onward, therefore no evolution of a reproductive barrier in sympatry. The development of reproductive isolation of the Big Bird lineage was not typical of allopatric, sympatric, or ecological speciation, although it comprised elements of all of them.

line from *fortis* weakened and eventually disappeared; they became absorbed into the population of *fortis*. Only one member of line *A*, 19800 (a male), survived the 2003–4 drought, and he died in 2008. He and his known descendants continued the trend of breeding only with *fortis* up to our last study of breeding in 2012. Thus lines *A* and *B* had different fates, and we suggest the reason for this was their difference in size (fig. 13.14).

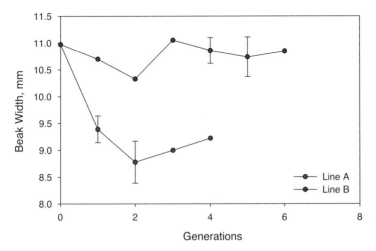

Fig. 13.14 Average beak width of birds hatched in successive generations in the *A* line and the *B* line (Big Bird lineage). The founder, 5110, is the sole member of generation 0 for both lines. Large average size persisted across generations in the *B* line, in part due to the high frequency of inbreeding, whereas average beak width of the *A* line fell in generation 1, owing to 5110 breeding with a small backcross female (BW = 8.2 mm), and remained at approximately that size for the following generations. The difference in generation 1 is pronounced. Measurements of 8 line-*A* birds (maximum 9.8 mm) all fall well below the measurement of the single line *B* bird (10.7 mm). Ninety-five-percent confidence limits on estimates of the means are shown with vertical lines for only those based on samples larger than 5 individuals. Sample sizes are 8, 10, 3, and 2 for the four generations of the *A* line and 1, 2, 2, 25, 8, and 2 for the *B* line. Sex is known for most but not all birds. Corrections for sexual dimorphism (Price 1984b) have not been applied; they are too small to account for the differences between lines.

Success of the Lineage So Far

The Big Bird lineage has persisted for more than 30 years since the arrival of its founder in 1981 (fig. 13.15). What determined its success? Intrinsic factors associated with the hybrid nature of the genome, conferring hybrid vigor, could be part of the answer. Extrinsic factors affecting survival are a second part.

INTRINSIC FACTORS

We have no direct measure of an intrinsic factor affecting vigor or robustness. If longevity can be considered a reflection of an intrinsic

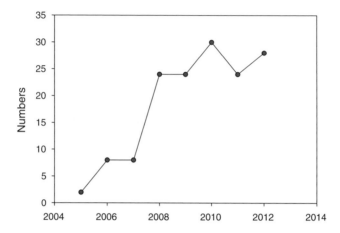

Fig. 13.15 Numbers of adult Big Birds after the 2003–4 drought and censused in January and February each year. The sex ratio was approximately equal among banded birds: 24 males and 21 females, summed across years. Included are an estimated 12–15 additional birds that were not banded but recognized individually by subtle plumage and beak-color features. Adult males outnumbered females throughout the growth of the population, not through differential production but through differential survival: the sex ratio of genotyped nestlings and fledglings was, if anything, female-biased (4 males : 8 females). In 2011 the number of breeding females was reduced to 6, but there were more in 2012 at our last census early in the breeding season.

property of vigor, as opposed to a response to extrinsic ecological opportunity (chapter 6), then Big Birds showed signs of vigor, starting with 5110 who lived for 13 years. Finches rarely live this long. Only 0.62% of *fortis* that hatched in 1976–98 and who survived their first year ($n = 1,445$) lived for 13 years or more. Moreover the Big Birds appear to have survived the severe drought of 2003–4 relatively well, even though this amounted to only two birds out of a known eight (25%) in the *B* line. Survival of banded birds of the other species was 18.5% *fortis* ($n = 173$), 25.4% *scandens* ($n = 110$), and 3.9% *magnirostris* ($n = 122$). The two drought-surviving Big Birds, 19798 and 19228, lived for 8–9 and 9–10 years respectively.

Genetic composition provides another reflection of vigor. The genome of 5110 is expected to be unusually heterozygous owing to its hybrid origin. In fact, surprisingly, it was unusually homozygous: 10 of the 14 autosomal microsatellite loci were homozygous. To put this in perspective only 0.66% of the total *fortis* sample ($n = 2,123$) were homozygous at 10 or more of the 14 loci. Low heterozygosity at the microsatellite loci does not necessarily contradict the notion of hybrid vigor. It may reflect genome-wide transgressive segregation, in which beneficial alleles are combined in homozygous form in the hybrid genome (Rieseberg, Widmer et al. 2003, Stelkens and Seehausen 2009). The result is a genotype and phenotype beyond the range of both parents and potentially with high fitness.

EXTRINSIC FACTORS

Success of the lineage after 2005 was enhanced by the heavy mortality experienced by the other species in the drought of 2003–4. More niche space for the Big Birds opened up with the morphological and ecological divergence of *fortis* and *magnirostris* (character displacement) in 2004–5 (chapter 7). As a result the Big Birds occupy ecological and morphological space in a zone between the other species (fig. 13.16). For example, in average beak width they are approximately equidistant from *fortis* (27.7% smaller), *scandens* (25.7% smaller), and *magnirostris* (37.2% larger). Lineage members have a distinctive generalist diet. They crack open *Tribulus* fruits (fig. 13.5), like *magnirostris* (fig. 6.10), feed on a variety of small seeds picked up from the ground, and feed on the seeds, nectar, and pollen of *Opuntia* (fig. 13.17). They differ from *magnirostris* in feeding on *Opuntia* flowers, they differ from *scandens* and most *fortis* in feeding on *Tribulus* seeds, and they differ from large *fortis* in efficiency when opening *Tribulus* fruits.

Future Prospects

It is unlikely that the drama we have witnessed will result in a long-lasting species. The small population may become extinct through the deleterious effects of inbreeding (e.g., Nieminen et al. 2001), such as those shown by *magnirostris* in the first few generations (chapter 6), and demographic stochasticity. Nevertheless it need not be doomed; rare interbreeding with *fortis* or *scandens* or even *magnirostris* could provide an escape from those effects. Two examples involving males that were apparently imprinted on another species hint at how this could happen. One male *fortis* (18570) sang the typical song of the lineage in 1993 and paired with a granddaughter of 5110 on the *A* line. Breeding success was unknown. One male *scandens* (21131) was recorded singing the song of the lineage in 2011 and bred with a female that looked like a *scandens* × *fortis* hybrid but was not captured. Genetic rescue, combined with exploitation of a unique, generalized niche, may enable the Big Bird lineage to flourish for quite some time.

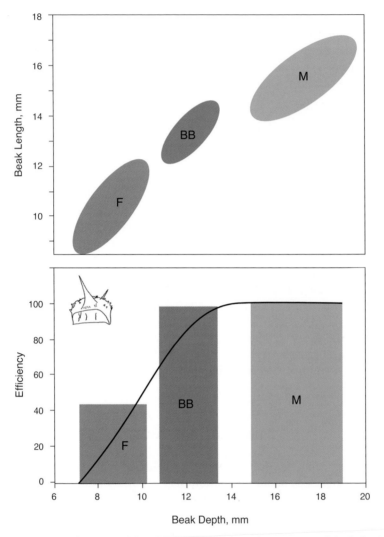

Fig. 13.16 Morphology and diet. **Upper**: Morphological separation of the hybrid (Big Bird) lineage (BB) from *fortis* (F) and *magnirostris* (M), illustrated with 95% density ellipses. **Lower**: A composite curve of feeding efficiency on the woody fruits of caltrop, *Tribulus cistoides*, in relation to beak depth of *fortis*, *magnirostris*, and the BB lineage, averaged from numerous timed observations. *G. scandens*, with different jaw musculature (Bowman 1961), does not crack caltrop fruits. Efficiency is defined as 1/*CT*: *CT* = time to crack the fruit in seconds (Grant 1981b). A minimum beak depth is needed to crack a fruit, and several *fortis* have smaller beaks than the minimum. *G. magnirostris* apparently do not vary in their efficiency in relation to beak depth or other dimensions. Their efficiency has been scaled to 100.

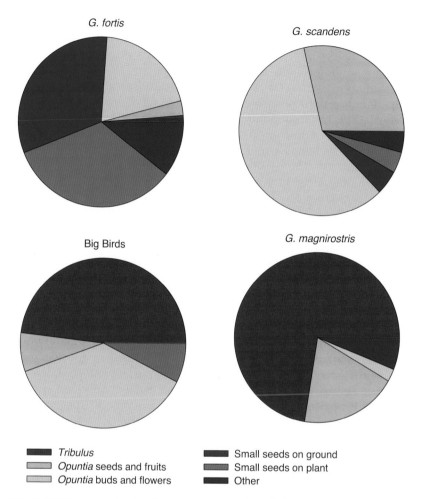

Fig. 13.17 Diets based on feeding observations of the following birds, one observation per bird, 2006–11: 190 *fortis*, 27 Big Birds, 140 *scandens*, and 84 *magnirostris* (see also fig. 6.11).

Future prospects are not necessarily bleak. They are discussed further in the next chapter.

Summary

The chapter discusses the possibility of speciation occurring through hybridization and the development of reproductive isolation of a hybrid lineage from the two species that gave rise to it. In 1981 a *fortis* × *scandens* hybrid male, probably a backcross to *fortis*, immigrated from the nearby large island of Santa Cruz. It was exceptionally large and sang a unique song. We followed its fate and the fate of most descendants for the next 30 years by making observations of breeding pairs, banding their offspring, and inferring relatedness by using a genetic marker. The immigrant initially bred with a hybrid female, giving rise to one line of descent in subsequent generations (line *A*). He later bred with *fortis* females, and one of them gave rise to a second line of descent (line *B*). The two lines experienced different fates, probably owing to differences in average body size. The *A* line (smaller birds) became reproductively absorbed into the *fortis* population, whereas the *B* line became reproductively isolated from the other species. Members of the *B* line were reduced to a single brother-sister pair by a severe drought and heavy mortality in 2003–4. From 2005 onward they and their offspring bred endogamously, and never with *fortis* or *scandens*. Ecological success and reproductive isolation of the lineage were enhanced by heavy mortality in the 2003–4 drought of their principal competitors, *magnirostris* and large-beaked members of the *fortis* population. These observations provide insight into speciation and hence into the origin of a new species. They show how a single diploid immigrant can start the process by breeding with a resident species; tolerance of the effects of inbreeding is needed to sustain and complete the process. The barrier to interbreeding arose behaviorally and without genetic change in sympatry. This is a rare example of speciation by introgressive hybridization: introgressive speciation.

PART 4

Syntheses

The Future of Finches on Daphne

Scientists cannot predict the future any better than
anyone else—even about their own field of research.

(Kendrew, 1966)

Now if I wanted to be one of those ponderous scientific
people and "let on" to prove . . . what will occur in the
far future by what has occurred in late years,
what an opportunity is here!

(Mark Twain 1883, pp. 172–173)

Introduction

OUR GOAL HAS BEEN TO GAIN INSIGHTS into the past by studying the present for long enough to capture significant change
(Grant and Grant 1996c, 2011b), to use the information to
extrapolate to the broader context of the Darwin's finch radiation
(Grant and Grant 2008a), and to look into the future and ask such
questions as what will the Daphne finch community be like at the end
of the century? Predictions are hazardous (appendix 10.1), as the cau-
tionary quotations above imply; nonetheless 40 years of experience are
a better basis for extrapolation than, say, 4 years.

The present is a guide to both the past and the future. Contemporary
study of finches identifies the most important environmental factors
that govern their fates. The Daphne study shows that fluctuations in
annual rainfall are more important than changes in temperature (Grant

et al. 2000), that droughts have more severe selective effects than abundant rain, and that the effects of any one drought are conditioned by preceding conditions (i.e., how past rainfall has affected the vegetation and food supply of the finches). It shows that two plant species are crucial for the survival of two of the finch species: *scandens* is dependent on *Opuntia* cactus, and *magnirostris* is dependent on *Tribulus*. Members of the Big Bird hybrid lineage are dependent on both. These facts are unlikely to change in the future.

The Past as Context of the Present

Daphne was formed perhaps as recently as 23,000 years ago as a volcanic satellite of Santiago (Geist et al. MS). It became an island about 15,000 years ago when a rapid rise in air temperature caused polar ice melting (Blard et al. 2007, Deschamps et al. 2012), initiating a rise in the level of seawater (box 14.1), and Daphne became separated from an enlarged Santa Cruz Island (fig. 2.4; Grant and Grant 1996b, Geist et al. MS). Continued rise of about 100 m in the next 9,000 years (Grant et al. 2012) reduced the height of the island to half (~ 120 m) and increased its isolation to about 8 km. As the island shrank in size, it may have lost species. In the light of finch communities elsewhere in the archipelago (Grant 1986) the species most likely to have been lost is *Camarhynchus parvulus* (Small Tree Finch).

The history of Daphne and other Galápagos islands has been one of repeated and often rapid climatic change. Conditions were either drier than now or as warm and wet but no wetter (Colinvaux 1972, Restrepo et al. 2012). Periods of one or more centuries in which El Niño events were common alternated with periods when they were apparently scarce or lacking (Anderson 1992, Moy et al. 2002, Riedinger et al. 2002, Rein et al. 2005). About 3,000 years ago an essentially modern climate replaced a drier one. The question we would like to answer is whether observations on Daphne for 40 years are an adequate representation of climatic swings and evolutionary events over centuries and millennia. How frequent are extremes beyond the range of our observations, and what were their effects on vegetation and the finches? Climatic fluctuations are known, whereas evolutionary events must be inferred. We concentrate mainly on rainfall.

The Galápagos coral core data spanning nearly 400 years provide an indirect record of sea-temperature fluctuations and associated rainfall (Dunbar et al. 1994). The record shows that it takes a long time for environmental variation to approach an asymptote (~170 years: Grant

Box 14.1 Mollusks as Markers

Fragments of four or five fossil bivalve mollusks (fig. B.14.1) substantiate the reconstruction of Daphne during the lowering of seawater. The fossils are a mixture of actual shell fragments and impressions in a carbonate rock, together with coral fragments and sea-urchin spines. They were found in sector 19 (fig. A.3.1) in 1991, at an estimated height above seawater of 10–20 m and 20 m upslope from the sea cliff. The mollusks lived in the intertidal, and the valves were washed into calmer waters where the rock was formed. Their age is not known. They have been tentatively identified by Matthew James (pers. comm.) as those of a contemporaneous Galápagos species, *Periglypha multicostata* (Sowerby 1835), which is known to occur as fossils on other Galápagos islands.

Fig. B.14.1 Fossil bivalve mollusks, tentatively identified as *Periglypha multicostata* (M. James, *in litt.*). It is evidence of submergence of Daphne's peripheral area.

and Grant 1996c). Even though conditions were warmer in some half centuries than others, and more variable in some than others, there is no evidence of a long-term trend or of large-amplitude changes preventing a close approach to an asymptote (fig. 14.1).

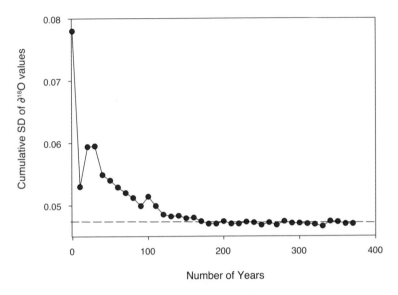

Fig. 14.1 Environmental variation through time. Stable oxygen-isotope ratios provide a measure of sea-surface temperatures in the Galápagos corals *Pavonia clavus* and *P. gigantea*, 1607–1981 (data from Dunbar et al. 1994). Original $\partial^{18}O$ values have been ln-transformed, and negative signs have been ignored: high values correspond to high sea-surface temperatures. Cumulative standard deviations (SD) from the first five years onward have been expressed as a function of years to show the rate at which an asymptote is approached. It took 170 years to reach approximate constancy: there is no sign of recent upward or downward trends. Variation in isotope values may also reflect variation in rainfall, as they are correlated with rainfall on Santa Cruz ($r = 0.551$) in the years 1965–81 (Grant and Grant 1996c).

What are the extremes in the record? There is no evidence of an El Niño event more severe than the 1982–83 one. Droughts could have strong effects on finch populations if they last for many years. However, short-term reversals between El Niño events of abundant rainfall and La Niña droughts, brought about by the Southern Oscillation of atmospheric pressure differences across the tropical and southern Pacific (Philander 1990), typically prevent severe climatic (drought) conditions from lasting long. In fact there is only one sequence of three successive years (1637–39) of cool sea-surface temperatures and hence low rainfall (Dunbar et al. 1994) in the Galápagos coral core record (Grant and Grant 1996c), and none longer. Ecological and evolutionary effects at this time were possibly more severe than those observed on Daphne in 2003–5. With this single exception, we conclude that finches were exposed to the same range of environmental conditions during the last 40 years as their predecessors experienced over the preceding 400 years.

The likelihood of more extreme conditions and effects increases as we go back further in time. Annual climatic fluctuations are revealed by a variety of proxy measures, including tree rings and pollen, diatoms, and spores from lake, ice, and coral cores (Philander 1990, Michaelsen and Thompson 1992, Flexor 1993, Cobb et al. 2003). These show once-in-a-millennium El Niño events of extraordinary severity occurring about 900 and 1,400 years ago (Michaelsen and Thompson 1992, Restrepo et al. 2012) as well as much earlier (McGlone et al. 1992). They might have had drastic effects upon *Opuntia* cactus and Cactus Finches, in the light of observations on Daphne during the 1982–83 El Niño event and aftermath. Extinction, recolonization, and strong evolutionary change are more likely on this time scale.

MERGE-AND-DIVERGE DYNAMICS

Ecological conditions in the terrestrial Galápagos environment depend on climatic conditions that in turn depend on oceanographic conditions (Grant et al. 2000). These conditions fluctuate on long timescales. Thirty-four years before our study began, cactus bushes were much more sparse on the inner slope of the crater (fig. 1.5) than at any time during our study. This indicates more arid conditions occurred then, and photographs in Beebe (1924) show the same.

Sea-surface temperatures fluctuate at approximately 25-year intervals (Chavez et al. 2003). Our study began at the end of a relatively cool period (Guilderson and Schrag 1998), and the Galápagos are perhaps now beginning to experience the next cool period. If so, a switch from wet to dry conditions, back to the conditions prevailing in the 1970s, may no longer allow hybrids to survive well. If this happens, the current trend of convergence will be reversed and the species will diverge. Thus, over the long term *fortis* and *scandens* on Daphne may alternately converge through hybridization and selection and diverge through selection alone. They may have been doing that for centuries, if not millennia.

The Present as a Guide to the Future

All else being equal, the past 40 years allow a projection into the future by simple extrapolation. However, the origin and flourishing of the hybrid lineage is a reminder of the inherent unpredictability of the biological system on Daphne. Compounding the uncertainty of the future is the prospect of climate change that is being experienced globally (Latif and Spellman 2009). It has given rise to concern about how rising temperatures, more severe droughts, and the increasing frequency of extreme weather events will influence natural communities, and uncertainty as to the consequences (McMichael 2012). Will populations alter reproductive schedules and activity patterns, or shift in latitudinal or altitudinal distribution (Davis et al. 2005, Gardner et al. 2009, Tingley et al. 2009, Colwell and Rangel 2010)? Small populations on islands, and in habitat islands on continents, may be especially vulnerable to the effects of increasing temperatures because they have no escape in space (e.g., Santisteban et al. 2012). Will populations have sufficient time and genetic variation to respond evolutionarily by adapting in physiology, morphology, or in other ways (Lynch and Lande 1993, Gienapp et al. 2008, Gomulkiewicz and Houle 2009, Dillon et al. 2010)? What will be the ecological and evolutionary consequences of altered environments (Parmesan 2006, Lavergne et al. 2009)? Uncertainties are so large that answers have to be full of mights and maybes.

GLOBAL WARMING AND GALÁPAGOS

Climate projections for the rest of the century indicate a moderate rise in temperatures for Galápagos of about 2°C–3°C, a rise of 3°C–5°C for the continent, a 1 m increase in sea level, and, as elsewhere, an increase in climate variability and extreme events (Solomon et al. 2007, Willis and MacDonald 2011). Possible geophysical effects have been registered already. A warming trend in tropical Pacific waters began in the mid-1970s (Guilderson and Schrag 1998, Zhang et al. 1998), associated with a reduced gradient of sea-surface temperatures across the Pacific and a shift in atmospheric convection from the east to central Pacific (Tokinaga et al. 2012). The altered spatial pattern is likely to become much more frequent with global warming (Yeh et al. 2009), and El Niño events may intensify. Thus according to current signs climatic conditions will not return this century to those prevailing before the 1970s.

If the global climate change is already affecting Galápagos climate, we might see a signal in the rainfall data, with a gradual increase in average rainfall and an increase in the variance. No such signal is apparent in figure 14.2. This figure shows no change after the first 10 years in the long-term average rainfall with successive annual additions to the cumulative total. The only possible signals we have experienced were an unusually long El Niño event (1991–93), even longer outside Galápagos waters (Trenberth et al. 1996), and exceptionally high tem-

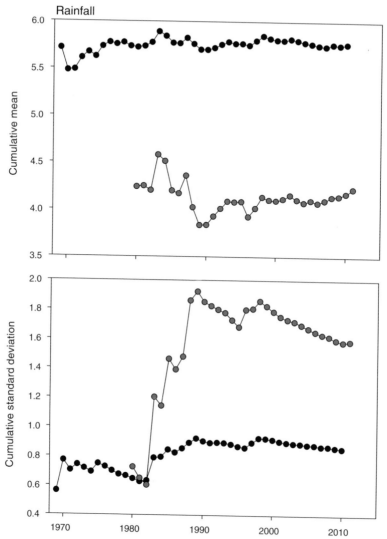

Fig. 14.2 Rainfall (mm) on Daphne (pale blue) and Santa Cruz (dark blue) plotted on a natural log scale as cumulative means (above) and standard deviations (below). The first point on each figure is the average for the first five years. By plotting numbers as a cumulative function of time, we show how they approach a quasi-equilibrial state. The figure shows that rain is lower and more annually variable on Daphne than at the southern coastal location on Santa Cruz; the north shore of Santa Cruz is drier and possibly more variable. Long-term averages are reached in a smaller number of years on Santa Cruz than on Daphne and remain approximately constant, with no obvious tendency to increase or decrease. Annual variation (standard deviation) tends to decline; extreme values have diminishing effects on long-term estimates.

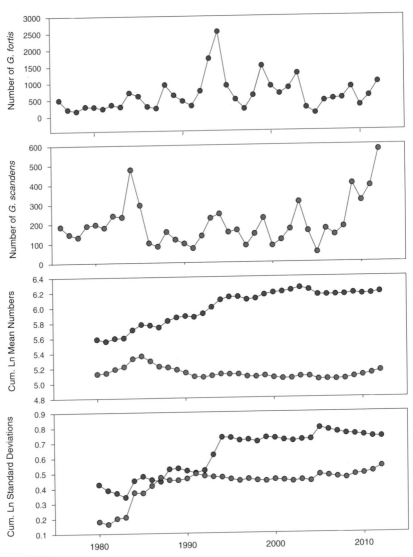

Fig. 14.3 Finch population sizes on Daphne. January 1 estimates are shown in the upper two panels for *fortis* (blue) and *scandens* (red). Annual averages and standard deviations on a natural log scale are plotted as cumulative functions of time in the two panels below, as in figure. 14.2. For both species approximately constant long-term estimates in the cumulative panels were reached in the early 1990s, in other words, after about 15 years. After that, large increases in population sizes as a result of El Niño conditions favorable for breeding had progressively smaller influence on the long-term estimation of averages and standard deviations. *G. fortis* were always more abundant than *scandens*, and fluctuated in numbers much more as a result of occasionally prolific breeding followed by population crashes in droughts.

peratures in the sea and air during the 1997–98 El Niño event, higher even than in the longer and more intense event of 1982–83 (Grant et al. 2000). High temperatures during the 1997–98 El Niño had no conspicuous effect upon the vegetation or the demography of the finches beyond the affected years (fig. 14.3). Long-term average numbers did not increase with successive additions of annual values, with one exception; *scandens* were more abundant in the last year (2012) than in any of the preceding 35 years.

Finch Futures

MEANS AND EXTREMES

Consequences of climate change in Galápagos are difficult to anticipate because there are two aspects to it: average temperature and rainfall and the variance of both. These could have opposite effects on vegetation, arthropods, and finch populations, so the outcome will depend on which one predominates. We focus on rainfall.

An increase in average rainfall will make the climate, vegetation, and finch dynamics on Daphne a little more similar to current conditions on Santa Cruz. Over the long term Daphne has low average rainfall with a high standard deviation, whereas a similar coastal location on the south coast of the much larger Santa Cruz Island has a high average rainfall but low standard deviation (fig. 14.2). Expressed as a coefficient of rainfall variation, Daphne rainfall variance is 60.9% of the mean whereas at the Santa Cruz locality the variance is only 12.6% of the (much higher) mean. The contrast in average rainfall on the two islands and the relative variation implies that population fluctuations are greater on Daphne (they have not been studied on Santa Cruz). Since natural selection occurs under extreme drought conditions causing high mortality on Daphne, it is likely that selection is rarer on Santa Cruz, where finches can move to higher altitudes and encounter different foods in a wetter climatic zone. In 18 of the 37 years in the period 1976–2012 annual rainfall on Daphne was lower than the *minimum* recorded at the coastal site on Santa Cruz in the same period. Therefore selection of the sort observed on Daphne may never have occurred on Santa Cruz during this period. In the future, Daphne may receive more rain, finch population sizes may increase and fluctuate above and below a higher average, and, importantly, selection due to mortality may occur less often.

A different set of consequences follows from the anticipated increase in frequency of extreme climatic events. If intense droughts become more frequent on Daphne, despite a general tendency for rainfall to increase, finches may be subject to more frequent or intense selection. Annual variation in population density (fig. 14.3) may increase, and so will the risk of extinction (Grant and Grant 1996c). Computer simulations show broad-scale vulnerability of bird populations to the effects of environmental stochasticity (Sæther et al. 2005). In a comparison of 13 species of birds Sæther et al. (2002) found that Daphne *scandens* and *fortis* ranked first and second in environmental stochasticity, and first and third in coefficients of variation in population size over 15 years.

These general considerations apply to all the finch species on Daphne. Some responses to changes in climate are species-specific.

G. FORTIS, SCANDENS, AND *FULIGINOSA*

Extrapolating beyond measurements, it is reasonable to suppose that populations will prosper in El Niños, suffer in droughts, and their fates will be determined by the relative strengths and durations of these opposing influences. An increase in average rainfall and predominant production of small seeds could foster a continued fusion of the *fortis* and *scandens* populations through introgressive hybridization. On the other hand the current trend of convergence might be reversed if intense droughts increase in frequency and finches become more dependent at those times on *Tribulus* and *Opuntia*. These circumstances would favor greater specialization of *fortis* on *Tribulus* fruits, which would now be more common, and *scandens* on *Opuntia*. *G. scandens* with short beaks as a result of introgressive hybridization would be at a selective disadvantage because they are likely to be relatively inefficient at exploiting *Opuntia* cactus flowers for pollen and nectar. Our reasoning presupposes the continued existence of *Opuntia* on Daphne, and in a warmer and wetter climate this may not be correct.

G. MAGNIROSTRIS

This species is dependent on *Tribulus cistoides*, and to a lesser extent the seeds of *Opuntia echios*. A similar situation exists on the island of Genovesa, where the dry-season diet of *magnirostris* is dominated by a single seed type, *Cordia lutea* (Grant and Grant 1989a). If *Tribulus* or *Cordia* are adversely affected by an altered climate, then *magnirostris* could disappear from one or the other island. Given the near extinction of *magnirostris* in 2005 when the *Tribulus* food supply was nearly exhausted, a small intensification of droughts might administer the coup de grâce for this species on Daphne. Introduction of a disease specific to *Tribulus* would have the same effect. In 1982 a fungus adversely affected *Tribulus* seed production (Grant and Grant 1993).

HYBRID LINEAGE

It is highly unlikely that we have witnessed the origin of a long-lasting species, but not impossible. The future of the Big Bird lineage will be determined by environmental fluctuations but also by effects of inbreeding, and by ecological and possibly reproductive interactions with other members of the finch community.

According to the relationship between species number and island area among Galápagos islands, Daphne should have two species (fig. 14.4). The establishment of *magnirostris* added a third species to the *fortis-scandens* community. The total becomes four when the lineage

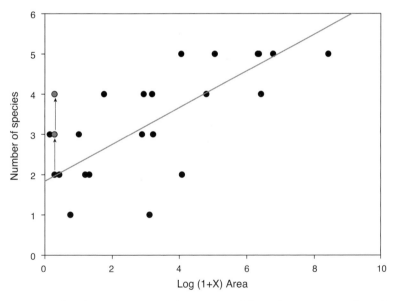

Fig. 14.4 Daphne has more species than predicted from the regression of number of species on the logarithm of area (km²) for 22 islands. The predicted number is 2. The actual number became 3 with the establishment of *magnirostris* (chapter 6) and is 4 with the inclusion of the hybrid lineage (connected with arrows). For the relationship without Daphne, $r = 0.712$, $p = 0.0002$; with Daphne included as a four-species island, $r = 0.641$, $p = 0.001$. This is a specific example of a global relationship: a short distance from nearest landmass modifies the way area influences the number of species on an island (Kalmar and Currie 2007). From Grant and Grant 1996c.

is counted as one species and five if the occasional *fuliginosa* is included as well (chapter 6). Ignoring *fuliginosa*, the Big Bird lineage has the poorest prospects for long-term persistence because it is much rarer than the other three species. The lineage increased to a maximum of approximately 30 individuals in 2010 (fig. 13.15). Fitnesses may be jeopardized by inbreeding. Effects of inbreeding on fitness and random loss of alleles may increase in time if the population remains small. *G. magnirostris* escaped this fate through repeated immigration (chapter 6), whereas no further immigration is known to have occurred in the hybrid lineage.

The lineage may have an alternative escape through breeding with *fortis*, *scandens*, or even *magnirostris*. In other species it has been found that genes introduced into inbred populations spread rapidly (Saccheri and Brakefield 2002). If members of the lineage breed with any of the other species, the outcome will depend on the direction of gene flow, the rates of introgression of genes, and the fates of those genes. For example, if a female from the lineage breeds with a male *fortis*, the offspring, having imprinted on the father, are likely to breed with *fortis*. In this case genes flow out of the lineage. However if a male from the lineage breeds with a female *fortis*, the offspring, having imprinted on the father's song, are likely to breed with members of the lineage, and genes will flow from *fortis* into the lineage. Repeated introgression of heterospecific alleles, even if infrequent, could nullify the effects of inbreeding in the lineage, restore mean fitness, and allow the population to increase. Since this would also result in the production of small offspring, it might increase the probability of further interspecific breeding in subsequent generations and a gradual fusion of the lineage with *fortis*, an example of reproductive absorption, as happened with line *A* (chapter 9).

Thus the Big Bird lineage may become extinct, through environmental and demographic stochasticity, deleterious effects of inbreeding, or introgression. On the other hand it may flourish by exploiting a unique, generalized niche. The critical factor determining the future of the lineage may be the rate of interbreeding with other species.

These speculative changes are likely to be dependent on the demographic dynamics of the other species in the community, on their future evolution, and on any change in ecological interactions with those species such as greater competition for food. For example, long-term coexistence of the lineage with the other species could be weakened by an increase in numbers of the principal competitors, the large-beaked members of the *fortis* population. The future of the Daphne community could be monitored every generation (chapter 4), that is, at five-year intervals, to follow the ecological and evolutionary fates of the new lineage and the other resident species. The community could provide a valuable window into long-term effects of climate change in a natural environment.

INVASIVE PLANT SPECIES AND DISEASE

The future of Daphne finches is uncertain for one final reason: introduced alien organisms, both plants and animals. The introduction of an invasive plant species could alter the dynamics of the plant community and the food supply of the finches. One such plant species, *Cleome viscosa*, has already arrived (fig. 14.5, appendix 1.1). It was introduced to neighboring Baltra, and has explosively spread on this and North Seymour Island. In the last decade we have found *Cleome* on Daphne around seabird nests on the lower plateau and adjacent outer slope (fig. 2.5) and removed them by hand to slow the expected increase in numbers and distribution. Several years of rainfall could make this

Fig. 14.5 Invasive plant. *Cleome viscosa*, a threat to the natural vegetation on Daphne and other islands.

monitoring and removal an impossible task. Some native plant species may be at risk.

Diseases such as avian malaria (*Plasmodium*) could enter the archipelago and devastate finches. A mosquito vector, *Culex quinquifasciatus*, has been present in the archipelago since 1985 (Whiteman et al. 2005). A similar threat is posed by West Nile virus, for which a possible vector (*Aedes taeniorhynchus*) is already present (Bataille et al. 2009a, 2009b).

Finches currently have two persecutors. In some years avipox is contracted by several finches (fig. 5.13), although on Daphne it has not affected survival (Huber et al. 2010). The parasitic fly *Philornis downsi*, known to be in the archipelago in 1964, is much more serious. It is now known from observations (Fessl and Tebbich 2002, Fessl et al. 2006, O'Connor et al. 2010) and experiments (Koop et al. 2011) to cause high nestling mortality of Darwin's finches and other bird species (Fessl and Tebbich 2002, Fessl et al. 2006, Koop et al. 2011). *Philornis* has been recorded intermittently on Daphne but is usually not there, being more prevalent in the moist habitats of high islands. This could change if droughts become scarcer and El Niño events become more frequent.

Genomes for the Future

Darwin set the evolution revolution in motion. The wheels are turning faster every year as a result of genomics. We began the study with no knowledge of how environmental variation caused finch evolution, and no knowledge of its genetic foundation. When we consider how much has been learned in 40 years, it becomes apparent that not only can we not answer all our questions; we cannot anticipate all questions themselves, as some will emerge only from knowledge gained in the future. For the immediate future, however, the crystal ball is clear. Genetic variation, within and among species, will be much better understood now that tools are available for studying whole genomes.

An exciting new era in finch biology began with the sequencing of the *magnirostris* genome in 2011 (Rands et al. 2013) and the *fortis* genome in the following year (Zhang et al. 2012). Genomes of the other species are being sequenced, and knowledge of the genetics of the adaptive radiation will be improved in three ways: in phylogenetic history, in the genetic changes accompanying speciation, and in the genetic basis of adaptive change. The foundation of such studies will be direct measurement of genetic properties that are the subject of long-term research of microorganism evolution in the laboratory (e.g., Blount et al. 2008) but are currently out of reach of a field study of finches (chapter 9): properties such as mutation, gene duplication, recombination, chromosomal inversions, pleiotropy, linkage, and their various frequencies. Genomic data will reveal what changes in genetic architecture take place during speciation and hint at the causes (Feder et al. 2012, Nosil and Feder 2012). For example, changes may involve selective sweeps, although they are likely to be fewer and more restricted than in genetically incompatible species such as the two species of *Ficedula* flycatchers in Scandinavia (Ellegren et al. 2012). Comparative genomics may also reveal genetic factors that are responsible for, and thus help to explain, the relatively rapid radiation of Darwin's finches and their Caribbean (Burns et al. 2002) and continental (Burns and Naoki 2004) relatives. The next and very large challenge will be to identify the functions of genes to answer questions of what they do and how their expression is controlled and regulated. A particularly promising area of investigation is the genetic basis of beak variation (fig. 12.2), where a start has already been made (Abzhanov et al. 2004, 2006, Mallarino et al. 2011, 2012).

One goal of our future studies is to investigate the genetic changes that take place in a population when it is subject to directional selection on beak size or shape. For a given selection regime how repeatable is the genetic response? Does each selection event result in unique changes in combinations of genes and their regulators stemming from redundancy in genetic networks? Are subsequent changes equally likely, or does one genetic change constrain or potentiate another? Can genetic limitations on phenotypic change be identified? A second goal of future studies is to determine the genetic consequences of hybridization and backcrossing. Hybridization can promote genomic reorganization through the disruption of DNA methylation and mobilization of transposable elements (Michalak 2009). The question we would like to answer is how much evolutionary novelty is generated by restructuring of the genome through introgressive hybridization (Fontdevila 2005)? When genes are exchanged between finch species and new genetic constitutions are thereby constructed, what combinations work well, and why, and what do not? From answers to these questions insight may emerge into the genetic factors that made the Big Bird lineage a success in the first few generations of inbreeding.

Whatever is discovered at genetic level will need to be interpreted in the light of environmental effects on phenotypes (fig. 4.1). This is where the Daphne study can make a contribution to the future.

Summary

The chapter describes the history of Daphne and the Galápagos, and uses information from the long-term study of the finches on Daphne to project into the future. Observations on Daphne for 40 years are an adequate representation of climatic swings and evolutionary events over centuries, as indicated by Galápagos coral core data spanning nearly 400 years that provide an indirect record of sea-temperature fluctuations and associated rainfall. Climate projections for the rest of the century indicate a moderate rise in temperatures for Galápagos of about 2°C–3°C, a 1 m increase in sea level, and, as elsewhere, an increase in climate variability and extreme events. The future of the finches depends on which of two climatic features prevail: an increase in average rainfall or an increase in extremes. If Daphne receives more rain, finch populations may increase in average size but fluctuate less and, importantly, selection due to mortality may occur less often. On the other hand if intense droughts become more frequent, more severe, or last longer, despite a general tendency for rainfall to increase, finches may be subject to more frequent or intense selection. We speculate on the future of the hybridizing species, and the Big Bird lineage, as well as *magnirostris*, in the face of anticipated changes. Unknown factors such as diseases and invasive plant species add to uncertainty about the future: they could have substantial effects upon the finch populations. A new era in understanding finch evolution has just begun with the sequencing of *magnirostris* and *fortis* genomes, and the chapter concludes with some remarks about what is likely to be learned from genomic research in the near future.

15

Themes and Issues

We add to what was learned before, raising the old
questions again and again, lifting them if we can toward
higher and higher ground, and ourselves with them. Why
are there so many kinds of animals, and why are we
among them? Probably we have been asking these
questions ever since we lived in caves.

(Weiner 1994, p. 293)

Biologists do not yet fully understand how the various
potential processes of evolutionary change actually
integrate over time to yield the patterns in which
species occupy morphospace.

(Brakefield and Jorion 2010, p. 93)

Introduction

FIGURE 15.1 SUMMARIES THE FOUR profound morphological
changes that took place in the Daphne finch community over 40
years. Two changes were additions, brought about by colonization (*magnirostris*) and speciation (Big Bird lineage). Two changes
were morphological transformations of *fortis* and *scandens* through
selection and introgressive hybridization.

From start to finish three themes dominated the study: interspecific
competition for resources, enhanced variation in continuously varying
traits like beak size, and the role these two play in species formation.

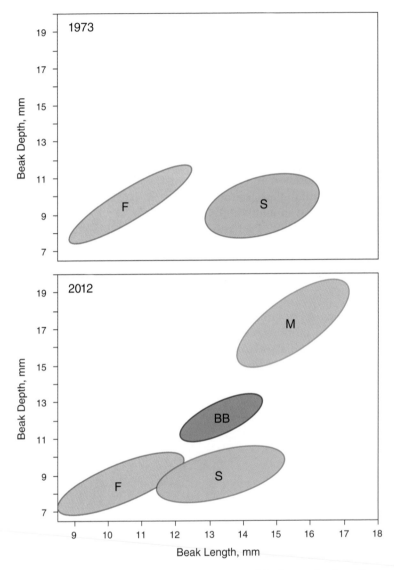

Fig. 15.1 Change in the morphological community of finches on Daphne. Between 1973 and 2012 *magnirostris* (M) colonized the island, a Big Bird lineage (BB) was formed, *fortis* (F) became smaller in beak depth, and *scandens* (S) became shorter in beak length through a combination of selection and hybridization. These dynamics provide a glimpse of evolution elsewhere on much longer time-scales.

Three themes emerged as the study progressed: natural selection as an observable, repeatable, and interpretable phenomenon with evolution-ary consequences; introgressive hybridization; and the role of both in species formation. In this chapter we synthesize what we have learned about speciation before discussing some of the issues raised by our findings, including the question of how generalizable they are.

Speciation, Selection, and Hybridization

A central problem in evolutionary biology is to explain how new species are formed. The world is biologically rich because it is filled with so many different kinds of species; therefore understanding speciation goes a long way toward explaining biodiversity. The Daphne study has yielded several important insights into a process that is rarely observed yet frequently discussed, debated, and necessarily inferred. In fact there is not one process but two. The first is divergence and lineage splitting: two species from one. This is part of the canonical (standard) allopatric model of speciation. The second is unequal genetic mixing of two species to form a third: three species from two—introgressive or reticulate speciation. Here we summarize and discuss our findings.

EVOLUTION

There is widespread agreement that speciation generally results from genetic differentiation of separate populations in allopatry (or parapa-try), whether the populations are genetically connected by occasional gene flow or not. This is followed in many cases by the coexistence of the differentiated populations in sympatry (Mayr 1963, Coyne and Orr 2004, Price 2008). Two obvious requirements for this to happen are genetic variation and ecological opportunity. The Daphne study provides evidence of both. It shows there is abundant polygenic varia-tion in the ecologically most important traits, beak size and shape. Moreover some of the genes involved in the development of different beak dimensions have been identified. These are expressed at the same or different stages of development but independently, and their independence helps to explain how evolutionary change takes place in beak shape as well as size (chapter 12). Thus the intrinsic potential for evolutionary change is high. Extrinsic potential, that is, ecological opportunity in the form of available resources, is very difficult to iden-tify before it is revealed by a species actually exploiting the resources, and yet it must have been present on Daphne to have allowed two colonizations to take place, the first by *magnirostris* during the extreme El Niño event of 1982–83 and the second by the originator of the hybrid lineage two years earlier. Since there are many islands in the archipelago differing in ecological conditions, there is a high extrinsic potential as well as high intrinsic potential for evolutionary change in allopatry. This is a major reason why Darwin's finch species have proliferated.

ECOLOGICAL IMPORTANCE OF FOOD: THE DAPHNE PERSPECTIVE

The four finch species on Daphne, together with related populations on other islands, illustrate the speciation problem and point to a solution. *G. fortis, fuliginosa, scandens*, and *magnirostris* each have several populations, respectively 17, 27, 15, and 12 (Grant 1986). Morphologically the 71 populations are variations on four discrete species themes (fig. B.1.2). The pattern in figure B.1.2 raises two questions. First, how is a new theme generated? Second, why are there no populations in the gaps between the means? We offer an answer to the first question here, and an answer to the second question in a section below (Size and Hybridization).

The multimodal nature of dry-season seed distributions (fig. 2.16) shows why there should be more than one granivorous finch species on large islands (Schluter and Grant 1984), but not how speciation takes place. Since the four species possess adaptations to feeding on different foods, the first question can be rephrased: how does the process of ecological divergence beyond the normal range of variation of populations of a species get started? How do finches encounter a new food type or types that potentiate speciation?

Insights from the Daphne study offer an answer by extrapolation from the observed to the unobserved. A new encounter occurs when finches arrive at an island that has a new food (fig. 1.1 and chapter 7) or when a new food unfamiliar to the finches arrives on the island. In the latter case a population of finches may go on fluctuating morphologically between more or less fixed limits that are set by the composition of the food supply and competition for it, as we observed on Daphne (chapter11), and then starts on a new evolutionary trajectory when by chance a new food arrives on the island. Character displacement may be a key element in the directional change by providing the initial impetus in shifting the population toward a new optimum. Mutation, recombination, and additional selection would all be involved in completing the process.

Paleoclimatic evidence indicates the climate was gradually becoming drier even as it fluctuated in the last million years (Grant and Grant 2008a). The arrival in the archipelago of drought-tolerant plants, including those with medium and large seeds such as *Opuntia* cactus, *Tribulus*, and *Cordia*, probably triggered speciation of the cactus and granivorous ground finches (Grant and Grant 2008a). These evolutionary dynamics echo the theme in chapter 11 of approximate stasis for long periods of time followed by relatively rapid evolutionary change and speciation.

BEHAVIORAL BARRIER TO INTERBREEDING

A third requirement for the sympatric phase of speciation is a barrier to the free exchange of genes. Genetic theories of speciation stress genetic incompatibilities when nascent species interbreed (Coyne and Orr 2004, Feder et al. 2012, Abbott et al. 2013) and largely ignore the contribution of learning to a behavioral barrier that prevents interbreeding. Darwin's finches do interbreed, on Daphne and on other islands (Grant and Grant 1989, 2008a, Grant et al. 2005), but rarely, and yet without loss of fitness when feeding conditions for hybrids and backcrosses are favorable. This shows that in the early stages of speciation, before genetic incompatibilities arise, populations are almost entirely isolated from each other by a premating behavioral barrier to interbreeding that is set up by learning through sexual imprinting; that is, the barrier is culturally inherited (chapters 6 and 8, and Grant and Grant 1996d, 1997a, 1998), and rarely breached.

An important question, therefore, is what constitutes the barrier to interbreeding, because identifying the constituents help us to understand how and why the barrier arose and how it functions. For many passerine bird species the answer would be plumage and courtship behavior (Prager and Wilson 1980, Price 2008), subject to sexual selection occurring predominantly through female choice. The answer for Darwin's finches is song and morphology. These function in interactions between males as well as between males and females. Species cues are learned early in life as revealed by experiments (Bowman 1983, Grant and Grant 2008a). Like beak morphology, songs diverge in allopatry (fig. 15.2), probably through a combination of influences of habitat structure on song transmission (Bowman 1983), copying errors accumulating through time (Grant and Grant 1996d), and changes caused by acoustical interference with songs of other finch species (Grant and Grant 2010c) occupying somewhat different parts of the total acoustical space (Nelson and Marler 1990). These topics have

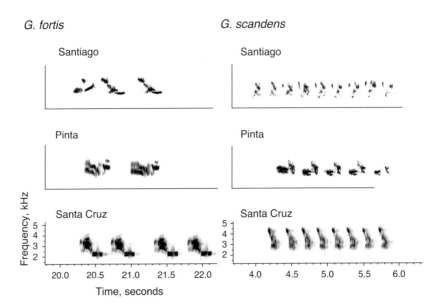

Fig. 15.2 Songs of *fortis* and *scandens* illustrate strong inter-island differences in note structure but species-specific similarity in temporal pattern. From Grant and Grant 2010c.

been reviewed recently: by Wilkins et al. (2013) for song divergence, Grether et al. (2009) for interspecific interference, and Verzijden et al. (2012) for song learning.

The coupling of song and morphology could be more than a matter of learning if song features were strictly determined (constrained) by beak morphology and musculature for biomechanical reasons (Podos 1997, 2001). Song would then be analogous to an inflexible trait that is genetically correlated with beak size. This attractively simple possibility has some support from an association between beak size and trill rate and frequency bandwidth on Santa Cruz Island (Huber and Podos 2006, Podos 2010), as well as from some studies of continental species (e.g., Badyaev et al. 2008). Evidence is mixed, however, both on Santa Cruz (Huber and Podos 2006) and outside Galápagos (Slabbekoorn and Smith 2004, Seddon 2005), sometimes lacking as on Daphne (Podos 1997), or even contrary to expectation (Huber and Podos 2006).

We doubt if strict determination of song by beaks plays a role in finch speciation, notwithstanding the original correlations between beaks and songs (Podos 2001) and results of some functional and biomechanical assays (Herrel et al. 2005, 2009). There are two reasons.

First, Bowman's (1983) extensive song recordings throughout the archipelago showed that in most finch populations individuals sing fast or slow variants of song, which he labeled as basic and derived respectively. The two intraspecific variants are not associated with differences in beak size, beak shape, or body size where this has been investigated, either in *fortis* and *scandens* on Daphne (Grant and Grant 1996d, 2009a, 2010c) or *conirostris* (Grant and Grant 1989) and *difficilis* (Grant and Grant 2002b) on Genovesa. Second, songs of some *Geospiza* species are not constrained, as shown by the fact that species with different beak sizes and shapes sing such precise copies of each other's songs that they elicit strong responses from the copied species (chapter 8). Moreover heterospecific singing is not restricted to Daphne, as birds are occasionally encountered elsewhere in the archipelago singing the song of another finch species with a different beak size and shape (Bowman 1983, Grant and Grant 1989).

SIZE AND HYBRIDIZATION

Occasional hybridization does more than reveal the absence of genetic incompatibilities; it shows in two ways that morphological cues are as important as song cues when they convey conflicting information about potential mates and species identity. First, singing each other's songs can lead to interbreeding when size differences are small, as between *fortis* and *fuliginosa* or *scandens*, but not when size differences are large, as they are between *magnirostris* and *fortis* on Daphne. Second, introgressive hybridization leads in opposite directions, either to the reversal or collapse of speciation (chapter 10) or to the formation of a new species (chapter 13). The outcome appears to depend on the magnitude of the size differences between the interacting species. This provides an answer to the long-standing question of gaps between clusters of species (Huxley 1938, 1940, Mayr 1942). A minimum morphological difference between species is set not only by competition for food (reviewed in Grant 1986) but by the propensity of similar species to hybridize and potentially fuse into one (chapter 10).

The importance of size differences is illustrated by the fates of two lines of descent in the hybrid lineage that can be attributed to a difference in size from generation 1 onward. Members of the *A* line were smaller on average than members of the *B* line in beak and body size, and resembled *fortis* more closely. Since males of both lines sang the same song as the immigrant male, the key difference between them was their size. The *A* line became absorbed into the *fortis* population through interbreeding, whereas the *B* line prospered as a reproductively isolated population. This is the clearest example of a single factor, morphology, tipping the balance toward either fission or fusion at the secondary sympatry stage.

PHYLOGENETIC IMPLICATIONS OF HYBRIDIZATION

We began the study with a simple conception of speciation, depicted in figure 1.2 as a process of divergence. In the light of experience on Daphne we should modify it to allow for reticulation (fig. 15.3), as exemplified by the exchange of genes between *fortis* on the one hand and *scandens* and *fuliginosa* on the other, and convergence as shown by *fortis* and *scandens*. Furthermore, rather than just a process of fission or fusion, repeated many times to give rise to many species from one, a third possibility should be added: interbreeding of two species contributes to the formation of a third. The genetic footprint seen in sympatric populations elsewhere in the archipelago (Grant et al. 2005) suggests that introgressive hybridization may have influenced, and even produced through introgressive speciation, some of the Darwin's finch species present today (Grant and Grant 2008a). Further tests need to be devised for detecting a genetic signature of past introgression and estimating when it occurred (e.g., Larsen et al. 2010, Rheindt and Edwards 2011).

EPHEMERALITY OF SPECIES

It is tempting to think of the Darwin's finch radiation as a simple process of accumulation of 14 diverse species from an original species, and that all species are known. Extinction, impossible to quantify in the absence of fossils, is acknowledged as a possibility in analyses of diversity and has to be estimated by making assumptions of its frequency

Fig. 15.3 (*facing page*) Darwin's first thoughts on phylogeny, from Notebook B written in 1837 (above), contrasted with how he might have depicted his thoughts with allowance for reticulation caused by introgressive hybridization (below, in red). In his famous sketch speciation is represented as a branching process. Some lines of descent become extinct: these lack terminal crossbars. In the lower panel species, and not just their genes, reticulate, and by interbreeding separate lineages may produce a new species (E). Reticulation is not a new idea (e.g., Huxley 1942). The upper illustration is reproduced with kind permission of the Syndics of Cambridge University Library, and the lower illustration is adapted from Grant and Grant 2010a.

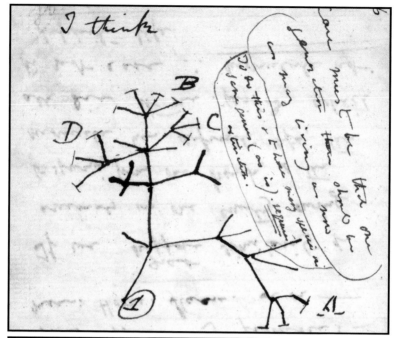

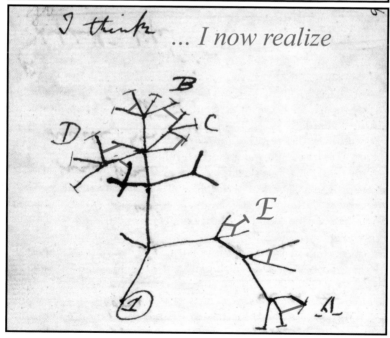

and temporal pattern (Ricklefs 2007, Glor 2010). Some distribution patterns of birds with a known phylogenetic history are best explained by invoking extinctions (Warren et al. 2006). The Daphne study suggests that species are likely to be arising, fusing, and collapsing more often than is generally appreciated, and unrecorded in either fossils or molecular phylogenies. For example, repeated hybridization with introgression may be thought of as repeated experimentation. Only some experiments yield results of evolutionary significance. If this view is correct, many episodes of speciation fail for every one that succeeds in reaching complete genetic isolation (Grant and Grant 2009a). Origination of ephemeral lineages will scarcely ever be observed, being rare and unpredictable in occurrence, cryptic, and local, and therefore underestimated. Long-term studies like the present one with marked individuals have the best chances of detecting and understanding them.

The 40 years spanned by our study is a short time in relation to the radiation, and yet we witnessed the origin of one incipient species by introgressive hybridization and, in the opposite direction, a process of fusion that is leading to a reduction from two species to one on Daphne. The hybrid lineage may be short-lived. We surely cannot have observed the only one in the history of the finches.

PREDICTABILITY AND EVOLVABILITY

There are predictable aspects of evolution—similar and repeated ecological, morphological, and genetic changes in similar environments (e.g., Hudson et al. 2011, Rutschmann et al. 2011, Jones et al. 2012, Zhen et al. 2012)—but also unpredictable aspects. Evolution of finches on Daphne displays both: it is genetically predictable but environmentally unpredictable.

Evolution is highly predictable when natural selection occurs because populations maintain a large amount of polygenic variation underlying body size and beak size and shape variation (chapters 3, 8, and 9). However, climatic fluctuations, which are the ultimate cause of natural selection through their effects on food supply (chapters 4 and 7), are unpredictable in their occurrence, duration, and severity, so the occurrence and direction of natural selection is unpredictable. Moreover a one-year drought may lead to strong natural selection, as in 1977 (chapter 4), whereas a two-year drought (e.g., 1988–89) may not (chapter 11). The factors governing the outcome are food types and quantity, as well as finch density at the beginning of a drought. Preceding conditions are therefore crucially important in determining whether natural selection will occur or not and, if it does, the direction it takes.

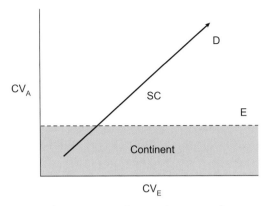

Fig. 15.4 The potential for evolution of a population is a function of both genetic (intrinsic) and environmental (extrinsic) variation. A line with arrowhead shows the direction of increasing potential. The coefficient of additive genetic variation (evolvability, chapter 3) is plotted against a similar index of environmental variation (chapter 1); both increase away from the origin. The environmental index is the long-term estimate of rainfall variance divided by the long-term mean, from figure 14.2. Expressed as a percentage of the mean, variation is five times greater on Daphne (60.9%) than on Santa Cruz (12.6%). Points on the graph refer to small (Enderby, E), medium (Daphne, D), and large (Santa Cruz, SC) islands. D and SC represent positions of two *fortis* populations, placed on the assumption that the unknown genetic variation is lower on Santa Cruz because introgressive hybridization is less frequent; this may not be correct (de León et al. 2010). The small island of Enderby has only a single population, of *fuliginosa*, and has been chosen because *fortis* does not occur alone anywhere. Its placement is hypothetical. The blue zone represents a band of conditions that might prevail in continental regions along an environmental gradient from stability ($CV_E \sim 0$), as in lowland tropical rain forests, to instability ($CV_E \sim 1$), as in deserts.

Thus the evolutionary potential of a population, as well as its predictability, is not just a function of genetic variation, or of environmental variation (including the social environment), but a product of the interaction between the two. A simple graph depicts the concept (fig. 15.4) and shows that the evolutionary potential varies among locations. Responses to environmental variation depend on the nature of the variation. The climatic environment may fluctuate as we have seen on Daphne and result in evolutionary change that fluctuates about a long-term mean within limits. The environment may also change to a new state, gradually through a change in those limits, or abruptly with the introduction of a novel food item, or new species.

A possible example of abrupt change on Daphne is the arrival of caltrop (*Tribulus*). *T. cistoides* is believed to be an adventive to the

New World (Svenson 1946, Porter 1967). A dozen relatives occur in the Old World, in Europe and Africa: one is food for camels in the Arabian desert (Thesiger 1959)! If caltrop is indeed an adventive, it arrived in Galápagos no earlier than 1535 on Fray de Berlanga's vessel, and on Daphne after that, perhaps attached to the feet of boobies (*Sula*) or gulls (*Larus*). *Tribulus* is part of the food niche of *fortis*, and a critical part of the niches of *magnirostris* and the hybrid lineage. A parallel example without any known or supposed human influence is the inter-island dispersal of seeds of *Opuntia* species and *Cordia lutea*, which are critical for the survival of *magnirostris* and *conirostris* on other islands (Grant and Grant 1982).

Overview

The value of Darwin's finches lies in what they reveal about the early stages of speciation in a young adaptive radiation, when ecological divergence has proceeded dramatically without the kind of genetic divergence that impairs fertility and viability (Grant and Grant 1997a). Daphne finches have shown us that speciation occurs by hesitant, wavering steps rather than by a large stride. It is pushed in one direction and then in another by selection, and pulled by hybridization in both: more a journey along a winding rural lane than an expressway. Experimental. Barriers to interbreeding are imperfect. Introgressive hybridization may cause them to decay and yet, paradoxically, it also provides the genetic variation for a new directional change. Recognizing the primary role of natural selection in their divergence, we could consider Darwin's finches to be an exemplar of ecological (Van Valen 1976, Schluter 1996, 2009, Nosil 2012) or ecogenetic speciation. Recognizing the primary role of learning in reproductive isolation, they could be considered an exemplar of ethological speciation (Dobzhansky 1941, Mayr 1942). In fact they are both.

In conclusion, speciation of Darwin's finches is ecologically driven by natural selection with reproductive isolation as an incidental effect (Fisher 1930, Dobzhansky 1940) or by-product of morphological divergence in allopatry (Grant 1986, Schluter 1996, 2000, Grant et al. 2000, Grant and Grant 2008a). It is also influenced by song, which constitutes an important element of the barrier to interbreeding by being culturally inherited and coupled with morphology through associative learning and not because it is determined by beak size. This may be a common form of speciation in birds because many species of birds learn their songs from conspecifics at an early age (Irwin and

Price 1999, ten Cate and Vos 1999). We think of speciation as an eco-behavioral-genetic process of evolutionary divergence (Grant and Grant 2008a). Genetic incompatibilities arise much later through mutation, and eventually the door to gene exchange closes forever. However, in this young radiation of Darwin's finches all species, apparently, still have their doors open (Grant 1986).

Summary

In this chapter we discuss the major themes of the study and some of the issues raised by our findings. The major themes are variation, natural selection, speciation, interspecific competition, and hybridization. The major issues are the causes of divergence of populations to the point of little or no gene exchange, the existence of morphological gaps between species, and the roles of song and beak size in speciation. We consider phylogenetic implications of hybridization, and the implication from the hybrid lineage that species may be arising and becoming extinct more frequently than is generally appreciated. Daphne finches show that speciation occurs by hesitant, wavering steps. Their evolution is genetically predictable but environmentally unpredictable, and so the evolutionary potential of a finch population, as well as its predictability, is a function of genetic variation, environmental variation, and their interaction. We end the chapter with an overview of speciation. Our main conclusion is that speciation of Darwin's finches is ecologically driven by natural selection, with reproductive isolation as an incidental effect or by-product of morphological and song divergence in allopatry.

Generalization

Perhaps—just perhaps . . . small populations, while teetering toward extinction and irrelevance, provide cauldrons of evolutionary novelty.

(Stern 2011, p. 218)

We can agree with Professor Popper that it does not matter for our purpose how generalisations are arrived at. The question which concerns us is what makes them acceptable.

(Ayer 2006, p. 19)

Generalizing When $N = 1$

WE STUDIED FINCHES ON DAPHNE: only one island! Are the evolutionary dynamics we uncovered just an interesting case history applicable only to Galápagos, or are they indicative of ubiquitous processes that are normally too subtle to be revealed in most field studies? In other words how representative are they of other island communities, in Galápagos or elsewhere, or of mainland birds, or of animals in general? Can Darwin's finches on Daphne be considered a model system of evolution in a fluctuating environment? Obtaining answers to questions of this sort require comparative studies of other species in other places (e.g., Price 2008, Bell 2010). Some general findings are obvious, however; for example, few scientists would doubt the generality of natural selection, and Thompson (2013), with echoes of Darwin (epigraphs of chapter 4), has argued that evolution by natural selection is "relentless," meaning pervasive and unceasing. On the other hand the importance of introgressive hybridization or interspecific competition is less well established and needs to be repeatedly assessed by measurement and experimentation (e.g., Arnold 1997, Dhondt 2012).

Daphne is small, and so are the finch populations; therefore we ask specifically to what extent can generalizations be made from studying finches with these features for a long time in one place (Grant and Grant 2010d, Billick and Price 2010)? We start with theory because it is general. Theory is explicit about genes and fitness but usually not about the intervening phenotype or the environment (fig. 4.1). Our study is explicit about the phenotype and environmentally driven fitness, but largely ignorant about the underlying genotypes. Acknowledging the difference, we use theory to interpret observations on Daphne.

THE SMALL POPULATION SYNDROME

Population genetics theory identifies several relevant genetic properties of populations that vary with population size (Crow and Kimura 1970). First, fewer mutations arise in small populations than in large ones because there are fewer genomes to mutate. Second, the probability that alleles will be lost by random genetic drift is higher in small populations. Third, as a result of both, small populations are expected to have a lower level of standing genetic variation. These three properties are mathematical consequences of small numbers of breeders (Price et al. 2010). When the reproductive success of breeders is heavily skewed, the genetically effective population size is even lower than the simple number of breeders, and the probability of genetic drift is correspondingly higher. The long-term genetically effective size is more influenced by low numbers than by high numbers, and this is important in a fluctuating environment such as Daphne. Other properties of small populations include high rates of inbreeding (Keller et al. 2006), diminished fitness, and loss of alleles; more rapid fixation of alleles, including ones with deleterious effects; reduced variation in epistatic interactions experienced by alleles (Mayr 1954); and possibly lower levels of pleiotropy (Stern 2011).

All things being equal, then, Daphne finch populations should vary less than related populations on large islands such as Santa Cruz and mainland populations. There is an additional ecological reason why this should be so. Habitats vary in plant composition along altitudinal transects on large islands such as Santa Cruz, and not on Daphne; therefore there is scope for some degree of genetic differentiation in different habitats on Santa Cruz (Kleindorfer et al. 2006, de Léon et al. 2010, Galligan et al. 2012). The expected difference in phenotypic (beak) variation of *fortis*, reflecting genetic variation, is in fact observed in a comparison of Daphne and Santa Cruz populations. However, several large islands support populations of *fortis* with lower variation than on Daphne (Grant et al. 1985). This shows that all things are not equal. The missing factor identified in the study of Daphne finches is introgression of alleles from other sympatric species.

Natural selection in small populations is sometimes described as being relatively inefficient because of the greater role of genetic drift in the loss or fixation of alleles. Selectively disadvantageous mutations are more likely to be fixed by genetic drift in small populations than in large ones. This may be happening the whole time on Daphne without us being aware of it, because we do not see the phenotypic effects of individual mutations; a single exception is described in the epilogue (chapter 17). Similarly, and for the same reason, we may fail to appreciate that stabilizing selection at the genetic level may be occurring the whole time, as typically assumed theoretically. Without denying both possibilities, we note that directional selection is occasionally strong on beak variation, and therefore not inefficient on this phenotype, and stabilizing selection on beak and body traits is apparently very rare. Moreover the Big Bird lineage did not lose any alleles at 16 microsatellite loci in three generations despite starting from a single inbred pair at the F_3 generation, and inbreeding did not result in the expected loss of variation, or the extinction of either *magnirostris* or the Big Bird lineage.

The small population paradigm of low variation may apply without qualification to populations that are essentially closed to heterospecific gene flow. The most likely examples in Galápagos are the solitary populations of *fuliginosa* on several arid islands that are smaller and botanically more depauperate than Daphne (Grant 1986) (fig. 15.4). These populations are presumably low in genetic variation, low in ecological opportunity, subject to stochastic fluctuations in numbers and liable to become extinct, and therefore short-lived.

THE MEDIUM POPULATION SYNDROME

Daphne is unique among islands in the Galápagos in size, degree of isolation, and vegetation. Smaller islands are arid and largely treeless, except for those close to large islands as North and South Plazas are in relation to Santa Cruz. Larger islands are generally richer floristically, except for the strongly isolated islands of Wolf and Darwin. Thus the intermediate characteristics of Daphne foster populations of intermediate characteristics. The island is large enough and floristically diverse enough to support three or more related species and not just one or two. Their populations are similar enough to allow hybridization, large enough that extinction risk is low, and large enough that effects of random genetic drift and inbreeding are minimal. All these factors point to a rich evolutionary potential that is less constrained genetically and environmentally than is usual in most small populations.

Generalizing from a unique island strains credulity. Nonetheless, from a deep temporal perspective Daphne is not alone: islands like Daphne existed in the past. Several islands acquired similar characteristics to Daphne when sea level fell and small islands became larger and higher (fig. 2.4), and new ones were formed by volcanic activity. Rises and falls in sea level occurred repeatedly, and islands separated and coalesced several times, in association with periods of glaciation. This happened as many as 10 times at 100,000 year intervals in the last million years, sufficient time for generation of the eight youngest species in the radiation and perhaps more. Their long-term survival would have been dependent upon colonization of other islands by dispersal or vicariance when coalesced islands separated (Poulakakis et al. 2012, Geist et al. MS). This broader perspective implies that favorable conditions for speciation occurred more in the past than in the present. Assuming the interpretation is relevant to speciation outside the archipelago, we suggest that the fission-fusion-replacement dynamics we see in the Galápagos occur in similar island settings with similar histories. Two examples involving hybridization are *Nesospiza* finches on the Tristan group of islands (Ryan et al. 1994, 2007) and *Zosterops* white-eyes on Reunion (Gill 1970, Warren et al. 2006). Similar dynamics may occur on continents in local, semi-isolated patches of inhomogeneous habitat, most likely at the margins of species distributions (see below). From this point of view Daphne, although singular, is a model system.

LARGE ISLANDS

If Daphne finches are unique, we do not expect hybridization on large islands. On the other hand hybridization may occur elsewhere in the archipelago but perhaps less frequently, in which case we should see signs of gene exchange between sympatric species on other islands. Those signs would be greater genetic similarity between species A and B where they occur together than where they occur on different islands. We made these comparisons between all pairs of related species, including the tree finches, on many islands, all larger than Daphne, and found overwhelming support for widespread gene exchange: species were more similar on average at microsatellite loci in sympatry than in allopatry (Grant et al. 2005). Therefore Daphne finches are not unique. Nevertheless introgression may have stronger effects on islands like Daphne because of its relatively small size.

In other respects Daphne finches are unusual. For example, they are probably subject to natural selection when food is scarce more frequently and intensely than are finches on large islands. Droughts are likely to be more severe on Daphne: the climatic evidence was discussed in the previous chapter. Furthermore Daphne lacks resident predators as well as the diversity of habitats present on large islands that may enable finches to escape the worst effects of food scarcity. On Daphne they have nowhere to go.

These comparisons are relevant to the debate on the roles of island size and isolation in evolution (MacArthur and Wilson 1967, Losos and Ricklefs 2010). Relatively small and well-isolated Galápagos islands have been considered a greater source of evolutionary novelty than large islands because they have a high proportion of endemic subspecies of finches (Lack 1945, 1947, Hamilton and Rubinoff 1967). Alternatively the high endemism might reflect long-term persistence rather than high rate of evolutionary divergence (Price et al. 2010, Rosindell and Phillimore 2011), and Cox (1990) has argued that the large central islands might have contributed more to the radiation. Evidence from Daphne and other islands (Petren et al. 2005) supports the former view.

BEYOND GALÁPAGOS

Where might the medium population syndrome apply to birds in continental regions? The answer is discontinuous, fragmented, or peripheral habitats. One example is glacially disturbed habitats at high latitudes (Weir and Schluter 2007). A second example is montane habitats in the Andes, especially during uplift when changes occurred at habitat margins (Burns and Naoki 2004, Fjeldså and Rahbeck 2006, Cadena et al. 2012). These include newly created patches of open habitat that witnessed rapid speciation in certain plant groups (Hughes and Eastwood 2006) and isolated cloud-forest ridges undergoing similar rapid speciation in plant species (Gentry 1989). A third example is dry and humid habitats in the neotropics that underwent cycles of expansion and contraction (Campagna et al. 2012, Jetz et al. 2012, Smith et al. 2012). A group of eleven species of capuchino seedeaters (*Sporophila*) in South America illustrate the role of habitat size in speciation. Their rapid diversification in the Pleistocene was associated with periodic contractions of savanna and forested habitat, so at times their populations must have been small. They occur in various sympatric associations, and like Darwin's finches they display genetic signatures of introgressive hybridization (Campagna et al. 2012). In several respects capuchinos may be the continental equivalent of Darwin's finches.

Our thesis, developed more extensively in a previous book (Grant and Grant 2008a), is that Darwin's finch evolution is no different in principle from the earliest stages of evolutionary divergence elsewhere, but differs in detail. The most important difference in detail is color. Darwin's finches are not colorful, whereas most of their continental tanager relatives are colorful and patterned, many outstandingly so (Burns 1997, Isler and Isler 1999). Sexual selection through female choice is likely to have played a much larger role in the diversification of continental species than in Darwin's finches. Speciation, and the rate of evolution of such traits, may be governed by the rate at which new mutations arise (Lanfear et al. 2010, Price et al. 2010), except when species hybridize. Little is actually known empirically about how the mutations then spread throughout the range of a species, either from the center or the margins where hybridization with a parapatric species might occur.

A second difference is environmental, and this may affect the rate of speciation. Consider the continental tanagers again. The current number of species is about 400 (Isler and Isler 1999), many are allopatric, and they evolved in roughly 12 MY (Cracraft and Barker 2009). Assuming uniform rates of speciation and no extinction, these figures translate into a doubling of species numbers approximately every 1.5 MY. This is about one-third of the rate of accumulation of Darwin's finch species (Grant and Grant 2008a) and is in keeping with data from other neotropical species that show a relatively slow rate of speciation (Price 2010, Weir and Price 2011). However, two groups of tanagers confound the trend with high rates of speciation. The first is a group of species in another archipelago, the Caribbean, that contain the closest relatives of Darwin's finches (Sato et al. 2001, Burns et al. 2002). The second group comprises 49 species in the genus *Tangara*. Their evolution began 6.5 MYA, but most speciation (species accumulation) was concentrated in the period 3.5–5.5 MYA (Burns and Naoki 2004) at a time when the Andes were rising (Coltorti and Ollier 2000). Thus their diversification rate is more comparable to the speciation rate of Darwin's finches, and was even higher in a concentrated period of their history. This is an important observation, since it shows that archipelagos are not necessarily the sites of fastest diversification, so their inhabitants are not necessarily exceptional.

THE SPECTER OF EXTINCTION, THE BIG UNKNOWN

We conclude the chapter with speculations about extinction. The 600 MY fossil record shows that almost all species that ever existed have become extinct (Simpson 1953). Are recent radiations exempt? Probably not. Our four decades of observations provide tantalizing hints of repeated speciation and extinction on a much shorter timescale. Thus any explanation of biological richness is incomplete without some understanding of the subtractive process of extinction and its effects on modern communities. Unfortunately almost nothing is known from direct evidence about natural extinction in recently radiating groups of organisms. The assumption we made in the preceding paragraph, that speciation rate is uniform and extinction is absent, is just that, an assumption. In the absence of fossils all we know is the minimum number of species that evolved in a group. The statistical challenge is to devise a model with extinction and speciation parameters that best fits the pattern of species accumulation through time based on molecular phylogenies (Ricklefs 2006, Glor 2010, Rabosky and Glor 2011). An example is the recent analysis of endemic vangids (Vangidae) on Madagascar. They display a standard pattern of species accumulating rapidly early in the radiation and slower later, and modeling shows the pattern is best explained by a decrease in the rate of speciation and not by an increase in the rate of extinction (Jønsson et al. 2012, Reddy et al. 2012).

Darwin's finches are a classical example of adaptive radiation (Schluter 2000, Grant 2013), yet they lack the classical signature of an early burst of species proliferation followed by a slower rate (fig. 16.1). In this they are not alone (Harmon et al. 2010, Derryberry et al. 2011). Modeling studies have shown that the early-burst signature is erased by extinctions if they are numerous enough (Rabosky and Lovette 2008). On these grounds alone there is evidence of finch extinctions. It appears at face value that the finch radiation started slowly (fig. 16.1), and this may not be correct. Additionally, regardless of whether Darwin's finches really exhibited an early burst or not, or whether they are still experiencing one, three lines of biological reasoning indicate extinction could have been prevalent in their history.

First, later-formed species may be more efficient at resource exploitation than some of the early-formed ones, and competitively replace them. Second, extinctions are to be expected because the Galápagos environment changed (Grant and Grant 2008a). The composition of the vegetation changed as the climate became progressively cooler and drier through Darwin's finch history. Some resources and ecological opportunities would have been eliminated as others were created, re-

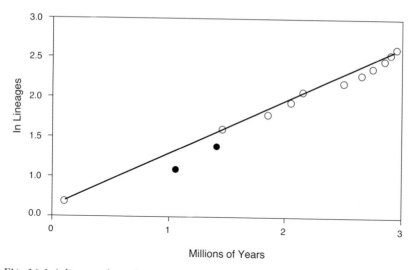

Fig. 16.1 A lineage-through-time plot of Darwin's finches shows a regular linear increase of species on a semilog plot after an apparent lag indicated by the two solid circles (Grant and Grant 2008a). The plot is constructed by taking a backward look at number of species from present to past and using molecular data to estimate time of species separations.

sulting in both extinction and speciation. The third reason is missing morphologies. The phylogenetically basal warbler finch (*Certhidea oli-vacea*) is very different from its closest relatives in continental South America and the Caribbean on the one hand, and very different from the next-oldest Darwin's finch species (*Geospiza difficilis* or *Platyspiza crassirostris*) on the other. The morphological transitions before and after *Certhidea* are so substantial that intermediate species must surely have existed and evolved into something else (pseudoextinction) or become extinct after generating one or more new species: the same reasoning from morphologically extreme species applies to the vangids mentioned above and to the Hawaiian honeycreepers (Pratt 2005). If our argument is correct, *Certhidea* may not be basal to the recon-structed phylogeny but a derived lineage from an ancestral species that has become extinct (Lack 1947, Petren et al. 1999b).

In combination these considerations suggest that modern Darwin's finch communities are the product of speciation and extinction that varied in time but not uniformly (fig. 16.2), perhaps co-occurring in pulses (Jackson 1995). Daphne has one such small community. Per-haps one day it will be possible to develop improved inferences about the past for all organisms, not just Darwin's finches, by reading evidence

Fig. 16.2 A graphical representation of the development of the Darwin's finch radiation. Species accumulate in a manner analogous to the MacArthur-Wilson (1967) immigration-extinction model of island biogeography (Grant 1984) and based on the same assumptions of density (diversity)-dependent processes and environmental saturation. There are three phases: (a) an early phase in which speciation, inevitably, exceeds extinction; (b) a middle phase of increasing speciation and extinction; and (c) a final phase when rates of speciation and extinction become equal and fall to a long-term equilibrial value. The logic of the last phase presupposes that a fixed, environmentally determined, long-term number of species in the archipelago has been reached. There is no evidence for the third phase in Darwin's finches, but there is for Caribbean *Anolis* lizards (Rabosky and Glor 2011), and evidence of an overshoot in species accumulation in a laboratory microcosm (Meyer et al. 2011). Extinction in the second phase could be responsible for an apparent

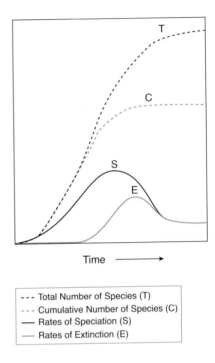

- - - Total Number of Species (T)
- - - Cumulative Number of Species (C)
—— Rates of Speciation (S)
—— Rates of Extinction (E)

lag in speciation at the beginning of the finch radiation (fig. 16.1), and high morphological disparity in surviving species, if some of the early formed species were outcompeted and replaced by later, more efficient, and morphologically different species (e.g., Ricklefs and Cox 1972). Quantitative models differ from this scheme by continuously varying speciation and extinction uniformly through time or setting extinction to zero throughout (e.g., Ricklefs 2006, Rabosky and Lovette 2008, Rabosky and Glor 2011).

of extinction as well as evolution into the genomes of extant species (Grant and Grant 2008a).

Summary

In this chapter we discuss the question of how generalizable are the results of studying finches on one island. The field observations are put in the context of theoretically expected evolution in small populations. Populations on Daphne do not conform entirely to expectation, for several reasons. Despite being small the island is floristically diverse enough to support three or more related species. Their populations are similar enough to allow hybridization, large enough that extinction risk is low, and large enough that effects of random genetic drift and inbreeding are minimal. We suggest that although Daphne is a unique island with unique biological properties, other islands of similar characteristics occurred in the archipelago in glacial periods. They may have been especially important platforms for generating new species. The same may be true in island-like situations on continents. We conclude by speculating about extinction, and suggest that future studies of avian genomes may help to reveal the signs of hitherto unknown species and extinctions. Finch populations must have frequently come and gone as new islands emerged and some were submerged.

Epilogue

It is now more than forty years that I have paid some
attention to the ornithology of this district, without being
able to exhaust the subject; new occurrences still arise as
long as any inquiries are kept alive.

(Letter XLIX, May 7, 1779: White 1789 [1877], p. 246)

The main difference [between short-term and long-term
research] is that with a prolonged study of a natural
system, the scientist is not in charge of the schedule.

(Edmondson 1991, p. 235)

Reflections on the Value
of Long-Term Studies

LONG-TERM PROCESSES require long-term study. A few such field
studies have contributed disproportionately to our knowledge of
how natural populations fluctuate in numbers and evolve (e.g.,
Anderson et al. 1991, Weider et al. 1997, Majerus 1998, Clegg et al.
2002, Beaugrand and Reid 2003, Packer et al. 2005, Baltensweiler et al.
2008, Ozgul et al. 2009, Armitage 2010, Clutton-Brock and Sheldon
2010, Husby et al. 2010, Qvarnström et al. 2010, van der Pol et al. 2010,
Hanski 2011, Alberts and Altmann 2012, Lebigre et al. 2012, Rijssel and
Witte 2013). In the experience of many field biologists the longer the
study persists, the greater the insights into the functioning of the study
system, and the greater the likelihood of making new discoveries, especially in strongly fluctuating environments (Strayer et al. 1986, Likens

et al. 1989, Edmondson 1991, Cody 1996, Billick and Price 2010). This is an argument for persistence and continuity: persistence to observe, and continuity to interpret, both ecological and evolutionary change. The argument is most cogent for studies of natural and unmanipulated populations whose members are individually identifiable.

In concluding this book we highlight with examples two ways in which benefits are gained from long-term investigations: from enlightenment that comes from the gradual accumulation of data over many years and, in contrast, from rare and unpredictable events.

LONG-TERM DYNAMICS OF A COLOR POLYMORPHISM

In 1978 we discovered a beak color polymorphism in nestlings (fig. 17.1): beaks were either pink or yellow (Grant et al. 1979). The frequency of the yellow morph was distinctly higher among *scandens* than *fortis* nestlings. These observations motivated us to continue recording colors to address questions of inheritance, long-term frequencies, and function, even though they had nothing to do with our primary research program. By continuing, we established that (a) the polymorphism was not a transient one but approximately stable in both *fortis* and *scandens* (fig. 17.2), (b) the frequency of the rarer yellow morph was always higher in *scandens* than in *fortis*, and (c) the morph frequencies of parents and offspring were consistent with a simple model of mendelian inheritance, yellow being recessive to pink (tables A.17.1 and A.17.2). Smaller samples of *magnirostris* showed much less stability in frequencies (fig. A.17.1 and table A.17.3).

A similar polymorphism with similar inheritance is well known in chickens (Bateson 1902, Dunn 1925, Hutt 1949). The responsible gene has been identified (Eriksson et al. 2008) and recently found to be the same gene in Darwin's finches (L. Andersson pers. comm.). Located on chromosome 24 (in chickens), it produces β-carotene deoxygenase 2 (BCO2), an enzyme that cleaves β-carotene into a colorless precursor of vitamin A (Walsh et al. 2011). A *cis*-acting regulatory, tissue-specific mutation inhibits this function in the skin but not in the liver or intestine (Eriksson et al. 2008), and ingested carotenes are deposited in the beak, skin, legs, and fat, where they give the bird a yellow appearance. Birds lacking the mutation appear pink because the blood supply can be seen.

What maintains the polymorphism in its current state, and why is the yellow frequency higher in *scandens* than *fortis*? The answer to the first question appears to be heterozygous advantage (fig. 17.3). The heterozygotes have a clear survival advantage over the yellow homozygotes in *fortis* in our combined data and in three cohorts out of five.

Fig. 17.1 Nestling beak color polymorphism. **Upper left**: *magnirostris* nestling, pink. **Upper right**: *magnirostris* nestling, yellow, from the same nest. **Lower left**: *scandens* fledgling, pink. **Lower right**: *scandens* fledgling, yellow. The yellow morph is also illustrated in figures 13.5 and 13.8.

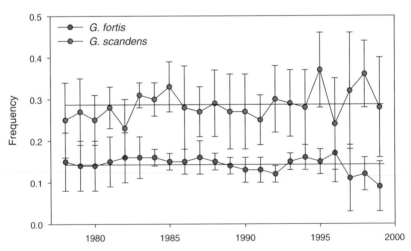

Fig. 17.2 Yellow-morph frequencies exhibit stability in *fortis* and *scandens*. Vertical bars are 95% confidence limits on frequency estimates. Sample sizes are in table A.17.3. After 1999 sample sizes fell below 20 until 2011. In that year the *fortis* frequency was 0.12 ± 0.15 ($n = 34$), and the *scandens* frequency was 0.35 ± 0.22 ($n = 17$). Frequencies of the yellow phenotype in *magnirostris* were much less stable (appendix 17.1), principally as a result of a population bottleneck in the severe drought of 2003–5. Horizontal lines indicate long-term averages.

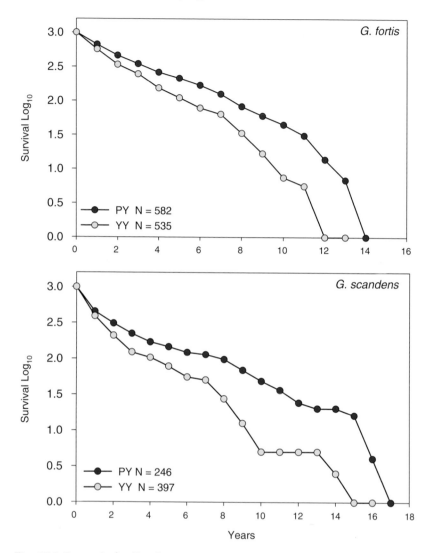

Fig. 17.3 Survival of yellow homozygotes and pink heterozygotes in five cohorts combined. The cohorts hatched in 1981, 1983, 1984, 1987, and 1991. Parents of heterozygotes were identified by having one yellow-morph parent and one pink-morph parent. Offspring with pink-morph fathers and mothers could be either heterozygotes or homozygotes (pink) and therefore are not included. The differences in survival between heterozygotes and yellow homozygotes were significant in both *fortis* ($\chi^2 = 19.517, p = 0.000, n = 582$) and *scandens* ($\chi^2 = 7.695, p = 0.0055, n = 246$) with a proportional hazards model. The survival curves of pink phenotypes of unknown genotype lie between the curves for heterozygotes and yellow homozygotes. This implies a relatively low survival of the pink homozygotes, and hence an advantage to heterozygotes, in comparison with both homozygote classes.

The heterozygotes also have the same advantage in *scandens* in our combined data and in two of the cohorts, but in one cohort the yellow homozygotes survived better than the heterozygotes.

The answer to the second question is not yet known, but there are two reasons for believing the higher frequency of the yellow morph in *scandens* is associated with exploitation of the carotene-rich pollen of *Opuntia* cactus (Grant and Grant 1981, Grant 1996). First, a fitness advantage to the yellow morph was experienced by the cohort of *scandens* produced in a year (1991) when cactus flowers were abundant during the nestling and fledgling phases. Cactus flowers were generally rare in 1981–87 when offspring of the other four cohorts were being fed by parents. Nevertheless first-year survival of the yellow morphs in the five cohorts was positively correlated with flower abundance ($r = 0.945$, $p = 0.005$), in contrast to a lack of correlation among the heterozygotes ($r = -0.100$, $p = 0.950$). The second reason is an association between yellow-morph frequency and cactus feeding in other populations of *Geospiza* species. On Wolf *difficilis* are uniquely monomorphic for the yellow morph (Grant et al. 1979) and are known to be dependent on cactus flowers during the breeding season. Likewise *conirostris* on Genovesa are strongly dependent on cactus flowers and have a high frequency of the yellow morph (Grant and Grant 1980b, 1989). All other finch populations that have been studied so far have lower frequencies of the yellow morph and are less dependent on cactus pollen.

The association with cactus feeding raises the question of why depositing carotenoids in tissues would be beneficial. Carotenoids have a signaling function in the context of courtship or territorial defense when deposited in feathers (McGraw et al. 2005, Hill 2007) or in the beaks of adults (Navarro et al. 2010, Walsh et al. 2011). However, there is no signaling value to the particular color of nestling beaks; parents do not feed the different-color morphs preferentially, according to our observations (Grant et al. 1979). Moreover the color can no longer be seen easily after the first week of life when melanin becomes deposited in the dermis of beaks and masks the underlying color.

A physiological value is more probable. Carotenoids function as a protective mechanism against oxidative damage by eliminating free-radical scavengers (Surai 2002, Hill and Johnson 2012) and by being the source of vitamin A. However, carotenoids and vitamin A are toxic in high concentrations (Russel 1999, Blomhoff 2001), leading, among other things, to impairment of skeletal muscles (Huggins et al. 2010). When Darwin's finches are feeding almost exclusively on *Opuntia* pollen, the high concentration of ingested carotenoids might be toxic and difficult for the liver to break down and excrete fast enough, or toxic when stored as vitamin A in high concentration in the duodenum (Olson 1984, Tajima et al. 2001). An ability to sequester carotenoids in the beak and other peripheral tissues and release them gradually later could be advantageous. Yellow-morph chickens become bleached when nutritionally stressed, implying storage, release, and utilization of the carotenoids (P. B. Siegel pers. comm.).

The toxic avoidance hypothesis could be tested in other species. Although the polymorphism appears to be rare in bird species (Grant et al. 1979), pairs of related species can be found with either pink or yellow beaks or legs (e.g., European thrushes, *Turdus ericetorum* and *T. merula*, and gulls, *Larus argentatus* and *L. fuscus*).

This example of a genetic polymorphism followed through time illustrates the value of studying long enough to answer basic questions. It enabled us to establish the stability of the polymorphism and the association with cactus flowering. The long term was also needed to both detect and quantify a perturbing factor: introgressive hybridization between species with different morph frequencies. The effect is small (table A.17.4), however, because hybrids and backcrosses are a minority of the breeding populations.

RARE EVENTS AND THEIR CONSEQUENCES

A second value of long-term studies is the detection of rare but strong events, coupled with an understanding of the long-lasting consequences for the environment, for the organisms that exploit it, and for the relationship between the two (Grant and Grant 2010d). Short-term studies are more likely to miss such events and fail to interpret, or misinterpret, the consequences. The best example is the critical El Niño event of 1982–83. If we had stopped before that year, after the tenth field season, we could have concluded, incorrectly, that droughts invariably select for large body and beak size, hybrids do not breed, and the island cannot support another species. If we had started after 1983, we would have missed the colonization by *magnirostris* and the *fortis* × *scandens* hybrid that gave rise to the hybrid lineage. Our appreciation of the ecological and evolutionary importance of annual environmental variation and hybridization would have been greatly diminished (Grant and Grant 1996c).

We are intrigued by the idea that a chance conjunction of rare events may be especially potent evolutionarily. For example, extinction of dinosaurs and many other organisms at the same time may have been caused by a combination of two rare phenomena, the impact of an asteroid or comet and extreme volcanism, and not by one of them alone (Keller 2012, Renne et al. 2013). Similarly, the current success of the hybrid lineage can be traced to two improbable and unrelated events in the 1980s. First, a *fortis-scandens* hybrid immigrated in 1981, and second, this was followed by the establishment of a breeding population of *magnirostris* at the end of 1982. Almost certainly an important factor in the success of both was the unusually favorable ecological conditions caused by the exceptionally prolonged, once-in-400-years El Niño event in 1982–83. There was no connection between the two colonization events. Twenty years later a connection was made when *magnirostris* determined the fate of the hybrid lineage by competitively eliminating most of the large members of the *fortis* population in the drought of 2003–5. Without the 1983 El Niño event and the population of *magnirostris* that it fostered, members of the hybrid lineage may not have survived beyond 2004. Rare events need not be exactly coincident to have profound and long-lasting effects, and the consequences may emerge after a long delay, as these examples illustrate. So although the environment determines the direction of evolution, there is an element of chance in what constitutes the environment at any one time; hence evolution is partly stochastic.

CHANGES IN PERSPECTIVE

With regard to the larger picture, results of studying finches on a single island for 40 years have changed our perspective on the broad context of adaptive radiation to a more dynamic, less unidirectional one. The results suggest a high degree of change and instability in the properties of populations and communities: species are repeatedly gained by fission and divergence, and lost by fusion and extinction (Grant and Grant 1996c, 2008a). The archipelago exhibits a similar instability through volcanism and changing sea-level and climates. Thus the radiation of finches is best viewed as the joint product of inherent (biological) and environmental dynamism, not marching inexorably to its current and fixed state in a constant and permanent environment, with ecological niches being sequentially filled along the way, but oscillating about a long-term trend of increasing diversity. The radiation cannot be fully understood without an appreciation of environmental change over short and long spans of time. Radiations of cichlid fishes in the African Great Lakes are other examples of recent and rapid radiations (Kocher 2004, Seehausen 2006, Santos and Salzburger 2012) (and much more prolific), and they appear to exhibit a similar dynamism. As in the finches, hybridization is believed to have been important early in the radiations (Seehausen 2004, Albertson and Kocher 2005) in an environment of profound tectonic and topographical change (Schwarzer et al. 2012).

Fig. 17.4 Finis. **Upper:** Field laboratory, kitchen, and cave for cool reflection (ch. 15, epigraph). **Lower:** Boots.

Long-term studies should build toward greater understanding and revelation and not simply repeat what has gone on before. We never reached an identifiable point of diminishing returns, or experienced a sense of completion. By the end of the study (fig. 17.4) the environment had changed (fig. 1.5), and the composition of the Daphne community of finches and the characteristics of the individual populations had been transformed (fig. 15.1). Change was continuing. In the last year (2012) *scandens* were more numerous, *fortis* beaks were more variable, *scandens* beaks were more blunt, and *fortis* and *scandens* were more similar morphologically than at any other time in the study. Questions remain, to be answered in the future even as more arise. Therefore one conclusion we draw after 40 years is the same as the conclusion we reached after 20 years: long-term studies of ecology and evolution should be pursued in an open-ended way because for many of them there is no logical end point (Grant and Grant 1996c, 2010d). Daphne finches have much more to teach us.

Coda

The cradle that is Daphne was once undersea,
Umbilicaled to Santiago.
She emerged from a deep subterranean sleep,
The year, an exceptional El Niño.

Daphne is a cradle that is now above sea.
She nurtured the birth of a finch
Compounded of genes from two species or more,
The beak much less than an inch.

The cradle that is Daphne will sink once again
Or explode and then it will be gone.
By then the finches will have flown elsewhere,
Continuing to evolve, on and on.

Appendixes

Appendix 1.1 Daphne Plants

TABLE A.1.1
Daphne plants

Plant species	Growth form	Seed index
Resident		
Bursera malacophylla Burseraceae	Tree	4.4
Croton scouleri Euphorbiaceae	Tree	1.3
*Opuntia echios** Cactaceae	Shrub	3.0
Abutilon depauperatum Malvaceae	Herb	1.6
Acalypha parvula Euphorbiaceae	Herb	1.1
Amaranthus squamulatus Amaranthaceae	Herb	•
*Amaranthus sclerantoides** Amaranthaceae	Herb	0.4
*Boerhaavia caribaea** Nyctaginaceae	Herb	0.8
*Boerhaavia erecta** Nyctaginaceae	Herb	0.9
Cacabus miersii Solanaceae	Herb	2.0
*Chamaesyce amplexicaulis** Euphorbiaceae	Herb	0.8
Chamaesyce punctulata Euphorbiaceae	Herb	1.1
Cleome viscosa Capparidaceae	Herb	•
Desmodium glabrum Leguminosae	Herb	2.2
*Heliotropium angiospermum** Boraginaceae	Herb	0.8
*Herissantia crispa** Malvaceae	Herb	1.4
Mentzelia aspera Loasaceae	Herb	1.7
Mollugo flavescens Molluginaceae	Herb	•
*Portulaca howelli** Portulacaceae	Herb	0.4
*Sesuvium edmonstonei** Aiozoaceae	Herb	0.7
*Sida salviifolia** Malvaceae	Herb	0.9
Tephrosia decumbens Leguminosae	Herb	3.1
*Tiquilia galapagoa** Boraginaceae	Herb	0.7
*Tribulus cistoides** Zygophyllaceae	Herb	9.8
Cuscuta gymnocarpa Convolvulaceae	Vine	•
Sarcostemma angustissima Asclepiadaceae	Vine	•
*Aristida subspicata** Gramineae	Grass	0.8
Cenchrus platyacanthus Gramineae	Grass	3.9
*Chloris virgata** Gramineae	Grass	•
*Eragrostis cilianensis** Gramineae	Grass	0.5
*Trichoneura lindleyana** Gramineae	Grass	0.8

TABLE **A.1.1**
(*continued*)

Plant species	Growth form	Seed index
Rarely recorded		
Brickellia diffusa[1] Compositae	Herb	•
Chamaesyce nummularia Euphorbiaceae	Herb	0.6
Cyperus anderssoni[2] Cyperaceae	Sedge	•
Cyperus confertus[2] Cyperaceae	Sedge	•
Desmodium procumbens[1] Leguminosae	Herb	•
Digitaria horizontalis[3] Gramineae	Grass	•
Eleusine indica[1] Gramineae	Grass	•
Galactia striata[3] Leguminosae	Herb	3.7
Gossypium barbadense[1] Malvaceae	Herb	4.1
Paspalum galapageium[1] Gramineae	Herb	•
Paspalum redundans[1] Gramineae	Grass	•
Phaseolus atropurpureus[2] Leguminosae	Herb	•
Physalis pubsecens Solanaceae	Herb	0.6
Porophyllum ruderale[3] Compositae	Herb	0.3
Rhynchosia minima[3] Leguminosae	Herb	3.6
Salvia sp.[1] Labiatae	Herb	•
Tiquilia darwinii[3] Boraginaceae	Herb	•
Trianthema portulacastrum Aiozoaceae	Herb	1.2
Fern[1] Polypodiaceae	Herb	•
Rare in numbers		
Alternanthera filifolia[E] Amaranthaceae	Herb	0.6
Castela galapageia Simaroubaceae	Tree	6.6
Cordia lutea Boraginaceae	Tree	13.2
Heliotropium curassavicum[E] Boraginaceae	Herb	•
Nolana galapagensis Nolanaceae	Herb	•
Scalesia crockeri[E] Compositae	Tree	0.7
Tournefortia psilostachya[E] Boraginaceae	Tree	1.7
Waltheria ovata Sterculiaceae	Shrub	0.5
Not consumed by finches		
Ipomoea linearifolia Convolvulaceae	Vine	3.8
Merremia aegyptica Convolvulaceae	Vine	7.1

Note: Species of plants, growth forms, and, where known, the size-hardness index of the seeds or fruits $(DH)^{1/2}$, are from Abbott et al. (1977); also box 2.1. Missing values, indicated by a dot, are all low. Finches subsist on seeds in the dry season until *Opuntia* begin flowering in October or November; 16 common seed species are indicated by *. One species (*Cleome viscosa*), a nonnative (fig. 14.5), invaded and became established in 1995, probably brought to the island by seabirds from Baltra, where it is now abundant. Four species represented by single individuals became extinct after several years (E). Plants rarely recorded include several encountered only one, two, or three times ([1], [2], [3]). The remaining species are rare residents. Although rare in numbers, *Castela galapageia* (one), *Cordia lutea* (one) and *Waltheria ovata* (four) were present in all years. *Cordia lutea* rarely produced fruits, in El Niño years. Names are from Wiggins and Porter (1971) and Porter (1977).

Appendix 1.2 Measurements of Finches

Six morphological traits were measured on adult birds (box 2.2), and estimates of the means and standard deviations are provided in table A.1.2. Only *fortis*, *scandens*, and *magnirostris* that were known to breed (at least one egg in a nest) were included. Many more that did not breed were also measured: some may have been immigrants, especially *magnirostris* and *fuliginosa*. The *fuligi-*

TABLE A.1.2
Means and standard deviations (SD) of four finch species breeding on Daphne

	Males			Females		
	N	Mean	SD	N	Mean	SD
G. fortis						
Weight	547	17.05	1.810	518	15.67	1.615
Wing	545	69.20	2.266	517	66.60	2.105
Tarsus	530	19.35	0.725	506	18.74	0.744
Beak length	547	11.02	0.752	518	10.72	0.744
Beak depth	547	9.59	0.830	518	9.19	0.795
Beak width	547	8.87	0.601	518	8.57	0.582
G. scandens						
Weight	235	21.17	1.688	212	14.04	1.798
Wing	233	74.21	2.184	214	8.95	2.249
Tarsus	223	21.06	0.764	201	8.45	0.743
Beak length	235	14.46	0.873	214	19.75	0.880
Beak depth	235	9.37	0.463	214	70.86	0.467
Beak width	235	8.82	0.417	214	20.42	0.396
G. magnirostris						
Weight	65	30.87	1.906	41	28.55	1.641
Wing	65	82.31	2.222	41	78.93	2.132
Tarsus	65	23.46	0.761	41	22.79	0.789
Beak length	65	15.55	0.682	41	15.30	0.531
Beak depth	65	17.47	0.871	41	16.67	0.675
Beak width	65	15.27	0.899	41	14.71	0.647
G. fuliginosa						
Weight	16	12.71	1.520	17	12.61	1.948
Wing	16	63.10	2.286	17	61.92	3.098
Tarsus	16	18.26	1.070	17	17.66	1.057
Beak length	16	9.01	0.735	17	8.85	0.677
Beak depth	16	7.07	0.693	17	6.95	0.611
Beak width	16	6.97	0.572	17	7.11	0.609

Note: Weight is in grams, and the other traits were measured to the nearest mm (wing) or 0.1 mm (the remainder). Measurements of the same bird in different years were repeatable (table 3.1) and averaged. N refers to sample size.

nosa samples included those known to breed and those present in the breeding seasons of 2000–2012, when breeding was suspected but not followed in detail.

Appendix 1.3 Other Species of Darwin's Finches

Two other species complete the summary of ground finches in the genus *Geospiza*. One, the Large Cactus Finch (*G. conirostris*), is geographically restricted to Genovesa in the northeast of the archipelago and to Española and its satellite Gardner in the southeast (fig. P.3). The other, the sharp-beaked ground finch (*G. difficilis*), is ecologically restricted though geographically more widespread. Populations on Santiago, Pinta, and Fernandina occupy *Zanthoxylum* forest at high elevations in the humid zone. Populations were formerly present in similar habitat on Santa Cruz and Floreana, and possibly Isabela. Populations on the northwestern islands of Darwin and Wolf bear a morphological resemblance to the Cactus Finch (*G. scandens*) and were once classified as such. The population on Genovesa resembles the Small Ground Finch (*G. fuliginosa*). The warbler finch, vegetarian finch, five species of tree finches, and a single species on Cocos Island (Costa Rica) are in different genera. They differ from each other in body size, size and shape of the beak, in plumage, and in song. Their evolution has been discussed in two other books (Grant 1986 [1999] and Grant and Grant 2008a).

Appendix 3.1 Mapping Breeding Locations

To facilitate study of nests and recording of observations of birds, we divided the island into 20 sectors (fig. A.3.1). Sector boundaries were marked with flagging, and bearings from north at these markers were determined by compass. Territories and nests were then mapped with reference to sector boundaries or markers, and distances between nests were either measured with a 50 m tape measure or estimated. Nests at relatively low density of breeders in February 1991 are depicted in figure A.3.2 to illustrate the wide distribution of nesting locations.

Appendix 3.2 Annual Changes in Measurements

Finches stop growing in their first year of life after about two months (Boag 1984). In the next year there are small tendencies to increase or decrease in size: from January–March of the year after hatching to the following year at the same time. Sixty-nine *fortis*, 18 *magnirostris*, and 16 *scandens* were measured (by P.R.G.) at both times. Average weight, wing length, and beak length increased in all three species, and average beak depth and beak width decreased in all three species, to approximately the same degree (fig. A.3.3). These trends are illustrated below with the *fortis* sample of measurements. Minor changes lower the repeatability of measurements (table 3.1).

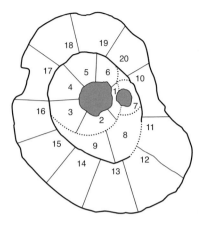

Fig. A.3.1 Daphne divided into 20 sectors for nesting studies. The gray areas are the crater floors.

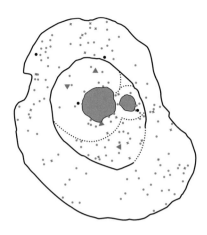

Fig. A.3.2 Distribution of nests in 1991. Symbols: *fortis*, blue circles; *scandens*, red crosses; *fuliginosa*, black circles; *magnirostris*, green triangles.

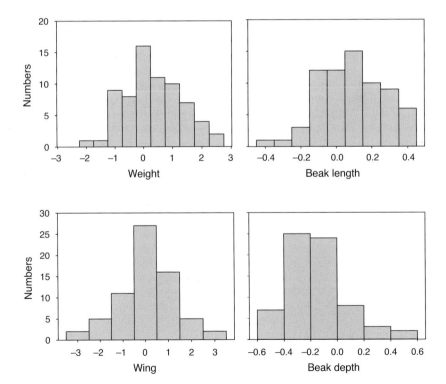

Fig. A.3.3 Measurements of 69 *fortis* individuals compared at the end of their first and second years. Differences are expressed relative to the first measurements.

Appendix 5.1. Extra-pair Mating

Male *fortis* and *scandens* gain extra fitness in terms of number of offspring produced through extra-pair mating (Grant and Grant 2011a). The benefit is gained early in life by *scandens* but is delayed in *fortis* until about year 6, when fitness losses from being cuckolded are exceeded on average by gains from their extra-pair mating.

In contrast *fortis* females do not produce more offspring through extra-pair mating. Moreover there are three reasons for believing they do not produce better offspring through extra-pair mating. First, extra-pair *fortis* young hatched in 1987 and 1991 did not become recruits to the breeding population more frequently than their nest mates sired by the social father. Sixty-eight nests had both extra-pair and within-pair offspring, and at least one recruit was produced from 30 of them. Recruitment of extra-pair offspring (56.6%, $n = 53$) was almost the same as recruitment of within-pair offspring (54.0%, $n = 87$).

Second, extra-pair *fortis* young hatched in the 68 nests lived shorter lives than their half sibs as often as they lived longer lives. Mean longevities of extra-pair young exceeded mean longevities of their half sibs in less than half of the comparisons (45.6%, $n = 68$ pairs). Longevity is an indirect indication of their lifetime reproductive success (figs. 5.10 and 5.11). Third, individuals with extra-pair fathers did not produce more offspring than their half sibs: when matched with their half sibs they were usually less successful. This applies to number of eggs, where those with extra-pair fathers have the higher number in a minority of cases (37.5% of 16 comparisons), number of successful clutches (34.5%), and number of fledglings (34.5%), including, in the case of males, those resulting from extra-pair matings. The mean number of fledglings produced by those with extra-pair fathers (4.1 ± 4.15 s.d., range 0–13) is less than their half-sibs with social fathers (5.8 ± 5.82, range 0–21). Recruitment of this generation is poorly known because not all pairs were known after 1993, but there is no indication of compensatory success of extra-pair young at this stage (see also Sardell et al. 2012a). Three of those with extra-pair fathers became recruits, whereas four with social fathers became recruits. Recognizing that lifetime reproductive data are lacking for the grand-offspring generation, we draw a cautious conclusion: there is no indication of an average fitness benefit gained by females from extra-pair mating.

We have considered other possible advantages of extra-pair mating, such as gaining a better social mate in the future, with a better territory and nest sites, but could find none (Grant and Grant 2011a). Nor could Sardell et al. (2012b) in a more comprehensive study of Song Sparrows (*Melopsiza melodia*) on Mandarte Island, Canada. Extra-pair mating within species of Darwin's finches appears to be opportunistic and stochastic to a large degree (compare Reyer et al. 1997). When preparing to form a clutch of eggs, a female needs energy and sperm, both of which may be obtained in part by feeding occasionally in another territory, typically a neighbor's, in the absence of her social mate. Radio tracking could help to pinpoint the location of extra-pair mating, home or away (Akçay et al. 2012), as observations are scarce. We have seen (fewer than 10 times) *fortis* and *scandens* females soliciting copulations from intruding males close to their own nests or close to a neighbor's nest. With one exception the intruder was chased away by the resident male or female. In summary, we suggest extra-pair mating is not so much a choice of a particular male by a female, unless she is attempting to change mates, as an incidental effect of being ready to mate in two circumstances: when feeding on or off territory in the absence of her social mate, and when close to her nest.

Frequencies of extra-pair mating are lower in the larger species (*magnirostris* and *scandens*), reflecting stronger attachment to, and defense of, nests and perhaps greater mate guarding. However, extra-pair mating may be exceptionally high in years of unusual sex ratios. In 1979 and 1980, when male *scandens* outnumbered females by 3:1, unpaired males fed nestlings in the nests of neighbors (Price et al. 1983). Possibly this reflected an unusually high frequency of extra-pair mating with neighbors. In later years, when extra-pair paternity could be determined and the sex ratio was closer to 1:1, feeding neighbors' offspring was not observed during more than 300 nest watches of 0.5–1.0h duration (Grant and Grant 2011a).

In other species elsewhere that are highly colored or ornamented and subject to strong sexual selection, there are fitness benefits to be gained by selecting particular extra-pair mates (Albrecht et al. 2009, MacFarlane et al. 2010, Gerlach et al. 2012).

Appendix 5.2 Visitors and Predation

Finches may remove an egg or kill a nestling in another finch's nest (fig. 5.13) and even eat the contents of an egg. We have witnessed these events rarely. Usually we only know that occasionally an egg or a nestling disappears from a nest. We have never seen a lizard (*Microlophus*), the only resident reptile, visiting a finch nest. Short-eared Owls (*Asio flammeus*) prey upon nestlings of finches on Genovesa (Grant and Grant 1989) but not on Daphne, perhaps because the protective spines of the cactus nest bushes are much stiffer and more difficult for owls to negotiate than they are on Genovesa. Most bizarre is the death of two nestlings squashed by an *Opuntia* fruit! Remarkably two other (*fortis*) nestlings survived to fledging. The suspected perpetrator was a male *scandens*. We recorded a total of five nests containing *Opuntia* fruits that must have been pushed inside the side entrance.

Once out of the nest, young finches as well as adults on Daphne are vulnerable to predation by owls, Great Blue Herons (*Ardea herodias*), and Great Egrets (*Casmerodias albus*) (Grant et al. 1975) (fig. 5.12). None of these are permanent residents on Daphne; they visit the island in small numbers for short periods. Other predatory birds that occasionally visit the island are Galápagos Hawks (*Buteo galapagoensis*), migratory Peregrines (*Falco peregrinus*), Lava Herons (*Butorides sundevalli*), Cattle Egrets (*Bubulcus ibis*), and Smooth-billed Anis (*Crotophaga ani*); the last two species have been introduced to Galápagos. The numbers of these and other nonpredatory occasional visitors are given in table A.5.1. The contents of owl pellets found throughout the island are given in table A.5.2. The frequency of pellets containing finch bones varied from 0 (1993, $n = 10$) to 33.6 (1995, $n = 327$), with an average in 23 years of 12.7 ± 0.093 s.d.

TABLE A.5.1

Avian visitors to Daphne

Species	Common name	Numbers	Years	Dates
Asio flammeus	Short-eared Owl	1–10	21	1973–2012
Buteo galapagoensis	Galápagos Hawk	1	17	1973–2011
Falco peregrinus	Peregrine	1–2	17	1973–2011
Ardea herodias	Great Blue Heron	1–2	18	1977–2011
Casmerodius albus	Great Egret	1–3	8	1975–2001
Butorides sundevalli	Lava Heron	1–2	13	1988–2012
Bubulcus ibis	Cattle Egret	1–20	9	1988–2012
Crotophaga ani	Smooth-billed Ani	1–3	7	1987–2011
Myiarchus magnirostris	Galápagos Flycatcher	1–3	14	1973–2011
Coccyzus melacorhyphus	Galápagos Cuckoo	1–5	6	1984–1998
Certhidea olivacea	Warbler Finch	1	2	1973–1976
Camarhynchus parvulus	Small Tree Finch	1	4	1984–2002
Camarhynchus psittacula	Large Tree Finch	1	2	1977–1984

Note: Numbers were recorded in 33 of the 40 years. In addition to resident ground finches, a few pairs of Galápagos martins (Progne modesta) and Yellow Warblers (Dendroica petechia), and Galápagos doves (Zenaida galapagoensis; 6–30 pairs) breed on the island. Two North American migrants, a Blackpoll Warbler (Dendroica striata; Boag and Ratcliff 1979) and possibly a Prothonotary Warbler (Protonotaria citrea; Petit and Tarvin 1990) have also been recorded.

Appendix 9.1 Variation and Mortality

G. scandens may be subject to stabilizing selection because it is a resource specialist. If so we should see a trend toward a reduction in variation in many years, with survivors being less variable than nonsurvivors. If, on the other hand, hybrids have a distinct advantage over nonhybrids, the opposite result would be expected because hybrids are added to one end of morphological frequency distributions and thereby broaden the range of variation. To examine these alternatives, we compared variances in the three principle components traits in each year with samples greater than five.

Seven differences in variance were significant ($p < 0.05$ Levene's test). These should be ignored because 5 are expected by chance from the total of 105 tests. This leaves only two: nonsurvivors varied more in beak size than survivors in 1986 ($p = 0.0023$) but less than them in 2008 ($p = 0.0095$, close to chance expectation). Thus there is little evidence of stabilizing or diversifying selection. An alternative procedure ignores statistical significance, pays attention only to the sign of the differences, and adds up the pluses and minuses separately. This yields no net trend. The nonsurvivors vary in body size more than the survivors in 19 of the 35 years, in beak size in 18 and in beak shape in 14.

A different argument leads to a different expectation. G. fortis might be subject more often to stabilizing selection than scandens because it has a higher equilibrial level of morphological variation due to introgression (chapter 9). However fortis, like scandens, shows little evidence of stabilizing selection,

TABLE A.5.2
Finches found dead on the ground, and the finch and other contents of owl pellets

Year	Dead finches	% Banded	Pellets	% Finches	% Seabirds
1973	•	•	49	8.2	59.2
1974	•	•	56	21.4	48.2
1987	46	100.0	•	•	•
1988	20	90.0	25	16.0	4.0
1989	23	100.0	•	•	•
1990	42	97.6	268	4.9	81.7
1991	24	100.0	142	3.5	83.8
1992	41	38.0	211	1.4	40.3
1993	3	100.0	10	0.0	70.0
1994	4	100.0	117	19.7	35.0
1995	20	40.0	327	33.6	56.3
1996	11	36.4	173	28.3	45.7
1997	8	25.0	94	19.1	64.9
1999	5	0.0	•	•	•
2000	11	9.1	12	16.7	75.0
2001	4	0.0	18	5.6	55.6
2002	18	0.0	42	21.4	33.3
2003	8	0.0	7	0.0	71.4
2004	105	12.4	63	20.6	63.5
2005	61	39.3	66	13.6	77.3
2006	7	57.1	81	8.6	72.8
2007	2	100.0	82	3.7	96.3
2008	15	20.0	98	11.0	78.6
2009	3	33.3	52	11.5	63.5
2010	20	25.0	144	17.4	69.4
2011	21	19.0	70	17.1	77.1
2012	6	0.0	14	64.3	35.7

Note: Seabirds include boobies *Sula*, tropicbirds *Phaethon*, petrels *Oceanodroma*, which nest on the island, and shearwaters *Puffinus*, which do not. Finches include all four resident species. Other prey items comprise shorebirds, warblers, doves, martins and cuckoos, locusts, scorpions, beetles, fish fragments, lizards, rats, and mice. Rats (1974–2011) and mice (1973–92) do not live on Daphne. They are caught by owls that hunt on Seymour, Baltra, and Santa Cruz and roost on Daphne (Grant et al. 1975).

with 4 differences in variance at $p < 0.05$ out of 111 comparisons and 1 at $p < 0.005$. Variation was as likely to be greater in the survivors as in the non-survivors, in body size (20:17), and in beak size and shape (both 19:18).

In both species survivors to year $x + 1$ did not vary significantly less, or more, than the total sample in year x, in any year and in any of the three traits. We conclude that variation rarely changes as a result of differential mortality.

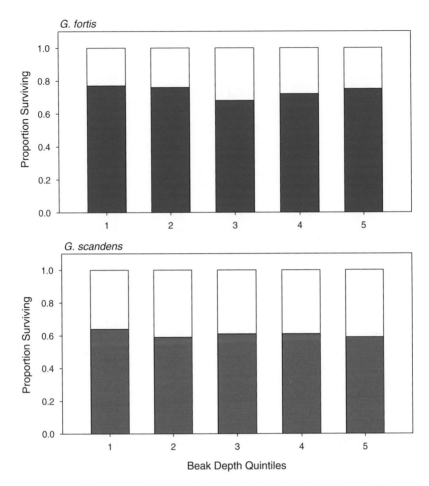

Fig. A.9.1 Uniform survival (solid bars) across five beak-depth classes with equal sample sizes. With directional selection survival of one extreme-size class is lower than the rest, and with stabilizing selection survival of both extreme classes is lower than the others. Starting sample sizes are 553 *fortis* (1989–90) and 350 *scandens* (1984–85). Robustness of uniformity was checked by repeating the analysis with 10 beak-size classes (deciles) instead of five quintiles. Size increases from left to right on the x-axis. Survival was more heterogeneous in both species, but survival was again not depressed in the extreme classes.

Figure A.9.1 illustrates uniform survival across beak size categories with large samples.

Surviving *magnirostris* varied more than nonsurvivors in beak shape in 1983 ($p < 0.05$) and 1988 ($p < 0.05$), but less than nonsurvivors in beak size in 1995 ($p < 0.0001$) and less than nonsurvivors in body size in 1995 ($p < 0.0001$).

These differences may be the result of emigration rather than local selection (see text).

Appendix 10.1 On the Dangers of Extrapolation

"The Mississippi between Cairo and New Orleans was twelve hundred and fifteen miles long one hundred and seventy-six years ago. In the space of this time the lower Mississippi has shortened itself by two hundred and forty-two miles. Now if I wanted to be one of those ponderous scientific people and "let on" to prove what had occurred in the remote past by what had occurred in a given time in the recent past, or what will occur in the far future by what has occurred in late years, what an opportunity is here! —

That [change] is an average of a trifle over one mile and a third per year. Therefore, any calm person, who is not blind or idiotic, can see that in the Old Oölitic Silurian Period, just a million years ago next November, the Lower Mississippi River was upwards of one million three hundred thousand miles long, and stuck out over the Gulf of Mexico like a fishing rod. And by the same token any person can see that seven hundred and forty-two years from now the Lower Mississippi will be only a mile and three quarters long, and Cairo and New Orleans will have joined their streets together, and be plodding comfortably under a single mayor and a mutual board of aldermen. There is something fascinating about science. One gets such wholesale returns of conjecture out of such trifling investment of facts" (Twain 1883, pp. 172–173).

Appendix 10.2 Plumage

Male ground finches acquire more black in their plumage with increasing molts, beginning at the head and extending over the whole body (illustrated in Grant 1986, 1990). The number of years it takes to pass from immature plumage (categories 0–3 in Grant 1986) to mature plumage (categories 4–5) varies from one year (2 scandens) to seven years (2 fortis). G. scandens ($n = 62$) acquire mature plumage more rapidly than fortis ($n = 245$; $\chi_6^2 = 80.533$, $p < 0.0001$). There is a deficiency in fortis taking two years to reach plumage maturity ($\chi_1^2 = 9.730$, $p < 0.005$) and a strong excess of scandens taking two years ($\chi_1^2 = 38.448$, $p < 0.0001$) over the joint expectations. Additionally there are more scandens than expected reaching plumage maturity in one year ($\chi_1^2 = 6.307$, $p < 0.02$) and less than expected reaching plumage maturity in three ($\chi_1^2 = 9.057$, $p < 0.005$) and four ($\chi_1^2 = 8.438$, $p < 0.01$) years. The mode is two years in scandens and three years in fortis. The mode in magnirostris is also two years (9 out of 19; ages 2 to 4 years only), like scandens.

Rates of maturation are environmentally sensitive in fortis but not in scandens. There are four fortis cohorts and three scandens cohorts. Sample sizes for fortis are 15 (cohort 1983), 17 (1984), 91 (1987), and 114 (1991), and for scandens they are 5 (1983), 35 (1987), and 22 (1991). Annual variation overall is strong in fortis ($\chi_{15}^2 = 133.526$, $p < 0.0001$) and weak in scandens ($\chi_6^2 = 12.748$, $p = 0.0472$). Annual variation in the well-represented 1987 and 1991 cohorts considered alone is pronounced in *fortis* ($\chi_5^2 = 93.697$, $p < 0.0001$) and absent in *scandens* ($\chi_3^2 = 0.834$, $p = 0.8413$). In these two years most *scandens* took only two years to reach plumage maturity: 83% of the 1987 cohort and 73% of the 1991 cohort. Most of the 1987 cohort (80%) of *fortis* took 4 years, whereas most of the 1991 cohort (88%) took three years. Similarly most of the 1983 cohort of *fortis* (55%) took four years, and most of the 1984 cohort (67%) took three years.

Annual variation in plumage maturation of *fortis* is explained by annual variation in feeding conditions. A drought occurred in 1985, and 1986 was also dry. *G. fortis* males that had not molted into mature plumage before those years did so in the El Niño year of 1987. This applies to both 1983 and 1984 cohorts; hence maturity was reached in a shorter time by the latter cohort. The other two cohorts, produced in 1987 and 1991, experienced different conditions in the following years. Two drought years and one dry year (1990) followed 1987, whereas two wet years followed 1991. Correspondingly most *fortis* males took longer to molt into mature plumage after 1987 than after 1991. In this species plumage maturation rate may also be adversely affected by adult density.

If there is an inherited tendency to mature at a particular rate under a particular set of conditions, and the genetic factors are different in the two species, the offspring of mixed matings should reflect their heterogeneous inheritance. We expect the first generation backcrosses in both directions to be more similar to the species to which they have backcrossed than to the other under a scheme of additive inheritance. However, if the trait is inherited with dominance the backcrosses in both directions should be most similar to one of the species, the one in which it is dominant. This can be tested with the backcross FFS sample that was produced in 1987 ($n = 9$) and compared with the contemporary sample of *fortis* and *scandens*. The SSF sample produced in 1991 ($n = 7$) is compared with the 1991 cohorts of *fortis* and *scandens*.

The tests reveal that FFS differ from *fortis* ($\chi_5^2 = 19.122$, $p = 0.0018$) but not from *scandens* ($\chi_4^2 = 6.521$, $p = 0.4350$), and SSF also differ from *fortis* ($\chi_5^2 = 8.862$, $p = 0.0119$) but not from *scandens* ($\chi_3^2 = 1.097$, $p = 0.7779$). The test criterion in each case is p less than 0.025, and the results are clear. Members of both backcross classes mature relatively early, like *scandens*, and unlike *fortis*. This provides evidence of an inherited difference between the species in rate of plumage maturation, and dominance in *scandens* relative to *fortis*. Strictly paternal inheritance can be ruled out by the fact that the FFS have a *fortis* father, yet differ from *fortis* and resemble *scandens*. The necessary caveat, however, is a caution about lack of statistical independence in the backcrosses. The two groups are independent, but several brothers are treated as independent, which is not strictly legitimate. The 9 FFS come from only four families and the 7 SSF come from three families.

Appendix 11.1 Samples of Measurements
for Selection Analyses

Samples for the morphological analyses of chapter 11 are listed here in two tables.

TABLE A.11.1
Samples of measurements for figs. 11.1, 11.2, and 11.10

Year	G. fortis		G. scandens		G. magnirostris	
	Survived	Died	Survived	Died	Survived	Died
1973	10	11	10	5	•	•
1974	73	31	36	10	•	•
1975	51	22	25	10	•	•
1976	98	105	33	14	•	•
1977	122	245	68	75	•	•
1978	98	27	61	7	•	•
1979	213	20	129	13	•	•
1980	219	44	119	10	•	•
1981	204	19	156	12	•	•
1982	205	84	166	24	•	•
1983	134	72	118	57	•	•
1984	266	154	179	121	8	23
1985	272	266	90	123	11	16
1986	201	71	78	12	•	•
1987	173	28	55	23	•	•
1988	553	182	115	18	44	10
1989	407	146	96	19	28	21
1990	272	134	71	25	7	21
1991	211	64	52	19	•	•
1992	327	114	72	21	•	•
1993	364	63	99	11	30	21
1994	160	229	99	36	35	36
1995	87	77	84	15	33	9
1996	38	50	50	39	15	19
1997	21	17	32	17	•	•
1998	22	8	17	14	•	•
1999	24	22	13	18	26	14
2000	21	3	24	0	20	6
2001	17	4	21	31	35	8
2002	73	42	36	3	57	18
2003	75	96	71	39	55	48
2004	38	37	30	41	7	71
2005	32	6	25	5	•	•
2006	100	39	59	11	13	8
2007	69	32	46	13	18	2
2008	82	42	75	13	21	6
2009	140	97	134	47	59	12
2010	100	40	111	46	55	41
2011	85	115	92	90	43	43

Note: Survival to 2012 is shown on the last line.

TABLE A.11.2
Samples of measured adults alive each year and used in figs. 11.3–11.5
and 11.7–11.9

Year	G. fortis	G. scandens	G. magnirostris
1972	30	15	0
1973	221	72	0
1974	108	53	0
1975	332	95	0
1976	933	236	43
1977	367	143	3
1978	290	166	0
1979	288	215	14
1980	270	187	39
1981	338	239	3
1982	293	213	7
1983	507	364	22
1984	749	350	15
1985	500	213	19
1986	274	90	15
1987	792	156	0
1988	792	156	40
1989	557	115	5
1990	413	96	0
1991	514	112	1
1992	561	136	12
1993	465	146	40
1994	401	135	41
1995	173	104	7
1996	98	89	1
1997	48	50	0
1998	102	45	2
1999	94	31	35
2000	71	13	0
2001	223	47	23
2002	212	113	40
2003	172	110	46
2004	76	71	23
2005	142	75	0
2006	143	70	16
2007	158	102	7
2008	210	191	9
2009	236	200	50
2010	235	223	37
2011	232	223	26
2012	117	122	59

Appendix 13.1 Identification of Breeders

Marker alleles enable us to reconstruct parts of the pedigree where observations of breeding pairs are lacking or one of a pair has not been identified (box 13.1). One of 5110's sons, 18350, hatched in 1992 and bred with an unknown female in 1997 or 1998. Two individuals are inferred to be their offspring because each of their genotypes matched the genotype of 18350 at all loci and no other contemporary male. Furthermore, like 18350, the two individuals were exceptionally large (fig. 13.7). One of them, 19669, bred with an unknown male in 2002; there was no successful breeding in 1999 and 2000 and scarcely any in 2001. Bird number 19669 was heterozygous at locus $Gf.11$ (183/-). Both members of the following generation, generation 4, were homozygous (183/183) at locus $Gf.11$; therefore their father must have carried the 183 allele. Given the scarcity of the 183 allele in the G. fortis population (~5%; Grant and Grant 2009a), it seems probable that he was a member of either the A line or the B line of descent from 5110. He was most likely to have been the one and only large male in shiny black plumage on the island that was repeatedly observed in 2000–2003 but never captured and measured. He sang the typical song of the lineage.

Two other facts point to an ancestry of this male that traces back through the B line to 17870. First, 18350 is inferred to have been his father because paternal alleles identified by the subtraction-exclusion technique (box 13.1) at seven loci present in the generation 3 brother-sister pair were entirely compatible with 18350's genotype. Second, he had another unusual genetic marker that enables us to reconstruct paternal ancestry. Both the brother and the sister of generation 3 possessed an allele (179) at locus $Gf.1$ that was not present in their mother's genotype, or grandfather's (18350), or in any of 37 genotyped individuals in the A line. The male probably inherited the 179 allele from an unknown daughter of 17870; the inferred sister, 19039, did not have this allele. Bird 17870 carried the 179 allele, and he died in 1995. Allele 179 is notably rare; its frequency in the population of G. fortis up to 1998 was estimated to be 0.16 ($n = 1,296$).

Appendix 17.1 Nestling Beak Color Polymorphism

Beak colors are either pink or yellow in all species (fig. 17.1). The polymorphism is governed by a single mendelian factor, with pink dominant to yellow (chapter 17). When both parents have the yellow morph, they should produce only yellow-morph offspring. With paternity confirmed there are only two exceptions in one family (table A.17.1), which we attribute to allelic dropout or an environmental effect of poor nutrition (L. Andersson pers. comm.). Where exceptions occur in the total samples, they may be attributed to misidentified paternity (table A.17.2).

The frequency of the yellow morph remained stable in fortis and scandens populations, subject to minor fluctuations (fig. 17.2). Morph frequencies in the third species, magnirostris, underwent a change in the late 1990s, possibly as a result of immigration, and a rapid change in the period 2004–9 (fig. A.17.1),

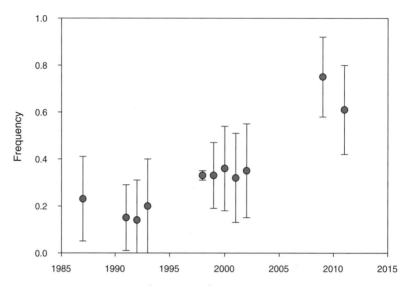

Fig. A.17.1 Variation in the frequency of the yellow morph in the *magnirostris* population. Increases in the frequency may have been due first to immigration and later to random change during and after a population bottleneck in 2005. Sample sizes are given in table A.17.3.

which we attribute to random effects of a population bottleneck in the drought of 2003–5 (chapter 8). Sample sizes are given in table A.17.3.

Despite approximate stability, frequencies are perturbed by introgressive hybridization between species with different morph frequencies. Table A.17.4 shows with aggregated data that the effect is small, because hybrids and back-crosses are a minority of the breeding populations. The table also shows the unexpected feature that the yellow-morph frequency is higher in the back-crosses than in the F_1 hybrids in all cases. Moreover hybrids produced by the interbreeding of *fortis* and *fuliginosa* combined with backcrosses to *fortis* have a higher frequency of the yellow morph than *fortis* ($G^2 = 6.028$, $p = 0.0141$). This is a conservative test because the expected frequency is not the *fortis* frequency but one intermediate between *fortis* and *fuliginosa* frequencies. Hy-brids produced by the interbreeding of *scandens* and *fortis* combined with backcrosses to *scandens* have a higher frequency of the yellow morph than *scandens* that borders on significance ($G^2 = 3.79$, $p = 0.0516$). Again the test is conservative. The frequency of the yellow morph in *fortis* × *scandens* hybrids and backcrosses to *fortis* is nominally higher than the *fortis* frequency, al-though statistically indistinguishable from it ($G^2 = 0.282$, $p = 0.5963$). The over-all pattern is suggestive of either nonrandom survival or nonrandom mating.

TABLE A.17.1 Inheritance of nestling beak colors

Parents	Families	Offspring pink	Offspring yellow
G. fortis			
Pink × pink	14	45	20
Pink × yellow	16	44	34
Yellow × yellow	1	2	5
Total	31	91	59
G. scandens			
Pink × pink	4	16	12
Pink × yellow	8	21	17
Yellow × yellow	0	0	0
Total	12	37	29

Note: Parents with beak colors identified when nestlings bred in various combinations; there is no evidence of assortative mating by nestling beak color, or differential reproductive success. Paternity was confirmed with microsatellite DNA in all cases. Frequencies of yellow morph offspring (39.3 for *fortis* and 43.9 for *scandens*) are unusually high when compared with the total samples (table A.17.2), probably an effect of small sample sizes.

TABLE A.17.2
Inheritance of nestling beak colors from the total samples including hybrids determined from pedigrees

Parents	Families	Offspring pink	Offspring yellow
G. fortis			
Pink × pink	286	1,358	110
Pink × yellow	86	256	123
Yellow × yellow	10	10	58
Total	382	1614	291
G. scandens			
Pink × pink	46	162	38
Pink × yellow	50	149	67
Yellow × yellow	9	11	33
Total	105	322	138

Note: Paternity was not checked by DNA in most cases. Paternity may not have been correctly identified in the yellow × yellow families that produced pink offspring. The percentages of yellow morphs are 15.2% for *fortis* and 30.0% for *scandens*.

TABLE A.17.3
Samples of birds of known beak color morph and known to be alive in each year used in figs. 17.2 and A.17.1

Year	G. fortis	G. scandens	G. magnirostris
1978	136	99	0
1979	148	131	0
1980	136	228	0
1981	350	269	0
1982	174	132	0
1983	1,133	827	8
1984	985	571	7
1985	483	230	2
1986	277	81	3
1987	1,240	228	26
1988	854	136	6
1989	569	99	6
1990	445	92	1
1991	1,039	208	41
1992	818	145	28
1993	655	117	20
1994	524	101	18
1995	221	101	10
1996	126	58	10
1997	55	47	10
1998	214	139	107
1999	90	57	43
2000	29	9	25
2001	24	8	22
2002	19	8	20
2003	16	7	17
2004	5	5	9
2005	2	2	0
2006	0	0	0
2007	0	0	0
2008	0	0	0
2009	5	6	28
2010	0	0	?
2011	34	17	23

TABLE A.17.4 Frequencies of beak color morphs in four species, and in F_1 hybrids and backcrosses (B_{1-5}) determined from pedigrees

Group	Species	Symbol	Pink	Yellow	Total	% Yellow
1	G. fuliginosa	f	9	0	9	0.0
	G. fortis	F	1,964	317	2,281	13.9
	G. scandens	S	492	194	686	28.3
	G. magnirostris	M	162	88	250	35.2
2	F_1	Ff	92	8	100	0.8
	B_1	FFf	60	18	78	23.1
	B_2–B_5	FFFf +	107	35	142	24.6
	F + hybrids		2,223	378	2,601	14.5
3	F_1	FS	9	1	10	10.0
	B_1	FFS	53	9	62	14.5
	B_2–B_5	FFFS +	29	7	36	19.4
	F + hybrids		2,055	334	2,389	14.0
4	F_1	SF	12	4	16	25.0
	B_1	SSF	21	15	36	41.7
	B_2–B_5	SSSF +	35	22	57	38.6
	S + hybrids		560	235	795	29.6

Note: All fortis × fuliginosa F_1 hybrids backcrossed to fortis (group 2), whereas fortis × scandens F_1 hybrids backcrossed to fortis (group 3) or to scandens (group 4). For the period 1978–98 in which the hybrids were produced, the percentages of yellow morphs were fortis 13.6% (n = 4,086), scandens 31.5% (n = 2,053). Paternity was not checked by DNA in most cases, as in table A.17.2.

Abbreviations

ANOVA—Analysis of variance.
BCO2—β-carotene deoxygenase 2 enzyme.
Bmp4—Bone morphogenetic protein 4.
CaM—Calmodulin.
CT—Cracking time; time taken to crack a seed.
CV—Coefficient of variation; 100 times the standard deviation divided by the
  mean. Subscripts denote additive genetic (A), environmental (E), pheno-
  typic (E), and residual (R).
D—Nei's genetic distance between two samples.
$DH^{1/2}$—Square root of depth times hardness of a seed.
EDTA—A preservative used to prevent bacterial breakdown of DNA in blood
  samples.
ENSO—The El Niño–Southern Oscillation phenomenon.
EPY—Extra-pair young.
Ff, FFff, FFf, FFFf, FS, FFSS, FSS, FSSS—Hybrids and backcrosses involving
  fortis (F), *fuliginosa* (f), and *scandens* (S). Ff or FFff and FS are F_1 hybrids,
  and the others are first- and second-generation backcrosses (B_1 and B_2). FS
  and FFSS are equivalent F_1 hybrids. The first initial of each refers to the spe-
  cies of the father. An equivalent series of symbols is used for hybrids whose
  father is *fuliginosa* (e.g., fF). For *scandens* see SF.
kHz—Kilohertz.
mg—Milligram.
Min—Minimum.
mm—Millimeters.
MY—Million years.
MYA—Million years ago.
p—Probability.
PC—Principal component; for example, PC 1 and PC 2 of a multivariate morphological analysis.
PCA—Principal components analysis.
Pmx—Premaxillary bone.
Pnc—Prenasal cartilage.
PVC—Polyvinyl chloride plastic for making leg bands for birds.
SD or s.d.—Standard deviation.
s.e.—Standard error of the mean.
SF, SSFF, SSF, and SSSF—Hybrids and backcrosses involving *scandens* (S)
  and *fortis* (F). SF and SSFF are F_1 hybrids, and the others are first- and
  second-generation backcrosses (B_1 and B_2).
yr—Year.
Z-linked—Gene on one of the sex chromosomes (Z).

Glossary

Acoustical interference—Obstruction of sound waves, for example by birds singing at the same time.

Adaptive landscape—A conceptual model of the way in which the fitness of organisms varies in relation to their traits in a heterogeneous environment, which is represented as a landscape with peaks and valleys.

Adaptive radiation—The rapid evolution from a common ancestor of several species that occupy different ecological niches.

Age structure of a population—The age composition of a population, represented as the number of individuals in each age class.

Alleles—Alternative forms of a gene at a particular locus.

Allometry—The way in which variation in one part of an organism changes in relation to variation in another; for example, the change in beak size in relation to body size.

Allopatry—Different geographical regions occupied by organisms.

Assignment test—Statistical test using genetic data to assign individuals to groups or populations on a probabilistic basis.

Asteroid—Extraterrestrial body. Impact of an asteroid on earth is believed to have contributed to the extinction of the dinosaurs.

Autosomes—Chromosomes other than the sex chromosomes.

Avipox—A disease of birds caused by a DNA virus (genus *Avipoxvirus*: Poxviridae) with pox symptoms.

Backcross—Offspring produced by the breeding of an F_1 hybrid with a member of one of the parental species. Designated by the symbol B, with subscript indicating the particular generation.

Barrier to interbreeding—Any feature of an individual that reduces the chances of its breeding with a member of another population.

β-carotene—A particular type of carotenoid with antioxidant function that can be converted to vitamin A.

β-catenin—A subunit of a protein complex that acts as an intracellular signal transducer. The signaling molecule is implicated in beak development.

Bone morphogenetic protein 4 (Bmp4)—A signaling molecule in several cellular functions including skeletal and beak development.

Brachial vein—Wing vein used for withdrawing a single drop of blood for DNA analysis.

Calmodulin—A calcium-binding protein present in all nucleated cells. It mediates a variety of cellular responses to calcium and participates in signaling pathways, including those involved in beak development.

Carotenoids—Chemical class of yellow and red pigments, manufactured by plants.

Character displacement—Enhancement of differences between coexisting species by natural selection. The characters that diverge may have ecological or reproductive functions, or both.

Character release—Morphological convergence on a species in its absence; the opposite of character displacement.

Chromosomal inversion—Inversion of a block of genes on a chromosome, resulting in a reversal in the order or sequence of genes in the block.

Chromosome—A DNA molecule in the nucleus or mitochondrion of a cell where genes reside. Nuclear chromosomes containing factors affecting sex are called sex chromosomes, and the remainder are autosomes. Sex chromosomes are identified by the symbols X and Y in most organisms, and as Z and W in birds.

Cis-acting mutation—Mutation that affects alleles on the same chromosome.

Close inbreeding—Breeding between kin or close relatives.

Coefficient of coancestry (ø)—A measure of the relatedness of a breeding pair of individuals.

Coefficient of variation—The standard deviation divided by the mean, and typically then multiplied by 100. Subscripts denote additive genetic (A), environmental (E), phenotypic (E), and residual (R).

Coevolution—Evolution of two or more reciprocally interacting species.

Cohort—All individuals of an age group, such as those born or hatched in 1975.

Colonization—Establishment of a new breeding population by immigrants.

Competition—The seeking of resources in short supply, such as food or mates.

Competitive exclusion—Competition that is sufficiently severe as to result in local extinction of one of the competitors.

Complementary epistasis—Epistasis in an offspring resulting from the masking or compensation of deleterious alleles at a locus in one parental genome by alleles at a different locus in the other parent's genome.

Confidence interval—The range of values on either side of an estimated mean within which the true mean lies at some specified level of statistical confidence, typically 95%.

Conspecific—Belonging to the same species, and applied to alleles, song, or individuals (see *heterospecific*).

Covariance matrix—A matrix of columns and rows of a set of variables, with quantitative measures of the variance of each variable and the covariation (covariances) of each pair of variables.

Cubic spline—Statistical function that connects a set of points, often irregular, on a bivariate plot.

Cultural drift—Random changes over time in song characteristics of a population.

Density dependence—Process that is dependent on population density, such as offspring growth or mortality.

Dermis—Skin tissue.

Despeciation—Speciation in reverse, when sister species exchange genes through introgressive hybridization, leading to the elimination of the differences between them.

Diatoms—Microscopic one-celled or colonial algae with silica walls.

Dickkopf3 (Dkk3)—A glycoprotein that acts as a signaling molecule implicated in beak development.

Discriminant function analysis—Statistical analysis for discriminating between two or more groups with a set of morphological variables. Maximum discrimination is achieved by a particular combination of the variables, which make different contributions to the discrimination.

Disruptive (diversifying) selection—Selective disadvantage of those individuals close to the mean relative to more extreme individuals.

Diversification—Evolutionary increase in the number and phenotypic diversity of species.

DNA—Deoxyribonucleic acid: the hereditary material of chromosomes.

DNA methylation—Process by which a small proportion of DNA becomes modified by addition of methyl groups, which lowers the rate of transcription.

Ecological opportunity—Availability of niches in an environment.

Ecological speciation—Speciation in which a primary role is assigned to ecology. *Ecogenetic speciation* refers to the same but with genetic involvement made explicit.

Eco-morphology—Ecologically significant morphological feature such as beak size.

El Niño—Occurrence of unusually warm surface waters in the Pacific Ocean, and associated heavy rains.

El Niño–Southern Oscillation—The El Niño phenomenon is closely linked to an oscillation of atmospheric pressure in the southern Pacific.

Electrophoresis—Method of separating and identifying proteins or DNAs by their mobility on a gel in an electric field according to their sizes and electrical charges.

Endemic species—A species that is restricted to one area and found nowhere else in the world.

Endogamy—Breeding confined to members of a group.

Epistasis—An interaction between genes at different loci producing an effect on phenotype or fitness.

Ethological speciation—Speciation in which a primary role is assigned to behavior associated with mate choice.

Evolution—Change in a population from one generation to another, in genetic composition (organic evolution). Also *cultural evolution*.

Extinction—Disappearance of a breeding population.

Extra-pair—Association between one member of a socially breeding pair and another individual that is not a member of the same pair. The extra-pair activities are copulation, mating, and fertilization resulting in extra-pair young.

F_1 hybrid—First-generation offspring produced by interbreeding species.

Fitness—Ability to survive and reproduce, measured by the number of offspring produced.

Fledging—Departure of a young bird from the nest in which it was raised.

Fledgling—A juvenile that has recently fledged from the nest and is unable to feed by itself.

Founder event—Establishment of a new population, typically by few individuals.

Free-radical scavengers—Chemical compounds that decompose reactive fragments of molecules produced as a by-product of metabolism.

Frequency bandwidth—The range of frequencies from lowest to highest, in birdsong for example, and measured in kilohertz (kHz).

Frequency dependence—Dependence of the fate of a group of individuals (or alleles) on whether they are relatively common or rare in the population.

Gamete—Single copy of a set of parental chromosomes in sperm or egg.

Gene—Unit of inheritance, typically at a single locus on a chromosome.

Gene conversion—A meiotic process of change in which one allele directs the conversion of a partner allele on the opposite chromosome into its own form. This can follow mismatched repair, where both mismatches are corrected to yield the same parental type.

Gene duplication—Generation of an extra copy of a locus usually by unequal crossing over at meiosis.

Gene flow—Addition of genes to a population through interbreeding with members of another population, usually of the same species.

Generalist diet—Eating a wide variety of foods.

Genetic absorption—The absorption of one population by another as a result of repeated interbreeding and unidirectional backcrossing. Also *Reproductive absorption*.

Genetic bottleneck—Reduction in number of alleles associated with a short-term reduction in population size, either in the founding of a new population or subsequently through genetic drift.

Genetic correlation—A quantitative association between two traits determined in part by shared genes.

Genetic drift—Random change in frequencies of alleles in a population.

Genetic incompatibility—Reduced fitness of hybrids arising from adverse interactions between maternal and paternal genomes during growth and development.

Genetic load—Reduction in the mean fitness of a population caused by deleterious alleles,

Genetic resistance—Constraining influence on the evolution of one trait caused by genetic correlations with other traits.

Genetically effective size (Ne)—The effective number of breeding individuals in a population taking account of the sampling variance of family sizes. It is less than the actual number of breeding individuals because it incorporates nonrandom distribution of family sizes and overlapping generations.

Genome—The complete set of chromosomes of an individual or from just one of its parents.

Genotype—Genetic constitution of an individual.

Genotype × environment correlation—Nonrandom distribution of genotypes among environments arising, for example, from a tendency of offspring and their parents (which share genotypes) to grow up in the same or similar environment. Referred to as *genotype by environment correlation*.

Genotype-environment interaction—Expression of an individual's genotype depends upon the environment experienced during growth and maturation.

Genus—A category in Linnean classification, one level above *species*.

Granivore—Seedeater.

Harmonic mean—The reciprocal of the arithmetic mean of the reciprocals of a set of numbers. Typically used when the average of rates is needed.

Heritability (h^2)—A measure of family resemblance as a result of shared genes.

Heterospecific—Belonging to another species, and applied to alleles, song, or individuals (see *conspecific*).

Heterozygosity—The state of being heterozygous.

Heterozygote—Different alleles at the same genetic locus, one on each of the paired chromosomes.

Homozygosity—The state of being homozygous.

Homozygote—Two copies of the same allele at a genetic locus.

Hubbs' principle—A relative scarcity of mates explains why a rare species hybridizes with another based on encounter rates. Invoked by the ichthyologist Carl Hubbs.

Hybridization—Interbreeding. Usually restricted to breeding between different species.

Imprinting—A rapid learning process during a short sensitive period early in life when a young animal learns to recognize the characteristics of an individual that it is attracted to and socializes with at that time. It leads to attachment to parents (filial imprinting) and to the use of learned information later in life in mate recognition and choice (sexual imprinting). Also used for an analogous process of learning features of the environment. See also *misimprinting*.

Inbreeding depression—Loss of fitness experienced by the offspring of kin-related parents, often as a result of an accumulation of deleterious recessive alleles in homozygous condition.

Introgression—Transfer of alleles from one species to another as a result of hybridization and backcrossing.

Isotope—The radioactive form of a chemical element such as carbon (C^{13}) or oxygen (O^{18}). Decay at a constant rate to the nonradioactive form of carbon can be used to date biological material.

La Niña—The opposite oceanographic condition of El Niño: unusually low sea-surface temperatures and scarce rainfall.

Least squares regression—A statistical method for estimating the relationship between two variables, when one (the predictor) can be used to predict the other. The relationship may be linear or it may be curvilinear (see *polynomial regression*).

Lethal equivalents—The number of deleterious genes whose cumulative effects equal that of one recessive lethal allele. Used as a measure of the fitness effects of inbreeding.

Lineage—General term for a line of descent in a pedigree, characterizing populations, individuals, or genes.

Linkage—The occurrence of two or more loci on the same chromosome.

Locus—Location of a gene on a chromosome.

Maternal effects—Effects of the maternal phenotype on offspring phenotype over and above the contribution made by nuclear genes.

Meiosis—The process of cell division in the formation of gametes (sperm and egg) that have half the number of chromosomes of the original cell. Recombination through chromosomal exchange takes place at this time.

Mendelian inheritance—Pattern of inheritance of a trait that conforms to a mechanism worked out by Gregor Mendel. The pattern is a statistical distribution of variation in a trait in successive generations of breeding that results from random combination of gametes and segregation of genes.

Mericarp—Component of the woody fruit of the plant caltrop, *Tribulus cistoides*.

Mesenchyme—Tissue of cells formed early in development of embryos. Neural crest–derived mesenchyme cells migrate to various parts of the embryo and give rise to bones of the skull and other structures.

Meta-analysis—A combined analysis of independent, and often heterogeneous, sets of data.

Microsatellites—Short pieces of DNA that usually have no coding function. They are composed of pairs, triplets, or quartets of nucleotides repeated a variable number of times.

Misimprinting—Imprinting on a member of another population: heterospecific imprinting.

Mist net—Nylon net with a fine mesh used to capture birds.

Molecular genetic marker—Alleles used as markers to keep track of an individual and its offspring.

Multicolinearity—The strong correlation between two variables that leaves insufficient independent variation in either to be associated reliably with a third variable.

Mutation—Alteration of a chromosome that is heritable.

Mutation-selection balance—A balance set by the opposing processes of mutation and selection against a deleterious allele.

Natural selection—A difference in survival or reproduction between members of a population due to a trait they possess or the expression of a trait. Some survive or reproduce better than others in a given environment as a result of the particular traits they have. Selection may be directional, in which individuals with traits at one extreme do best; disruptive or diversifying, in which individuals at both extremes do best; or stabilizing, in which intermediates do best at the expense of those at both extremes.

Neanderthals—An extinct species of the genus *Homo*, closely related to modern humans.

Neotropics—One of eight ecological zones dividing the earth's surface. It includes the tropical terrestrial regions of the Americas and the South American temperate zone.

Neural crest—A band of dorsal cells along the neural tube of developing embryos that migrates extensively to produce a variety of differentiated cell types including components of the central nervous system.

Niche—The part of the environment that is used by organisms.

Nuclear DNA—DNA that resides in the nucleus of a cell.

Nucleotides—The building blocks of DNA.

Oxidative damage—Damage produced by a free radical, that is, a molecule with unpaired valence electrons with a dangling bond from the oxygen. Antioxidants terminate these damaging reactions by becoming oxidized themselves.

Parapatry — Adjacent geographical areas occupied by populations; contiguous allopatry.

Partial regression coefficient — Statistical relationship between one of a set of predictor variables and a predicted variable, with the effects of other members of the predictor set held constant.

Peak shift — The learning of an exaggerated form of a favored signal or part of a signal in a direction away from a negative stimulus.

Phenotype — Structural or functional properties of an organism.

Phylogeny — A grouping of populations on the basis of their relationships, usually genetic, that reflects the pattern of descent from an ancestor, that is, their evolutionary history.

Pleiotropy — Effects on two or more traits of a single gene.

Poisson distribution — A discrete frequency distribution of randomly occurring events characterized by a variance that is equal to the mean.

Polygenic variation — Variation in a trait influenced by many interacting genes each with usually small effect.

Polymorphism — Two or more phenotypes in a population.

Polynomial regression — A statistical method for estimating a nonlinear relationship between two variables when one variable is predicted by linear and higher-order terms (squared, cubic, etc.) of the other.

Polyploid species — Species having extra sets of chromosomes.

Population — A group of interbreeding individuals and their offspring.

Predation — Killing and eating. Hawks and owls kill and eat finches. Finches occasionally kill each other but do not eat each other. They may prey upon each others' eggs, however, although this also is rare.

Principal components analysis — A statistical method of analyzing combinations of variables, for example, beak length, depth, and width, to produce a synthetic measure of beak size or shape.

Propagule — Individuals that colonize an island and initiate a new population.

Random genetic drift — See *genetic drift*.

Recessive allele — An allele that is only expressed in homozygous condition.

Recombination — Exchange of parts between DNA molecules or chromosomes through crossing over at meiosis. Sometimes referred to as genetic recombination.

Recruit — An individual surviving long enough to breed; it is recruited to the breeding component of the population.

Regression — see least squares regression.

Reproductive absorption — The absorption of one population by another as a result of repeated interbreeding and unidirectional backcrossing. Also *Genetic absorption*.

Reproductive barrier — A barrier that impedes mating. The barrier may be behavioral, or different timing of reproductive activity, or occupation of different habitats.

Reproductive isolation — Lack or scarcity of interbreeding between populations.

Resting metabolic rate—The amount of energy an individual expends when at rest.

Reticulation—A network of relationships between species as a result of introgression.

Seed handling—Should be seed billing. Ability of a bird to use its beak to deal with seeds: their procurement, mandibulation, crushing, and kernel extraction.

Selection differential—The combined direct effects of selection on a phenotypic trait together with the indirect effects of selection on traits that are correlated with it. Measured as the difference between the mean of a populations before selection and the mean of the individuals that become the parents of the next generation.

Selection gradient—The direct effect of selection on each of a set of correlated traits.

Selective sweep—The spread of a mutation throughout the entire population.

Sex chromosomes—Chromosomes associated with determining the difference in sex.

Sex-linked genes—Genes located on the sex chromosomes.

Sex ratio—The ratio of males to females.

Sexual imprinting—Early experience of parents affecting mate choice of offspring later in their life.

Sexual selection—A difference in average success in obtaining a mate or mates between individuals with a particular phenotype and those with other phenotypes. The phenotypic traits may be deployed in encounters between members of the same sex (intrasexual selection) or opposite sex (intersexual selection).

Signaling molecule—Molecule that activates others in biochemical pathways within cells and in cell-to-cell communication.

Sister species—Two species that are more related to each other than either is to another.

Sonogram—A sound spectrogram. Graphical representation of song characteristics (frequency or amplitude) as a function of time.

Speciation—The process of species formation; the evolution of reproductive isolation.

Species—A population or group of related breeding individuals that differ from other populations by certain criteria. The criteria include degree of genetic difference, whether the populations interbreed or not, and the consequences of interbreeding. The relative importance of these criteria is debated.

Stabilizing selection—Maintenance of an unchanging average phenotypic trait in a population as a result of extreme individuals having relatively low fitness: see *natural selection*.

Standard deviation (SD or s.d.)—A statistical measure of the degree to which measurements in a sample are dispersed on either side of the mean.

Standard error of the mean (s.e.)—A statistical measure of the uncertainty in the estimate of the mean of a trait.

Stochasticity—Random fluctuations in the properties of populations such as age and sex structure (demographic stochasticity) or in the environment such as climatic extremes with effects on population dynamics (environmental stochasticity).

Storage effect—Cumulative effect of past introgression on phenotypic variation that is not entirely reflected in recent introgression. Ecologists use a similar term to mean past demographic effects on present demography.

Sympatry—A geographical region jointly occupied by two or more populations.

Taxon—The most general term for a unit of classification, such as population or species. The plural is taxa.

Taxonomy—The science of classification. Groups are known as taxa (singular *taxon*).

Tectonic activity—Activity, such as seismic or volcanic, involving movement of the tectonic plates that comprise the earth's crust.

Transforming growth factor β II receptor (TGFβIIr)—A signaling molecule implicated in beak development.

Transgressive segregation—Hybrid offspring with genotypes and phenotypes beyond the ranges of both parents.

Transposable element (transposon)—A DNA sequence that can move from one location in the genome to another.

Trill rate—The rate at which individual song notes are repeated.

Variance—A measure of variation, and equal to the standard deviation squared.

Vector—A carrier and transmitter of disease.

Volcanism—Volcanic activity.

X-axis—Horizontal axis of a bivariate plot.

Z-linked genes—Genes on one of the sex chromosomes (Z).

References

Numbers in brackets after each reference indicate the page(s) on which it is cited in the text.

Abbott, I., L. K. Abbott, and P. R. Grant. 1977. Comparative ecology of Galápagos Ground Finches (*Geospiza* Gould): evaluation of the importance of floristic diversity and interspecific competition. *Ecol. Monogr.* 47: 151–184. [22–24, 29, 33, 179, 322]

Abbott, R., D. Albach, S. Ansell, J. W. Arntzen, S.J.E. Baird, and 34 others. 2013. Hybridization and speciation. *J. Evol. Biol.* 26: 229–246. [245, 291]

Abzhanov, A., W. P. Kuo, C. Hartmann, B. R. Grant, P. R. Grant, and C. J. Tabin. 2006. The calmodulin pathway and evolution of elongated beak morphology in Darwin's finches. *Nature* 442: 563–567. [235, 285]

Abzhanov, A., M. Protas, B. R. Grant, P. R. Grant, and C. J. Tabin. 2004. Bmp4 and morphological variation of beaks in Darwin's finches. *Science* 305: 1462–1465. [234, 235, 285]

Adams, J. R., L. M. Vucetich, P. W. Hedrick, R. O. Peterson, and J. A. Vucetich. 2011. Genomic sweep and potential genetic rescue during limiting environmental conditions in an isolated wolf population. *Proc. R. Soc. B* 278: 3336–3344. [119]

Agrawal, A. F., and J. R. Stinchcombe. 2009. How much do genetic covariances alter the rate of adaptation? *Proc. R. Soc. B* 271: 1183–1191. [75]

Akçay, C., W. A. Searcy, S. E. Campbell, V. A. Reed, C. N. Templeton, K. M. Hardwick, and M. D. Beecher. 2012. Who initiates extrapair mating in song sparrows? *Behav. Ecol.* 23: 44–50. [326]

Alberts, S. C., and J. Altmann. 2012. The Amboseli Baboon Research Project: themes of continuity and change. In P. Kappelar and D. P. Watt, eds., *Long-term field studies of primates*, 261–288. Springer Verlag, New York, NY. [310]

Albertson, R. C., and T. D. Kocher. 2005. Genetic architecture sets limits on transgressive segregation in hybrid cichlid fishes. *Evolution* 59: 686–690. [316]

Albrecht, T., M. Vinkler, J. Schnitzer, R. Polákavá, P. Munclinger, and J. Bryja. 2009. Extra-pair fertilizations contribute to selection on secondary male ornamentation in a socially monogamous passerine. *J. Evol. Biol.* 22: 2020–2030. [327]

Alerstam, T., B. Ebenman, M. Sylvén, S. Tamm, and S. Ulfstrand. 1978. Hybridization as an agent of competition between two bird allospecies: *Ficedula albicollis* and *F. hypoleuca* on the island of Gotland in the Baltic. *Oikos* 31: 326–331. [184]

Amadon, D. 1950. The Hawaiian honeycreepers (Aves, Drepaniidae). *Bull. Amer. Mus. Nat. Hist.* 95: 151–262. [5]

Anderson, R. Y. 1992. Long-term changes in the frequency of El Niño events. In H. F. Diaz and V. Markgraf, eds., *El Niño: Historical and Paleoclimatic Aspects of the Southern Oscillation*, 193–200. Cambridge University Press, Cambridge, U.K. [272]

Anderson, W. W., J. Arnold, D. G. Baldwin, A. T. Beckenbach, C. J. Brown, S. H. Bryant, J. A. Coyne, L. G. Harshman, W. B. Heed, and D. E. Jeffry. 1991. Four decades of inversion polymorphisms in *Drosphila pseudoobscura*. *Proc. Natl. Acad. Sci. USA* 88: 10367–10371. [310]

Araújo, M., D. I. Bolnick, and C. A. Layman. 2011. The ecological causes of individual specialisation. *Ecol. Lett.* 14: 948–958. [139]

Armitage, K. B. 2010. Individual fitness, social behavior, and population dynamics of Yellow-bellied Marmots. In I. Billick and M. V. Price, eds., *Ecology of Place: Contributions of Place-Based Research to Ecological Understanding*, 134–154. University of Chicago Press, Chicago, IL. [310]

Arnold, M. L. 1997. *Natural hybridization and evolution*. Oxford University Press, Oxford, U.K. [301]

Arnold, M. L., and A. Meyer. 2006. Natural hybridization in primates: one evolutionary mechanism. *Zoology* 109: 261–276. [245]

Ashmole, N. P. 1963. The regulation of numbers of tropical oceanic birds. *Ibis* 103b: 458–473. [79]

Ayer, A. J. 2006. *Probability and Evidence*. Columbia University Press, New York, NY. [300]

Badyaev, A. V. 2010. The beak of the other finch: coevolution of genetic covariance structure and developmental modularity during adaptive evolution. *Philos. Trans. R. Soc. B* 365: 1111–1126. [236, 237]

Badyaev, A. V. 2011. How do precise adaptive features arise in development? Examples with evolution of context-specific sex ratios and perfect beaks. *Auk* 128: 467–474. [236]

Badyaev, A. V., G. E. Hill, and L. A. Whittingham. 2001. The evolution of sexual size dimorphism in the House Finch III: Developmental basis. *Evolution* 55: 176–189. [237]

Badyaev, A. V., R. L. Young, K. P. Oh, and C. Addison. 2008. Evolution on a local scale: developmental, functional, and genetic bases of divergence in bill form and associated changes in song structure between adjacent habitats. *Evolution* 62: 1951–1964. [236, 292]

Baker, M. C., and A.E.M. Baker. 1990. Reproductive behavior of female buntings: isolating mechanisms in a hybridizing pair of species. *Evolution* 44: 332–338. [153]

Baltensweiler, W., U. M. Weber, and P. Cherubini. 2008. Tracing the influence of larch-bud-moth insect outbreaks and weather conditions on larch tree-ring growth in Engadine (Switzerland). *Oikos* 117: 161–172. [310]

Bapst, D. W., P. C. Bullock, M. J. Melchin, H. D. Sheets, and C. E. Mitchell. 2012. Graptoloid diversity and disparity became decoupled during the Ordovician mass extinction. *Proc. Natl. Acad. Sci. USA* 109: 3428–3433. [206]

Barrier, M., R. H. Robichaux, and M. Purugganen. 2001. Accelerated regulatory gene evolution in an adaptive radiation. *Proc. Natl. Acad. Sci. USA* 98: 10208–10213. [234]

Barton, N. H. 1990. Pleiotropic models of quantitative variation. *Genetics* 124: 773–782. [167]

Barton, N. H. 2001. The role of hybridization in evolution. *Mol. Ecol.* 10: 551–568. [167]

Barton, N., and B. Charlesworth. 1984. Genetic revolutions, founder effects, and speciation. *Ann. Rev. Ecol. Syst.* 15: 133–164. [105]

Bataille, A., A. A. Cunningham, V. Cedeño, M. Cruz, G. Eastwood, D. M. Fonseca, C. E. Causton, R. Azuero, J. Loayza, J. D. Cruz Martinez, and S. J. Goodman. 2009a. Evidence for regular ongoing introductions of mosquito disease vectors into the Galápagos Islands. *Proc. R. Soc. B* 276: 3769–3775. [284]

Bataille, A., A. A. Cunningham, V. Cedeño, L. Patiño, A. Constantinou, L. Kramer, and S. J. Goodman. 2009b. Natural colonization and adaptation of a mosquito species in Galápagos and its implications for disease threats to endemic wildlife. *Proc. Natl. Acad. Sci. USA* 106: 10230–10235. [284]

Bateson, W. 1902. Experiments with poultry. *Reports of the Evolution Commission of the Royal Society* 1: 87–124. [311]

Beaugrand, G., and P. C. Reid. 2003. Long-term changes in phytoplankton, zooplankton, and salmon related to climate change. *Global Change Biology* 9: 801–817. [310]

Beebe, W. 1924. *Galapagos: world's end.* Putnams, New York. [13, 94, 105, 124, 275]

Behm, J. E., A. R. Ives, and J. W. Boughman. 2010. Breakdown in postmating isolation and the collapse of a species pair through hybridization. *Amer. Nat.* 175: 11–26. [183, 184]

Bell, G. 2010. Fluctuating selection: the perpetual renewal of adaptation in variable environments. *Philos. Trans. R. Soc. B* 365: 87–97. [56, 228, 300]

Bell, M. A., and S. A. Foster. 1994. *The Evolutionary Biology of the Threespine Stickleback.* Oxford University Press, Oxford, U.K. [206]

Bell, M. A., M. P. Travis, and D. M. Blouw. 2006. Inferring natural selection in the fossil record. *Paleobiology* 32: 562–576. [206]

Benkman, C. W. 1999. The selection mosaic and diversifying coevolution between crossbills and lodgepole pine. *Amer. Nat.* 153: S75–S91. [99]

Benkman, C. W., J. W. Smith, M. Maier, L. Hansen, and M. Talluto. 2013. Consistency and variation in phenotypic selection exerted by a community of seed predators. *Evolution* 67: 157–169. [226]

Billick, I., and Price, M. V., eds. 2010. *Ecology of Place: Contributions of Place-Based Research to Ecological Understanding.* University of Chicago Press, Chicago, IL. [310, 311]

Billing, A. M., A. M. Lee, S. Skjelseth, A. A. Borg, M. C. Hale, J. Slate, H. Pärn, T. H. Ringsby, B-E. Sæther, and H. Jensen. 2012. Evidence of inbreeding depression but not inbreeding avoidance in a natural house sparrow population. *Mol. Ecol.* 21: 1487–1499. [97]

Bischoff, H-J., and N. Clayton. 1991. Stabilization of sexual preferences by sexual experience in male zebra finches *Taeniopygia guttata castanotis*. *Behaviour* 118: 144–155. [153]

Björklund, M. 1993. Phenotypic variation of growth trajectories in finches. *Evolution* 47: 506–514. [234]

Björklund, M., A. Husby, and L. Gustafsson. 2013. Rapid and unpredictable changes in the G-matrix in a natural bird population over 25 years. *J. Evol. Biol.* 26: 1–13. [231]

Blard, P-H., J. Lavé, R. Pik, P. Wagnon, and D, Bourlès. 2007. Persistence of full glacial conditions in the central Pacific until 15,000 years ago. *Nature* 449: 591–594. [20, 272]

Blomhoff, R. 2001. Vitamin A and carotenoid toxicity. *Food Nutr. Bull.* 22: 320–334. [314]

Blount, Z. D., C. Z. Borland, and R. E. Lenski. 2008. Historical contingency and the evolution of a key innovation in an experimental population of *Escherischia coli*. *Proc. Natl. Acad. Sci. USA* 105: 7899–7906. [285]

Boag, P. T. 1983. The heritability of external morphology in Darwin's ground finches (*Geospiza*) on Isla Daphne Major, Galápagos. *Evolution* 37: 877–894. [44, 45, 63, 132, 133, 232]

Boag, P. T. 1984. Growth and allometry of external morphology in Darwin's Ground Finches (*Geospiza*) on Isla Daphne Major, Galápagos. *J. Zool. Lond.* 204: 413–441. [43, 99, 224, 236, 237, 324]

Boag, P. T., and P. R. Grant. 1978. Heritability of external morphology in Darwin's Finches. *Nature* 274: 793–794. [44, 45]

Boag, P. T., and P. R. Grant. 1981. Intense natural selection in a population of Darwin's finches (Geospizinae) in the Galápagos. *Science* 214: 82–85. [57–59, 61, 73]

Boag, P. T., and P. R. Grant. 1984a. The classical case of character release: Darwin's Finches (*Geospiza*) on Isla Daphne Major, Galápagos. *Biol. J. Linn. Soc.* 22: 243–287. [20, 22, 23, 26–30, 33, 59, 126, 127, 130, 203]

Boag, P. T., and P. R. Grant. 1984b. Darwin's Finches (*Geospiza*) on Isla Daphne Major, Galápagos: breeding and feeding ecology in a climatically variable environment. *Ecol. Monogr.* 54: 463–489. [23, 30, 42, 57, 61, 80, 123, 132, 140, 167, 209]

Boag, P. T., and L. M. Ratcliffe. 1979. First record of the Blackpoll Warbler for the Galápagos. *Condor* 81: 218–219. [328]

Bock, C. E., and L. W. Lepthien. 1976. Population growth in the cattle egret. *Auk* 93: 164–166. [104]

Bolnick, D. I., T. Ingram, W. E. Stutz, L. K. Snowberg, O. L. Lau, and J. S. Paull. 2010. Ecological release from interspecific competition leads to decoupled changes in population and individual niche width. *Proc. R. Soc. B* 277: 1789–1797. [140]

Bolnick, D. I., R. Svanbeck, J. A. Fordyce, L. H. Yang, J. M. Davis, C. D. Hulsey, and M. L. Forister. 2003. The ecology of individuals: incidence and implications of individual specialization. *Amer. Nat.* 161: 1–28. [139]

Bowman, R. I. 1961. Morphological differentiation and adaptation in the Galápagos finches. *Univ. Calif. Publs. Zool.* 58: 1–302. [18, 30, 211, 266]

Bowman, R. I. 1983. The evolution of song in Darwin's finches. In R. I. Bowman, M. Berson, and A. E. Leviton, eds., *Patterns of Evolution in Galápagos Organisms*, 237–537. Amer. Assoc. Adv. Sci., Pacific Division, San Francisco, CA. [115, 143, 148, 250, 291, 293]

Brakefield, P. M., and M. Jorion. 2010. The flexibility of butterfly wing color patterns and evolution in morphospace. In P. R. Grant and B. R. Grant, eds., *In Search of the Causes of Evolution: From Field Observations to Mechanisms*, 93–114. Princeton University Press, Princeton, NJ. [287]

Brede, N., C. Sandrock, D. Straile, P. Spaak, T. Jankowski, B. Streit, and K. Schwenk. 2009. The impact of human-made ecological changes on the genetic architecture of *Daphnia* species. *Proc. Natl. Acad. Sci. USA* 106: 4058–4063. [184]

Brekke, P., P. M. Bennett, J. Wang, N. Pettorelli, and J. G. Ewen. 2010. Sensitive males: Inbreeding depression in an endangered bird. *Proc. R. Soc. B* 277: 3677–3684. [108]

Brelsford, A., B. Milá, and D. E. Irwin. 2011. Hybrid origin of Audubon's Warbler. *Mol. Ecol.* 20: 2380–2389. [245]

Bronowski, J. 1973. *The Ascent of Man*. Little, Brown and Co., Boston, MA. [205]

Brouwhuis, S., A. Charmantier, S. Verhulst, and B. C. Sheldon. 2010. Transgenerational effects on aging in a wild bird population. *J. Evol. Biol.* 23: 636–642. [77]

Brower, A.V.Z. 2013. Introgression of wing pattern alleles and speciation via homoploid hybridization in *Heliconius* butterflies: a review of evidence from the genome. *Proc. R. Soc. B* 280: 20122302. [246]

Broyles, S. B. 2002. Hybrid bridges to gene flow. *Evolution* 56: 1943–1956. [169]

Bullini, L. 1994. Origin and evolution of animal hybrid species. *Trends Ecol. Evol.* 9: 422–426. [246]

Bürger, R. 1999. Evolution of genetic variability and the advantage of sex and recombination in changing environments. *Genetics* 153: 1055–1069. [167]

Burns, K. J. 1997. Molecular systematics of tanagers (Thraupinae): evolution and biogeography of a diverse radiation of neotropical birds. *Mol. Phylogenet. Evol.* 8: 334–348. [6, 305]

Burns, K. J., S. J. Hackett, and N. K. Klein. 2002. Phylogenetic relationships and morphological diversity on Darwin's finches and their relatives. *Evolution* 56: 1240–1256. [285, 305]

Burns, K. J., and K. Naoki. 2004. Molecular phylogenetics and biogeography of Neotropical tanagers in the genus *Tangara*. *Mol. Phylogenet. Evol.* 32: 838–854. [285, 304]

Caballero, A., and P. D. Keightley. 1994. A pleiotropic nonadditive model of variation in quantitative genetics. *Genetics* 138: 883–900. [52,172]

Cade, T. J. 1983. Hybridization and gene exchange among birds in relation to conservation. In C. M. Schonewald-Cox, S. M. Chambers, B. MacBryde, and W. L. Thomas, eds., *Genetics and Conservation*, 288–309. Benjamin/Cummings, Menlo Park, CA. [139]

Cadena, C. D., K. H. Kozak, J. P. Gómez, J. L. Parra, C. M. McCain, and 10 others. 2012. Latitude, elevational climatic zonation and speciation in New World vertebrates. *Proc. R. Soc. B* 279: 194–201. [304]

Calsbeek, R., and T. B. Smith. 2007. Probing the adaptive landscape using experimental islands: density-dependent natural selection on lizard body size. *Evolution* 61: 1052–1061. [78]

Campagna, L., P. Benites, S. C. Lougheed, D. A. Lijtmaer, A. S. Di Giacomo, M. D. Eaton, and P. L. Tubaro. 2012. Rapid phenotypic evolution during incipient speciation in a continental avian radiation. *Proc. R. Soc. B* 279: 1847–1856. [304, 305]

Campàs, O., R. Mallarino, A. Herrel, A. Abzhanov, and M. P. Brenner. 2010. Scaling and shear transformations capture beak shape variation in Darwin's finches. *Proc. Natl. Acad. Sci. USA* 107: 3356–3360. [234]

Carney, S. E., K. A. Gardner, and L. H. Rieseberg. 2000. Evolutionary changes over the fifty-year history of a hybrid population of sunflowers (*Helianthus*). *Evolution* 54: 462–474. [184]

Carson, H. L., F. E. Clayton, and H. D. Stalker. 1967. Karyotypic stability and speciation in Hawaiian *Drosophila*. *Proc. Natl. Acad. Sci. USA* 57: 1280–1285. [5]

Carson, H. L., and A. R. Templeton. 1984. Genetic revolutions in relation to speciation phenomena: the founding of new populations. *Ann. Rev. Ecol. Syst.* 15: 97–131. [105]

Cavalli-Sforza, L. L., and F. Cavalli-Sforza. 1995. *The Great Human Diasporas: The History of Diversity and Evolution*. Addison-Wesley, New York, NY. [77]

Charlesworth, B. 1980. *Evolution in Age-Structured Populations*. Cambridge University Press, Cambridge, U.K. [77]

Charlesworth, B. 1987. The heritability of fitness. In J. W. Bradbury and M. B. Andersson, eds., *Sexual selection: testing the alternatives*, 21–40. John Wiley and Sons, New York, NY. [97]

Charlesworth, D., and B. Charlesworth. 1987. Inbreeding depression and its evolutionary consequences. *Ann. Rev. Ecol. Syst.* 18: 237–268. [107]

Charnov, E. L., R. Warne, and M. Moses. 2007. Lifetime reproductive effort. *Amer. Nat.* 170: E129–E142. [89]

Chavez, F. P., J. Ryan, S. E. Lucha-Cota, and M. Ñiguin. 2003. From anchovies to sardines and back: multidecadal change in the Pacific Ocean. *Science* 299: 217–221. [275]

Chen, D., M. A. Cane, A. Kaplan, S. E. Zebiak, and D. Huang. 2004. Predictability of El Niño over the past 140 years. *Nature* 428: 733–736. [5]

Chiba, S. 1993. Modern and historical evidence for natural hybridization between sympatric species in *Mandarina* (Pulmonata: Camaenidae). *Evolution* 47: 1539–1556. [177]

Clarke, B. C., M. S. Johnson, and J. Murray. 1998. How "molecular leakage" can mislead us about island speciation. In P. R. Grant, ed., *Evolution on Islands*, 181–195. Oxford University Press, Oxford, U.K. [169, 183, 203]

Clayton. N. S. 1990. The effects of cross-fostering on assortative mating between Zebra finch subspecies. *Anim. Behav.* 40: 1102–1110. [152]

Clegg, S. M., S. M. Degnan, C. Moritz, A. Estoup, J. Kikkawa, and I.P.F. Owens. 2002. Microevolution in island forms: the roles of drift and directional selection in morphological divergence of a passerine bird. *Evolution* 56: 2090–2099. [310]

Clutton-Brock, T., and B. C. Sheldon. 2010. Individuals and populations: the role of long-term, individual-based studies of animals in ecology and evolutionary biology. *Trends. Ecol. Evol.* 25: 562–573. [310]

Cobb, K. M., C. D. Charles, H. Cheng, and R. L. Edwards. 2003. El Niño/Southern Oscillation and tropical Pacific climate during the last millennium. *Nature* 424: 271–276. [275]

Cody, M. L. 1996. Introduction to Long-Term Community Ecological Studies. In M. L. Cody and J. A. Smallwood, eds., *Long-Term Community Ecological Studies*, 1–15. Academic Press, San Diego, CA. [205]

Coen, E. 2012. *Cells to Civilizations. The principles of change that shape life.* Princeton University Press, Princeton, NJ. [41]

Colinvaux, P. A. 1972. Climate and the Galápagos Islands. *Nature* 240: 17–20. [272]

Coltorti, M., and C. D. Ollier. 2000. Geomorphic and tectonic evolution of the Ecuadorian Andes. *Geomorphology* 32: 1–19. [305]

Colwell, R. K., and T. F. Rangel. 2010. A stochastic evolutionary model for range shifts and richness on tropical elevational gradients under Quaternary glacial cycles. *Philos. Trans. R. Soc. B* 365: 3695–3707. [276]

Conant, S. 1988. Geographic variation in the Laysan finch (*Telyspiza cantans*). *Evol. Ecology* 2: 270–282. [104]

Connor, J. K. 2012. Quantitative genetic approaches to evolutionary constraint: how useful? *Evolution* 66: 3313–3320. [213, 231]

Covas, R. 2012. Evolution of reproductive life histories in island birds worldwide. *Proc. R. Soc. B* 279: 1531–1537. [79]

Cox, G. 1990. Centres of speciation and ecological differentiation in the Galápagos bird fauna. *Evol. Ecol.* 4: 130–142. [304]

Coyne, J. A., and H. A. Orr. 2004. *Speciation*. Sinauer, Sunderland, MA. [39, 229, 230, 246, 289, 291]

Cracraft, J., and F. K. Barker 2009. Passerine birds (Passerifomes). In S. B. Hedges and S. Kumar, eds., *The Timetree of Life*, 423–431. Oxford University Press, New York, NY. [6, 305]

Crow, J. L., and M. Kimura. 1970. *An Introduction to Population Genetic Theory*. Harper and Row, New York, NY. [107, 301]

Darwin, C. 1859. *On the Origin of Species by Means of Natural Selection, or preservation of favoured races in the struggle for life*. J. Murray, London, U.K. [xxvii, 3, 4, 55, 103, 205, 245]

Davis, M. B., R. G. Shaw, and J. R. Etterson. 2005. Evolutionary responses to changing climate. *Ecology* 86: 1704–1714. [276]

Dawley, R. M., and J. P. Bogert, eds. 1989. *Evolution and Ecology of Unisexual vertebrates*. Bulletin of the New York State Museum, Albany, NY. [246]

De Jong, G. 1994. The fitness of fitness concepts and the description of natural selection. *Quart. Rev. Biol.* 69: 3–29. [77]

de León, L. F., E. Bermingham, J. Podos, and A. P. Hendry. 2010. Divergence with gene flow as facilitated by ecological differences: within-island variation in Darwin's finches. *Philos. Trans. R. Soc.* 365: 1041–1052. [297, 302]

De León, L. F., J.A.M. Raeymaekers, E. Bermingham, J. Podos, A. Herrel, and A. P. Hendry. 2011. Exploring possible human influences on the evolution of Darwin's finches. *Evolution* 65: 2258–2272. [30]

de Queiroz, K. 1998. The general lineage concept of species, species criteria, and the process of speciation. In D. J. Howard and S. H. Berlocher, eds., *Endless Forms. Species and Speciation*, 57–75. Oxford University Press, New York, NY. [199, 261]

de Queiroz, K. 2011. Branches in the lines of descent: Charles Darwin and the evolution of the species concept. *Biol. J. Linn. Soc.* 103: 19–35. [261]

Derryberry, E. P., S. Claramunt, G. Derryberry, R. T. Chesser, J. Cracraft, A. Aleixo, J. Pérez-Emán, J. V. Remsen, Jr., and R. T. Brumfield. 2011. Lineage diversification and morphological evolution in a large-scale continental radiation: the neotropical ovenbirds and woodcreepers (Aves: Furnariidae). *Evolution* 65: 2973–2986. [306]

Deschamps, P., N. Durand, E. Bard, B. Hamelin, G. Camoin, A. Thomas, G. M. Henderson, J. Okuno, and Y. Yokoyama. 2012. Ice-sheet collapse and sea-level rise in the Bølling warming 14,600 years ago. *Nature* 483: 559–564. [272]

Dhondt, A. A. 2012. *Interspecific Competition in Birds*. Oxford University Press, Oxford, U.K. [125, 301]

Diamond, J. 1974. Colonization of exploded volcanic islands by birds: the super-tramp strategy. *Science* 184: 803–806. [104]

Dieckmann, U., and M. Doebeli. 1999. On the origin of species by sympatric speciation. *Nature* 400: 354–357. [262]

Dillon, M. E., G. Wang, and R. B. Huey. 2010. Global metabolic impacts of recent climate warming. *Nature* 467: 704–706. [276]

Dobzhansky, T. 1940. Speciation as a stage in evolutionary divergence. *Amer. Nat.* 74: 302–321. [298]

Dobzhansky, T. 1941. *Genetics and the Origin of Species*, 2nd ed. Columbia University Press, New York, NY. [183, 298]

Dobzhansky, T. 1970. *Genetics of the Evolutionary Process*. Columbia University Press, New York, NY. [27]

Dunbar, R. B., G. M. Wellington, M. W. Colgan, and P. W. Glynn. 1994. Eastern Pacific sea surface temperature since 1600 A.D. The O^{18} record and climate variability in Galápagos corals. *Paleoceanography* 9: 291–315. [272, 274]

Dunn, L. C. 1925. The genetic relation of some shank colors of the domestic fowl. *Anat. Record* 31: 343–344. [311]

Edmondson, W. T. 1991. *The uses of Ecology: Lake Washington and beyond*. University of Washington Press, Seattle, WA. [310, 311]

Eldredge, N., and S. J. Gould. 1972. Punctuated equilibria: an alternative to phyletic gradualism. In T.J.M. Schopf, ed., *Models in Paleobiology*, 82–115. Freeman and Cooper, San Francisco, CA. [205]

Eldredge, N., et al. 2005. The dynamics of evolutionary stasis. *Paleobiology* 31: 133–145. [205]

Ellegren, H., L. Smed, R. Burri, P. I. Olason, N. Backström, T. Kawakami, A. Künster, H. Mäkinen, K. Nadachowska-Brzyska, A. Qvarnström, S. Uebbing, and J.B.W. Wolf. 2012. The genomic landscape of species divergence in *Ficedula* flycatchers. *Nature* 491: 756–760. [285]

Endler, J. A. 1986. *Natural Selection in the Wild*. Princeton University Press, Princeton, NJ. [167]

Engen, S., T. H. Ringsby, B-E. Sæther, R. Lande, H. Jensen, M. Lillegård, and H.

Ellegren. 2007. Effective size of fluctuating populations with two sexes and overlapping generations. *Evolution* 61: 1873–1885. [83]

Eriksson, J., G. Larson, U. Gunnarsson, B. Bed'hom, M. Tixier-Boichard, L. Strömstedt, D. Wright, A. Jungerius, A. Vereijken, E. Randi, P. Jensen, and L. Andersson. 2008. Identification of the Yellow Skin gene reveals a hybrid origin of the domestic chicken. *PLoS Genetics* 4: e1000010. [311]

Fairbairn, D. J., and R. F. Preziosi. 1996. Sexual selection and the evolution of sexual size dimorphism in the water strider, *Aquarius remigis*. *Evolution* 50: 1549–1559. [211]

Falconer, D. G., and T.F.C. Mackay. 1995. *Introduction to Quantitative Genetics*, 2nd ed. Longman, Harlow, U.K. [41, 43, 65, 107, 132]

Farrington, H., and K. Petren. 2011. A century of genetic change and metapopulation dynamics in the Galápagos warbler finch (*Certhidea*). *Evolution* 65: 3148–3161. [104, 174]

Feder, J. L., S. P. Egan, and P. Nosil. 2012. The genomics of speciation-with-gene-flow. *Trends Genetics* 28: 342–350. [285, 291]

Fessl, B., S. Kleindorfer, and S. Tebbich. 2006. An experimental study on the effects of an introduced parasite in Darwin's finches. *Biol. Conserv.* 127: 55–61. [284]

Fessl, B., and S. Tebbich. 2002. *Philornis downsi*—a recently discovered parasite on the Galápagos archipelago—a threat for Darwin's finches. *Ibis* 144: 445–451. [284]

Fisher, R. A. 1930. *The Genetical Theory of Natural Selection*. Oxford University Press, Oxford, U.K. [298]

Fjeldså, J., and C. Rahbek. 2006. Diversification of tanagers, a species rich bird group, from lowlands to montane regions of South America. *Integr. Comp. Biol.* 46: 72–81. [304]

Flexor, J-M. 1993. Southern oscillation signal in South American paleoclimatic data of the last 7000 years. *Quart. Res.* 39: 338–346. [275]

Fontdevila, A. 2005. Hybrid genome evolution by transposition. *Cytogenet. Genome Res.* 110: 49–55. [285]

Fryer, G., and T. D. Iles. 1972. *The Cichlid Fishes of the Great Lakes of Africa*. Oliver and Boyd, Edinburgh, U.K. [5]

Galligan, T. H., S. C. Donnellan, F. J. Sulloway, A. J. Fitch, T. Bertozzi, and S. Kleindorfer. 2012. Panmixia supports divergence with gene flow in Darwin's small ground finch, *Geospiza fuliginosa*, on Santa Cruz, Galápagos Islands. *Mol. Ecol.* 21: 2106–2115. [302]

Garant, D., L.E.B. Kruuk, T. A. Wilkin, R. H. McCleery, and B. C. Sheldon. 2005. Evolution driven by differential dispersal within a wild bird population. *Nature* 433: 60–65. [78]

Garcia-Gonzalez, F., L. W. Simmons, J. L. Tomkins, J. S. Kotiaho, and J. P. Evans. 2012. Comparing evolvabilities: common errors surrounding the calculation and use of coefficients of additive genetic variation. *Evolution* 66: 2341–2349. [52]

Gardner, J. L., R. Heinsohn, and L. Joseph. 2009. Shifting latitudinal clines in avian body size correlate with global warming in Australian passerines. *Proc. R. Soc. B* 276: 3845–3852. [276]

Garnett, S. G., A. C. Williams, and R. W. Hindmarsh. 1992. Island colonization after possible reconnaissance by the yellow-bellied sunbird *Nectarinia jugularis*. *Emu* 91: 185–186. [106]

Gee, J. M. 2003. How a hybrid zone is maintained: behavioral mechanisms of interbreeding between California and Gambel's quail (*Callipepla californica* and *C. gambelii*). *Evolution* 57: 2407–2415. [153]

Geist, D., H. Snell, H. Snell, C. Goddard, and M. Kurz. MS. Paleogeography of the Galápagos Islands and biogeographical implications. [20, 272, 303]

Genbrugge, A., A-S. Heyde, D. Adriaens, M. Boone, L. van Hoorebeke, J. Dirckx, P. Aerts, J. Podos, and A. Herrel. 2011. Ontogeny of the cranial skeleton in a Darwin's finch (*Geospiza fortis*). *J. Anat.* 219:115–131. [237]

Gentry, A. H. 1989. Speciation in tropical forests. In L. B. Holm-Nielsen and H. Balsev, eds., *Tropical Forests. Botanical Dynamics, Speciation and Diversity*, 113–134. Academic Press, New York, NY. [304]

Gerlach, N. M., J. W. McGlothlin, P. G. Parker, and E. D. Ketterson. 2012. Promiscuous mating produces offspring with higher lifetime fitness. *Proc. R. Soc. B* 279: 860–866. [327]

Gibbs, H. L. 1988. Heritability and selection on clutch size in Darwin's medium ground finches (*Geospiza fortis*). *Evolution* 42: 750–762. [81, 97, 99]

Gibbs, H. L. 1990. Cultural evolution of male song types in Darwin's medium ground finches, *Geospiza fortis*. *Anim. Behav.* 39: 253–263. [250, 251]

Gibbs, H. L., and P. R. Grant. 1987a. Ecological consequences of an exceptionally strong El Niño event on Darwin's finches. *Ecology* 68: 1735–1746. [67, 71, 105]

Gibbs, H. L., and P. R. Grant. 1987b. Oscillating selection on Darwin's finches. *Nature* 327: 511–513. [68, 72–74]

Gibbs, H. L., and P. R. Grant. 1987c. Adult survivorship in Darwin's ground finches (*Geospiza*) populations in a variable environment. *J. Anim. Ecol.* 56: 797–813. [123]

Gibbs, H. L., and P. R. Grant. 1989. Inbreeding in Darwin's medium ground finch (*Geospiza fortis*). *Evolution* 43: 1273–1284. [96, 161]

Gibbs, H. L., P. R. Grant, and J. Weiland. 1984. Breeding of Darwin's Finches at an unusually early age in an El Niño year. *Auk* 101: 872–874. [81]

Gienapp, P., J. Teplitsky, J. S. Aho, J. A. Mills, and J. Merilä. 2008. Climate change and evolution: disentangling environmental and genetic responses. *Mol. Ecol.* 17: 167–178. [276]

Gifford, E. W. 1919. Field notes on the land birds of the Galápagos Islands and Cocos Island, Costa Rica. *Proc. Calif. Acad. Sci. Series* 4, 2: 189–258. [13]

Givnish, T., and K. Sytsma, eds. 1997. *Molecular Evolution and Adaptive Radiation*. Cambridge University Press, Cambridge, U.K. [5]

Glor, R. E. 2010. Phylogenetic insights on adaptive radiation. *Annu. Rev. Ecol. Evol. Syst.* 41: 251–270. [296, 306]

Gompertz, Z., J. A. Fordyce, M. L. Forister, A. M. Shapiro, and C. C. Nice. 2006. Homoploid hybrid speciation in an extreme habitat. *Science* 314:1923–1925. [246]

Gomulkiewicz, R., and D. Houle. 2009. Demographic and genetic constraints on evolution. *Amer. Nat.* 174: E218–E219. [276]

Grant, B. R. 1985. Selection on bill characters in a population of Darwin's finches: *Geospiza conirostris* on Isla Genovesa, Galapagos. *Evolution* 39: 523–532. [75]

Grant, B. R. 1990. The significance of subadult plumage in Darwin's finches, *Geospiza fortis. Behav. Ecol.* 1: 161–170. [201, 331]

Grant, B. R. 1996. Pollen digestion by Darwin's Finches and its importance for early breeding. *Ecology* 77: 489–499. [81, 209, 314]

Grant, B. R., and P. R. Grant. 1981. Exploitation *of Opuntia* cactus by birds on the Galápagos. *Oecologia* 49: 179–187. [54, 81, 124, 211, 216, 314]

Grant, B. R., and P. R. Grant. 1982. Niche shifts and competition in Darwin's finches: *Geospiza conirostris* and congeners. *Evolution* 36: 637–657. [99, 298]

Grant, B. R., and P. R. Grant. 1983. Fission and fusion in a population of Darwin's finches: an example of the value of studying individuals in ecology. *Oikos* 41: 530–547. [164]

Grant, B. R., and P. R. Grant. 1989. *Evolutionary dynamics of a natural population: the large cactus finch of the Galápagos.* University of Chicago Press, Chicago, IL. [xxx, 75, 79, 96, 107, 167, 184, 201, 232, 280, 291, 293, 327]

Grant, B. R., and P. R. Grant. 1993. Evolution of Darwin's Finches caused by a rare climatic event. *Proc. R. Soc. B* 251: 111–117. [65, 68, 71, 72, 157, 281]

Grant, B. R., and P. R. Grant. 1996b. High survival of Darwin's finch hybrids: effects of beak morphology and diets. *Ecology* 77: 500–509. [35, 88, 122, 157, 158, 178]

Grant, B. R., and P. R. Grant. 1996d. Cultural inheritance of song and its role in the evolution of Darwin's finches. *Evolution* 50: 2471–2487. [115, 141, 144, 146–152, 250, 251, 291]

Grant, B. R., and P. R. Grant. 1998. Hybridization and speciation in Darwin's finches: the role of sexual imprinting on a culturally transmitted trait. In D. J. Howard and S. H. Berlocher, eds., *Endless Forms: Species and Speciation,* 404–422. Oxford University Press, New York, NY. [141, 148, 151, 291]

Grant, B. R., and P. R. Grant. 2002b. Simulating secondary contact in allopatric speciation: an empirical test of premating isolation. *Biol. J. Linn. Soc.* 76: 545–556. [293]

Grant, B. R., and P. R. Grant. 2008b. Fission and fusion of Darwin's finch populations. *Philos. Trans. R. Soc. B* 363: 2821–2829. [157, 159, 183]

Grant, B. R., and P. R. Grant. 2010c. Songs of Darwin's finches diverge when a new species enters the community. *Proc. Natl. Acad. Sci. USA* 107: 20156–20161. [147, 241–243, 291, 292]

Grant, K. M., E. J. Rohling, M. Bar-Matthews, A. Ayalon, M. Medina-Elizalde, C. Bronk Ramsey, C. Satow, and A. P. Roberts. 2012. Rapid coupling between ice volume and polar temperature over the past 150,000 years. *Nature* 491: 744–747. [272]

Grant, P. R. 1972. Convergent and divergent character displacement. *Biol. J. Linn. Soc.* 4: 39–68. [20]

Grant, P. R. 1981a. Speciation and the adaptive radiation of Darwin's finches. *Amer. Sci.* 69: 653–663. [8, 9]

Grant, P. R. 1981b. The feeding of Darwin's finches on *Tribulus cistoides* (L.) seeds. *Anim. Behav.* 29: 785–793. [30, 37, 59, 66, 118, 126, 127, 130, 266]

Grant, P. R. 1981c. Patterns of growth in Darwin's finches. *Proc. R. Soc. Lond. B* 212: 403–432. [230, 233, 236, 237]

Grant, P. R. 1982. Variation in the size and shape of Darwin's finch eggs. *Auk* 99: 15–23. [89]

Grant, P. R. 1983. Inheritance of size and shape in a population of Darwin's finches, *Geospiza conirostris*. *Proc. R. Soc. Lond. B* 220: 219–236. [230]

Grant, P. R. 1984. Recent research on the evolution of land birds on the Galápagos. *Biol. J. Linn Soc.* 21: 113–136. [308]

Grant, P. R. 1986. *Ecology and Evolution of Darwin's Finches* (2nd ed. 1999). Princeton University Press, Princeton, NJ. [xxx, 4, 6, 9, 19, 27, 61, 79, 126, 166, 184, 230, 272, 290, 293, 298, 299, 302, 324, 331]

Grant, P. R. 1993. Hybridization of Darwin's finches on Isla Daphne Major, Galápagos. *Philos. Trans. R. Soc. B* 340: 127–139. [142, 169, 170, 172]

Grant, P. R. 2013. Adaptive Radiation. In J. B. Losos, ed., *The Princeton Guide to Evolution*, 559–566. Princeton University Press, Princeton, NJ. [4, 306]

Grant, P. R., I. Abbott, D. Schluter, R. L. Curry, and L. K. Abbott. 1985. Variation in the size and shape of Darwin's finches. *Biol. J. Linn. Soc.* 25: 1–39. [7, 18, 48, 139, 240, 302]

Grant, P. R., and P. T. Boag. 1980. Rainfall on the Galápagos and the demography of Darwin's finches. *Auk* 97: 227–244. [78]

Grant, P. R., P. T. Boag, and D. Schluter. 1979. A bill color polymorphism in young Darwin's finches. *Auk* 96: 800–802. [311, 314, 315]

Grant, P. R., and B. R. Grant. 1980a. Annual variation in finch numbers, foraging and food supply on Isla Daphne Major, Galápagos. *Oceologia* 46: 56–62. [22, 24]

Grant, P. R., and B. R. Grant. 1980b. The breeding and feeding characteristics of Darwin's Finches on Isla Genovesa, Galápagos. *Ecol. Monogr.* 50: 381–410. [79, 314]

Grant, P. R., and B. R. Grant. 1992a. Demography and the genetically effective sizes of two populations of Darwin's finches. *Ecology* 73: 766–784. [83, 86, 90, 172]

Grant, P. R., and B. R. Grant 1992b. Hybridization of bird species. *Science* 256: 193–197. [141, 160, 161, 166, 245]

Grant, P. R., and B. R. Grant. 1994. Phenotypic and genetic effects of hybridization in Darwin's finches. *Evolution* 48: 297–316. [45, 65, 66, 115, 133, 169, 174, 176–178, 235]

Grant, P. R., and B. R. Grant. 1995a. Predicting microevolutionary responses to directional selection on heritable variation. *Evolution* 49: 241–251. [62–65, 68, 71, 73, 132, 133]

Grant, P. R., and B. R. Grant. 1995b. The founding of a new population of Darwin's finches. *Evolution* 49: 229–240. [105, 107–112, 115, 116, 132, 272]

Grant, P. R., and B. R. Grant. 1996a. Speciation and hybridization in island birds. *Philos. Trans. R. Soc. B* 351: 765–772. [20, 21]

Grant, P. R., and B. R. Grant. 1996c. Finch communities in a climatically fluctuating environment. In M. L. Cody and J. A. Smallwood, eds., *Long-term studies of vertebrate communities*, 343–390. Academic Press, San Diego, CA. [43, 86, 99, 105, 117, 123, 125, 271, 273, 274, 280, 281, 315, 316, 318]

Grant, P. R., and B. R. Grant. 1997a. Hybridization, sexual imprinting, and mate choice. *Amer. Nat.* 149: 1–28. [141–143, 148, 153, 155, 291]

Grant, P. R., and B. R. Grant. 1997b. Mating patterns of Darwin's finch hybrids determined by song and morphology. *Biol. J. Linn. Soc.* 60: 317–343. [163]

Grant, P. R., and B. R. Grant. 2000a. Quantitative genetic variation in populations of Darwin's finches. In T. A. Mousseau, B. Sinervo, and J. Endler, eds., *Adaptive Genetic Variation in the Wild*, 3–41. Oxford University Press, Oxford, U.K. [43, 44, 46, 48–53, 139, 168, 171, 176, 177, 231–233]

Grant, P. R., and B. R. Grant. 2000b. Non-random fitness variation in two populations of Darwin's finches. *Proc. R. Soc. B* 267: 131–138. [89, 93, 97, 98]

Grant, P. R., and B. R. Grant. 2002a. Unpredictable evolution in a 30-year study of Darwin's finches. *Science* 296: 707–711. [132, 195, 206, 209]

Grant, P. R., and B. R. Grant. 2006. Evolution of character displacement in Darwin's finches. *Science* 313: 224–226. [122, 126–130]

Grant, P. R., and B. R. Grant. 2008a. *How and Why Species Multiply: The Radiation of Darwin's Finches*. Princeton University Press, Princeton, NJ. [xxviii–xxx, 6, 8, 10, 21, 23, 27, 29, 32, 33, 35, 69, 130, 143, 144, 146, 160, 164, 229, 238, 239, 271, 291, 294, 298, 299, 305–308, 316, 324]

Grant, P. R., and B. R. Grant. 2008c. Pedigrees, assortative mating and speciation in Darwin's finches. *Proc. R. Soc. B* 275: 661–668. [148, 247, 252]

Grant, P. R., and B. R. Grant. 2009a. The secondary contact phase of allopatric speciation in Darwin's finches. *Proc. Natl. Acad. Sci. USA* 106: 20141–20148. [247, 251, 252, 258, 296, 335]

Grant, P. R., and B. R. Grant. 2009b. Sympatric speciation, immigration, and hybridization in island birds. In J. B. Losos and R. E. Ricklefs, eds., *The theory of island biogeography revisited*, 326–357. Princeton University Press, Princeton. NJ. [262]

Grant, P. R., and B. R. Grant. 2010a. Natural selection, speciation and Darwin's finches. *Proc. Calif. Acad. Sci*, ser. 4, 61, suppl. 2:245–260. [64, 294]

Grant, P. R., and B. R. Grant. 2010b. Conspecific versus heterospecific gene exchange between populations of Darwin's finches. *Philos. Trans. R. Soc. B* 365: 1065–1076. [114, 159, 170, 174, 176, 183, 258]

Grant, P. R., and B. R. Grant. 2010d. Ecological insights into the causes of an adaptive radiation from long-term field studies of Darwin's Finches. In I. Billick and M. V. Price, eds., *Ecology of Place: Contributions of Place-Based Research to Ecological Understanding*, 109–133. University of Chicago Press, Chicago, IL. [301, 315, 318]

Grant, P. R., and B. R. Grant. 2011a. Causes of lifetime fitness of Darwin's finches in a fluctuating environment. *Proc. Natl. Acad. Sci. USA* 108: 674–679. [47, 50, 89–94, 326, 327]

Grant, P. R., and B. R. Grant, eds. 2011b. *In Search of the Causes of Evolution: From Field Observations to Mechanisms*. Princeton University Press, Princeton, NJ. [27]

Grant, P. R., B. R. Grant, and A. Abzhanov. 2006. A developing paradigm for the development of bird beaks. *Biol. J. Linn. Soc.* 88: 17–22. [236]

Grant, P. R., B. R. Grant, L. F. Keller, J. A. Markert, and K. Petren. 2003. Inbreeding and interbreeding in Darwin's finches. *Evolution* 57: 2911–2916. [161, 162]

Grant P. R., B. R. Grant, L. F. Keller, and K. Petren. 2000. Effects of El Niño events on Darwin's finch productivity. *Ecology* 81: 2442–2457. [42, 83–86, 88, 271, 272, 275, 279, 298]

Grant, P. R., B. R. Grant, J. A. Markert, L. F. Keller, and K. Petren. 2004. Convergent evolution of Darwin's finches caused by introgressive hybridization and selection. *Evolution* 58: 1588–1599. [183]

Grant, P. R., B. R. Grant, and K. Petren. 2000b. The allopatric phase of speciation: the sharp-beaked ground finch (*Geospiza difficilis*) on the Galápagos islands. *Biol. J. Linn. Soc.* 69: 287–317. [115]

Grant, P. R., B. R. Grant, and K. Petren. 2001. A population founded by a single pair of individuals: establishment, expansion, and evolution. *Genetica* 112–113: 359–382. [107, 109–113, 116]

Grant, P. R., B. R. Grant, and K. Petren. 2005. Hybridization in the recent past. *Amer. Nat.* 166: 56–67. [176, 246, 291, 294, 304]

Grant, P. R., B. R. Grant, J.N.M. Smith, I. J. Abbott, and L. K. Abbott. 1976. Darwin's finches: population variation and natural selection. *Proc. Natl. Acad. Sci. USA* 73: 257–261. [30, 37, 54, 56, 57, 140]

Grant, P. R., and T. D. Price. 1981. Population variation in continuously varying traits as an ecological genetics problem. *Amer. Zool.* 21: 795–811. [52, 58, 140, 167]

Grant, P. R., T. D. Price, and H. Snell. 1980. The exploration of Isla Daphne Minor. *Noticias de Galápagos* No. 31: 22–27. [xxviii, 12]

Grant, P. R., J.N.M. Smith, B. R. Grant, I. Abbott, and L. K. Abbott. 1975. Finch numbers, owl predation and plant dispersal on Isla Daphne Major, Galápagos. *Oecologia* 19: 239–257. [327, 329]

Greenberg, R., V. Cadena, R. M. Danner, and G. Tattersall. 2012. Heat loss may explain bill size differences between birds occupying different habitats. *PLoS ONE* 7: e40933. [59, 73, 118, 128]

Grether, G. F., N. Losin, C. N. Anderson, and K. Okamoto. 2009. The role of interspecific interference competition in character displacement and the evolution of competitor recognition. *Biol. Revs.* 84: 617–635. [241, 243, 292]

Guilderson, T. P., and D. Schrag. 1998. Abrupt shift in subsurface temperatures in the tropical Pacific associated with changes in El Niño. *Science* 281: 240–243. [275, 276]

Gustafsson, L., and J. Merilä. 1994. Foster parent experiment reveals no genotype-environment correlation in the external morphology of *Ficedula albicollis*, the collared flycatcher. *Heredity* 73: 124–129. [78]

Haldane, J.B.S. 1937. *Adventures of a Biologist*. Harper and Bros., New York, NY. [122]

Hamann, O. 2001. Demographic studies of three indigenous stand-forming plant taxa (*Scalesia, Opuntia,* and *Bursera*) in the Galápagos Islands, Ecuador. *Biodivers. Conserv.* 10: 223–250. [124]

Hamilton, T. H., and I. Rubinoff. 1967. On predicting insular variation in endemism and sympatry for the Darwin's finches in the Galápagos archipelago. *Amer. Nat.* 101: 161–171. [304]

Handel, C. M., L. M. Pajot, S. M. Matsuoka, C. Van Hemert, J. Terenzi, S. L. Talbot, B. M. Mulcahy, C. U. Meteyer, and K. A. Trust. 2010. Epizootic of beak deformities among wild birds in Alaska: an emerging disease in North America? *Auk* 127: 882–898. [237]

Hanski, I. 2011. Eco-evolutionary spatial dynamics in the Glanville fritillary butterfly. *Proc. Natl. Acad. Sci. USA* 108: 14397–14404. [310]

Harmon, L. J., J. B. Losos, T. J. Davies, R. G. Gillespie, J. L. Gittleman, and 14 others. 2010. Early bursts of body size and shape evolution are rare in comparative data. *Evolution* 64: 2385–2396. [306]

Harris, M. P. 1973. The Galápagos avifauna. *Condor* 75: 265–278. [13, 37, 105, 140]

Harris, M. P. 1974. *A Field Guide to Birds of Galapagos*. Collins, London, U.K. [13]

Harrison, R. G. 1993. Hybrids and hybrid zones: historical perspective. In R. G. Harrison, ed., *Hybrid Zones and the Evolutionary Process*, Oxford University Press, New York, NY. [183]

Harrison, R. 1998. Linking evolutionary pattern and process. In D. J. Howard and S. H. Berlocher, eds., *Endless Forms: Species and Speciation*, 19–31. Oxford University Press, New York, NY. [199]

Harrison, R. G. 2012. The language of speciation. *Evolution* 66: 3643–3657. [262]

Hau, M., M. Wikelski, H. Gwinner, and E. Gwinner. 2004. Timing of reproduction in a Darwin's finch: temporal opportunism under spatial constraints. *Oikos* 196: 489–500. [78]

Helbig, A. J., A. G. Knox, D. T. Parkin, G. Sangster, and M. Collinson. 2002. Guidelines for assigning species rank. *Ibis* 144: 518–525. [199]

Hendry, A. P., S. K. Huber, L. De Léon, A. Herrel, and J. Podos. 2009. Disruptive selection in a bimodal population of Darwin's finches. *Proc. R. Soc. B* 276: 753–759. [262]

Hennig, W. 1966. *Phylogenetic Systematics*. University of Illinois Press, Urbana, IL. [27]

Hermansen, J. S., A. A. Saether, T. O. Elgvin, T. Borge, E. Hjelle, and G.-P. Saetre. 2011. Hybrid speciation in sparrows I: Phenotypic intermediacy, genetic admixture and barriers to gene flow. *Mol. Ecol.* 20: 3812–3822. [246]

Herrel, A., J. Podos, S. K. Huber, and A. P. Hendry. 2005. Evolution of bite force in Darwin's finches: a role for head width. *J. Evol. Biol.* 18: 669–675. [30, 292]

Herrel, A., J. Podos, B. Vanhooydonck, and A. P. Hendry. 2009. Force-velocity trade-off in Darwin's finch jaw function: a biomechanical basis for ecological speciation? *Funct. Ecol.* 23: 119–125. [30, 292]

Hersch, E. I., and P. C. Phillips. 2004. Power and potential bias in field studies of natural selection. *Evolution* 58: 479–485. [61]

Hill, G. E. 2007. Melanins and carotenoids as feather colorants and signals. In B.G.M. Jamieson, ed., *Reproductive Biology and Phylogeny of Birds*, vol. 6B. Science Publishers, Enfield, U.K. [314]

Hill, G. E., and J. D. Johnson. 2012. The vitamin A-redox hypothesis: a biochemical basis for honest signaling via carotenoid pigmentation. *Amer. Nat.* 180: E127–E150. [314]

Hill, W. G. 2010. Understanding and using quantitative genetic variation. *Proc. R. Soc. B* 365: 73–85. [167, 169]

Hill, W. G., and M. Kirkpatrick. 2010. What animal breeding has taught us about evolution. *Ann. Rev. Ecol. Evol. Syst.* 41: 1–19. [167, 169, 231]

Hoekstra, H. E., J. M. Hoekstra, D. Berrigan, S. N. Vignieri, A. Hoang, C. E. Hill, P. Beerli, and J. G. Kingsolver. 2001. Strength and tempo of directional selection in the wild. *Proc. Natl. Acad. Sci. USA* 98: 9157–9160. [128]

Hou, Z., B. Sket, C. Fiser, and S. Li. 2011. Eocene habitat shift from saline to freshwater promoted Tethyan amphipod diversification. *Proc. Natl. Acad. Sci. USA* 108: 14533–14538. [206]

Houle, D. 1989. The maintenance of polygenic variation in finite populations. *Evolution* 43: 1767–1780. [52, 172]

Houle, D. 1992. Comparing evolvability and variability of quantitative traits. *Genetics* 130: 195–204. [52]

Hubbs, C. L. 1955. Hybridization of fish species in nature. *Syst. Zool.* 4: 1–20. [141, 142]

Hubby, J. L., and R. C. Lewontin. 1966. A molecular approach to the study of genic heterozygosity in natural populations. I. The number of alleles at different loci in *Drosophila pseudobscura*. *Genetics* 54: 577–594. [xxvii]

Huber, S. K., L. F. De Léon, A. P. Hendry, E. Bermingham, and J. Podos. 2007. Reproductive isolation of sympatric morphs in a population of Darwin's finches. *Proc. R. Soc. B* 274: 1709–1714. [176, 262]

Huber, S. K., J. P. Owen, J. A. Koop, M. O. King, P. R, Grant, B. R. Grant, and D. H. Clayton. 2010. Ecoimmunity in Darwin's finches: invasive parasites trigger acquired immunity in the medium ground finch (*Geospiza fortis*). *PloSOne* 5: e8605. [79, 284]

Huber, S. K., and J. Podos. 2006. Beak morphology and song features covary in a population of Darwin's finches. *Biol. J. Linn. Soc.* 88: 489–498. [241, 250, 292]

Hudson, A. G., P. Vonlanthen, and O. Seehausen. 2011. Rapid parallel adaptive radiations from a single hybridogenic ancestral population. *Proc. R. Soc. B* 278: 58–66. [246, 296]

Huggins, K. A., K. J. Navara, M. T. Mendonça, and G. E. Hill. 2010. Detrimental effects of carotenoid pigments: the dark side of bright coloration. *Naturwissenschaften* 97: 637–644. [314]

Hughes, C., and R. Eastwood. 2006. Island radiation on a continental scale: exceptional rates of plant diversification after uplift of the Andes. *Proc. Natl. Acad. Sci. USA* 103: 10034–10039. [304]

Husby, A., D. H. Nussey, M. E. Visser, A. J. Wilson, B. C. Sheldon, and L.E.B. Kruuk. 2010. Contrasting patterns of phenotypic plasticity in reproductive traits in two Great Tit (*Parus major*) populations. *Evolution* 64: 2221–2237. [310]

Hutt, F. B. 1949. *Genetics of the fowl*. McGraw-Hill, New York, NY. [311]

Huxley, J. S. 1938. Species formation and geographical isolation. *Proc. Linn. Soc. Lond.* 150: 253–254. [229, 293]

Huxley, J. 1940. *The New Systematics.* Clarendon Press, Oxford, U.K. [293]

Huxley, J. 1942. *Evolution. The Modern Synthesis.* Allen & Unwin, London, UK. [246, 294]

Irwin, D. E., and T. D. Price. 1999. Sexual imprinting, learning and speciation. *Heredity* 82: 347–354. [148, 298]

Isler, M. L., and P. R. Isler. 1999. *The Tanagers.* Smithsonian Institution Press, Washington, DC. [6, 305]

Ito, H. C., and U. Dieckmann. 2007. A new mechanism for recurrent adaptive radiations. *Amer. Nat.* 170: E96–E111. [262]

Jackson, J.B.C. 1995. Constancy and change of life in the sea. In J. H. Lawton and R. M. May, eds., *Extinction Rates*, 45–54. Oxford University Press, Oxford, U.K. [307]

Jain, R., M. C. Rivera, and J. A. Lake. 1999. Horizontal gene transfer among genomes: the complexity hypothesis. *Proc. Natl. Acad. Sci. USA* 96: 3801–3806. [245]

Jensen, H., B-E. Sæther, T. H. Ringsby, J. Tufto, S. C. Griffiths, and H. Ellegren. 2004. Lifetime reproductive success in relation to morphology in the house sparrow *Passer domesticus. J. Anim. Ecol.* 73: 599–611. [77, 78]

Jetz, W., G. H. Thomas, J. B. Joy, K. Hartmann, and A. O. Mooers. 2012. The global diversity of birds in space and time. *Nature* 491: 444–448. [304]

Jo, N. 1983. Karyotypic analysis of Darwin's finches. In R. I. Bowman, M. Berson, and A. E. Leviton, eds., *Patterns of Evolution in Galápagos Organisms*, 201–217. Amer. Assoc. Adv. Sci., Pacific Division, San Francisco, CA. [188]

Johnson, K. P., F. R. Adler, and J. L. Cherry. 2000. Genetic and phylogenetic consequences of island biogeography. *Evolution* 54: 387–396. [115]

Jones, F. C., M. G. Grabherr. Y. F. Chan, P. Russell, E. Mauceli, and more than 20 others. 2012. The genomic basis of adaptive evolution in threespine sticklebacks. *Nature* 484: 55–61. [296]

Jønsson, K. A., P-H. Fabre, S. A. Fritz, R. A. Etienne, B. Haegeman, and 13 others. 2012. Ecological and evolutionary determinants for the adaptive radiation of the Madagascan vangas. *Proc. Natl. Acad. Sci. USA* 109: 6620–6625. [306]

Kalmar, A., and D. J. Currie. 2007. A unified model of species richness on islands and continents. *Ecology* 88: 1309–1321. [281]

Keck, B. P., and T. J. Near. 2010. Geographical and temporal aspects of mitochondrial replacement in *Nothonotus* darters (Teleostei: Percidae: Etheostomatinae). *Evolution* 64: 1410–1428. [169]

Keightley, P. D. 1998. Genetic basis of response to 50 generations of selection on body weight in inbred mice. *Genetics* 148: 1931–1939. [231]

Keller, B., J. Wolinska, M. Manca, and P. Spaak. 2008. Spatial, environmental and anthropogenic effects on the taxon composition of hybridizing *Daphnia. Philos. Trans. R. Soc. B* 363: 2943–2952. [184]

Keller, G. 2012. The Cretaceous-Tertiary Mass Extinction, Chicxulub Impact, and Deccan Volcanism. In J. A. Talent, ed., *Earth and Life*, International Year of Planet Earth, 759–793. Springer Science+Business Media B.V., Dordrecht, Netherlands. [315]

Keller, L. F., and P. Arcese. 1998. No evidence of inbreeding avoidance in a natural population of song sparrows (*Melospiza melodia*). *Amer. Nat.* 152: 380–392. [97]

Keller, L. F., P. R. Grant, B. R. Grant, and K. Petren. 2001. Heritability of morphological traits in Darwin's finches: misidentified paternity and maternal effects. *Heredity* 87: 325–336. [46, 47, 49, 50, 132]

Keller, L. F., P. R. Grant, B. R. Grant, and K. Petren. 2002. Environmental conditions affect the magnitude of inbreeding depression in survival of Darwin's finches. *Evolution* 56: 1229–1239. [96, 107, 109, 161, 163]

Keller, L. F., A. B. Marr, and J. M. Reid. 2006. The genetic consequences of small population size: inbreeding and loss of genetic variation. In J. N. M. Smith, L. F. Keller, A. B. Marr, and P. Arcese, eds., *Conservation and Biology of Small Populations: The Song Sparrows of Mandarte Island*, 113–137. Oxford University Press, Oxford, U.K. [108, 301]

Kendrew, J. C. 1966. *The thread of life: an introduction to molecular biology.* Harvard University Press, Cambridge, MA. [271]

Kim, M., M-L. Cui, P. Cubas, A. Gillies, K. Lea, M. A. Chapman, R. J. Abbott, and E. Coen. 2008. Regulatory genes control a key morphological and ecological trait transferred between species. *Science* 322: 1116–1119. [234]

Kingsolver, J. G., and S. E. Diamond. 2011. Phenotypic selection in natural populations: what limits directional selection? *Amer. Nat.* 177: 346–357. [92, 128]

Kingsolver, J. G., H. E. Hoekstra, J. M. Hoekstra, D. Berrigan, S. N. Vignieri, C. E. Hill, A. Hoang, P. Gilbert, and P. Beerli. 2001. The strength of phenotypic selection in natural populations. *Amer. Nat.* 157: 245–261. [128, 132]

Kirchman, J. J. 2012. Speciation of flightless rails on islands: a DNA-based phylogeny of the typical rails of the Pacific. *Auk* 129: 56–69. [238]

Kleindorfer, S. 2007. The ecology of clutch size variation in Darwin's small ground finch *Geospiza fuliginosa*: comparison between lowland and highland habitats. *Ibis* 149: 730–741. [79]

Kleindorfer. S., T. W. Chapman, H. Winkler, and F. J. Sulloway. 2006. Adaptive divergence in contiguous populations of Darwin's small ground finch (*Geospiza fuliginosa*). *Evol. Ecol. Res.* 8: 357–372. [302]

Kocher, T. D. 2004. Adaptive evolution and explosive speciation: the cichlid fish model. *Nature Revs. Genet.* 5: 288–298. [316]

Komdeur, J. 2003. Daughters on request: about helpers and egg sexes in the Seychelles warbler. *Proc. R. Soc. B.* 270: 3–11. [78]

Koop, J. A., S. K. Huber, S. M. Laverty, and D. H. Clayton. 2011. Experimental demonstration of the fitness consequences of an introduced parasite of Darwin's finches. *PLoS One* 6: e19706. [284]

Krakower, D. G. 1996. Natal dispersal in two species of Darwin's finches: the medium ground finch *Geospiza fortis* and the cactus ground finch *Geopsiza scandens*. Unpubl. BA thesis, Princeton University, Princeton, NJ. [49, 97]

Kruuk, L.E.B., T. H. Clutton-Brock, J. Slate, J. M. Pemberton, S. Brotherstone, and F. E. Guiness. 2000. Heritability of fitness in a wild mammal population. *Proc. Natl. Acad. Sci. USA* 97: 698–703. [97]

Kruuk, L.E.B., B. C. Sheldon, and J. Merilä. 2002. Severe inbreeding depression in collared flycatchers (*Ficedula albicollis*). *Proc. R. Soc. B* 269: 1581–1589. [108]

Kruuk, L.E.B., J. Slate, and A. J. Wilson. 2008. New answers for old questions: the evolutionary quantitative genetics of wild animal populations. *Annu. Rev. Ecol. Evol. Syst.* 39: 525–548. [43, 65]

Kunte, K., C. Shea, M. L. Aardema, J. M. Scribner, T. E. Juenger, L. E. Gilbert, and M. R. Kronforst. 2011. Sex chromosome mosaicism and hybrid speciation among tiger swallowtail butterflies. *PLoS Genet.* 7: e1002274. [246]

Lack, D. 1944. Ecological aspects of speciation in birds. *Ibis* 86: 260–286. [238]

Lack, D. 1945. The Galápagos finches (Geospizinae): A study in variation. *Occas. Pap. Calif. Acad. Sci.* 21: 1–159. [5–7, 13, 40, 105, 139, 304]

Lack, D. 1947. *Darwin's Finches.* Cambridge University Press, Cambridge, U.K. [xxx, 4, 5, 7, 17, 19, 105, 134, 138, 139, 304, 307]

Lande, R. 1976. The maintenance of genetic variability by mutation in a polygenic character with linked loci. *Genet. Res., Cambridge* 26: 221–235. [167]

Lande, R. 1979. Quantitative genetic analysis of multivariate evolution, applied to brain:body size allometry. *Evolution* 33: 402–416. [62, 65, 230, 232]

Lande, R. 1993. Risks of population extinction from demographic and environmental stochasticity and random catastrophes. *Amer. Nat.* 142: 911–927. [104, 276]

Lande, R., and S. Arnold. 1983. The measurement of selection on correlated characters. *Evolution* 37: 1210–1226. [61, 62]

Lanfear, R., S.Y.W. Ho, D. Love, and L. Bromham. 2010. Mutation rate is linked to diversification in birds. *Proc. Natl. Acad. Sci. USA* 107: 20423–20428. [305]

Larsen, P. A., M. R. Marchán-Rivadeneira, and R. J. Baker. 2010. Natural hybridization generates mammalian lineage with species characteristics. *Proc. Natl. Acad. Sci. USA* 107: 11447–11452. [246, 294]

Larsson, K., P. Forslund, L. Gustafsson, and B. S. Ebbinge. 1988. From the high arctic to the Baltic: the successful establishment of a barnacle goose *Branta leucopsis* population on Gotland, Sweden. *Ornis Scandinavica* 19: 182–189. [104]

Larsson, K., H. P. van der Jeugd, I. T. van der Veen, and P. Forslund. 1998. Body size declines despite positive directional selection on heritable size traits in a Barnacle Goose population. *Evolution* 52: 1169–1184. [213]

Latif, M., and G. M. Spellman. 2009. Tipping elements in Earth Systems Special Feature: El Niño/Southern Oscillation response to global warming. *Proc. Natl. Acad. Sci. USA* 106: 20578–20583. [276]

Lavergne, S., N. Mouquet, W. Tuiller, and O. Ronce. 2010. Biodiversity and climate change: integrating evolutionary and ecological responses of species and communities. *Annu. Rev. Ecol. Evol. Syst.* 41: 321–350. [276]

Lebige, C., P. Arcese, R. J. Sardell, L. F. Keller, and J. M. Reid. 2012. Extra-pair paternity and the variance in male fitness in song sparrows (*Melospiza melodia*). *Evolution* 66: 3111–3129. [310]

Lessells, C. M., and P. T. Boag. 1987. Unrepeatable repeatabilities: a common mistake. *Auk* 104: 116–121. [45]

Lewontin, R. C., and L. C. Birch. 1966. Hybridization as a source of variation for adaptation to new environments. *Evolution* 20: 315–336. [166, 177]

Likens, G. E. 1989. *Long-term Studies in Ecology: Approaches and Alternatives*. Springer-Verlag, New York and Berlin. [310]

Losos, J. B., and R. E. Ricklefs, eds. 2010. *The Theory of Island Biogeography revisited*. Princeton University Press, Princeton, NJ. [304]

Luther, D. A. 2009. The influence of the acoustic community on songs of birds in a neotropical rain forest. *Behav. Ecol.* 20: 864–871. [241]

Luther, D. A, and R. H. Wiley. 2009. Production and perception of communicatory signals in a noisy environment. *Biol. Lett.* 5: 183–187. [241]

Lynch, M., and W. G. Hill. 1986. Phenotypic evolution by neutral mutation. *Evolution* 40: 915–935. [52]

Lynch, M., and R. Lande. 1993. Evolution and extinction in response to environmental change. In P. M. Kareiva, J. G. Kingsolver, and R. B. Huey, eds. *Biotic Interactions and Global Change*, 234–250. Sinauer, Sunderland, MA. [276]

Lynch, M., and J. B. Walsh 1998. *Genetics and Analysis of Quantitative Traits*. Sinauer, Sunderland, MA. [43, 132]

MacArthur, R. H. 1972. *Geographical Ecology: Patterns in the Distribution of Species*. Harper and Row, New York, NY. [103]

MacArthur, R. H., and E. O. Wilson. 1967. *The Theory of Island Biogeography*. Princeton University Press, Princeton, NJ. [304, 308]

MacFarlane, M., M. R. Evans, K. A. Feldheim, M. Tréault, R.C.K. Bowie, and M. I. Cherry. 2010. Long tails matter in sugarbirds—positively for extrapair but negatively for within-pair fertilization success. *Behav. Ecol.* 21: 26–32. [327]

MacMillen, R. E. 1990. Water economy of granivorous birds: a predictive model. *Condor* 92: 379–392. [73]

Majerus, M.E.N. 1998. *Melanisms: Evolution in Action*. Oxford University Press, Oxford, U.K. [310]

Mallarino, R., O. Campàs, J. Fritz, K. J. Burns, O. G. Weeks, M. P. Brenner, and A. Abzhanov. 2012. Closely related bird species demonstrate flexibility between beak morphology and underlying developmental programs. *Proc. Natl. Acad. Sci. USA* 109: 16222–16227. [236, 285]

Mallarino, R., P. R. Grant, B. R. Grant, A. Herrel, W. P. Kuo, and A. Abzhanov. 2011. Two developmental modules establish 3D beak-shape variation in Darwin's finches. *Proc. Natl. Acad. Sci. USA*. 198: 4057–4062. [235, 236, 285]

Mallet, J. 2005. Hybridization as an invasion of the genome. *Trends Ecol. Evol.* 20: 229–237. [245]

Mallet, J. 2007. Hybrid speciation. *Nature* 446: 279–283. [246]

Marchant, S. 1958. The birds of the Santa Elena peninsula, S. W. Ecuador. *Ibis* 100: 349–387. [79]

Marchant, S. 1959. The breeding season in SW Ecuador. *Ibis* 101: 139–152. [79]

Marchant, S. 1960. The breeding of some SW Ecuadorean birds. *Ibis* 102: 349–381, 584–599. [79]

Markert, J. A., P. R. Grant, B. R. Grant, L. F. Keller, J. L. Coombs, and K. Petren. 2004. Neutral locus heterozygosity, inbreeding, and survival in Darwin's ground finches (*Geospiza fortis* and *G. scandens*). *Heredity* 92: 305–315. [96, 161, 162, 198]

Marquiss, M., and R. Rae. 2002. Ecological differentiation in relation to bill size amongst sympatric, genetically undifferentiated crossbills *Loxia* spp. *Ibis* 144: 494–508. [238]

Marr, A. B., L. F. Keller, and P. Arcese. 2002. Heterosis and outbreeding depression in descendants of natural immigrants to an inbred population of Song Sparrows (*Melospiza melodia*). *Evolution* 56: 131–142. [119]

Martin, P. R., and T. E. Martin. 2001. Ecological and fitness consequences of species coexistence: a removal experiment with wood warblers. *Ecology* 82: 189–206. [125]

Martin, T. E. 2004. Avian life-history evolution has an eminent past: does it have a bright future? *Auk* 121: 289–301. [79]

Mathys, B., and J. L. Lockwood. 2009. Rapid evolution of great kiskadees on Bermuda: an assessment of the Island Rule to predict the direction of contemporary evolution within exotic vertebrates. *J. Biogeog.* 36: 2204–2211. [104]

Matyjasiak, P. 2005. Birds associate species specific acoustic and visual cues: recognition of heterospecific rivals by male blackcaps. *Behav. Ecol.* 16: 467–471. [148]

Mavárez, J., C. A. Salazar, E. Bermingham, C. Salcedo, C. D. Jiggins, and M. Linares. 2006. Speciation by hybridization in *Heliconius* butterflies. *Nature* 441:868–871. [246]

Mayr, E. 1942. *Systematics and the Origin of Species*. Columbia University Press, New York, NY. [293, 298]

Mayr, E. 1954. Changes of genetic environment and evolution. In J. S. Huxley, A. Hardy, and E. B. Ford, eds., *Evolution as a process*, 157–180. Allen and Unwin, London, U.K. [105, 301]

Mayr, E. 1963. *Animal Species and Evolution*. Belknap Press, Harvard University, Cambridge, MA. [139, 141, 289]

Mayr, E. 1970. Populations, Species, and Evolution. Harvard Univ. Press, Cambridge, MA. [xxvii]

McCarthy, E. M. 2006. *Handbook of Avian Hybrids of the World*. Oxford University Press, New York, NY. [166]

McDade, L. 1990. Hybrids and phylogenetic systematics I. Patterns of character expression in hybrids and their implications for cladistic analysis. *Evolution* 44: 1685–1700. [177]

McGlone, M. S., K. P. Kershaw, and V. Markgraf. 1992. El Niño/Southern Oscillation. In H. F. Diaz and V. Markgraf, eds., *El Niño: Historical and Paleoclimatic Aspects of the Southern Oscillation*, 435–462. Cambridge University Press, Cambridge, U.K. [275]

McGraw, K. J., G. E. Hill, and R. S. Parker. 2005. The physiological costs of being colorful: nutritional control of carotenoid utilization in the American goldfinch (*Carduelis tristis*). *Anim. Behav.* 69: 653–660. [314]

McMichael, A. J. 2012. Insights from past millennia into climatic impacts on human health and survival. *Proc. Natl. Acad. Sci. USA* 109: 4730–4737. [276]

Medawar, P. 1991. *The Threat and the Glory. Reflections on Science and Society.* Oxford University Press, Oxford, U.K. (Includes the quoted 1963 chapter Is the scientific paper a fraud?) [17, 39]

Merilä, J., L.E.B. Kruuk , and B. C. Sheldon. 2001. Cryptic evolution in a wild bird population. *Nature* 412: 76–79. [213]

Merilä, J., and B. C. Sheldon. 2000. Lifetime reproductive success and heritability in nature. *Amer. Nat.* 155: 301–310. [97]

Merilä, J., and B. C. Sheldon. 2001. Avian quantitative genetics. *Current Ornithology* 16: 179–255. [43, 48]

Merilä, J., B. C. Sheldon, and L.E.B. Kruuk. 2001. Explaining stasis: microevolutionary studies in natural populations. *Genetica* 112–113: 199–222. [213]

Merrell, D. J. 1994. *The Adaptive Seascape: The mechanism of evolution.* University of Minnesota Press, Minneapolis, MN. [134]

Meyer, J. R., S. E. Schoustra, J. Lachapelle, and R. Kassen. 2011. Overshooting dynamics in a model adaptive radiation. *Proc. R. Soc. B* 278: 392–398. [308]

Michaelsen, J., and L. G. Thompson. 1992. A comparison of proxy records of El Niño/Southern Oscillation. In H. F. Diaz, and V. Markgraf, eds. *El Niño: Historical and Paleoclimatic Aspects of the Southern Oscillation*, 323–348. Cambridge University Press, Cambridge, U.K. [275]

Michalak, P. 2009. Epigenetic, transposon and small RNA determinants of hybrid dysfunctions. *Heredity* 102: 45–50. [285]

Milá, B., J. E. McCormick, G. Castañeda, R. K. Wayne, and T. B. Smith. 2007. Recent postglacial range expansion drives the rapid diversification of a songbird lineage in the genus *Junco. Proc. R. Soc. B* 274: 2653–2660. [238]

Milá, B., D.P.L. Toews, T. B. Smith, and R. K. Wayne. 2011. A cryptic contact zone between divergent mitochondrial DNA lineages in southwestern North America supports past introgressive hybridization in the yellow-rumped warbler complex (Aves: *Dendroica coronata*). *Biol. J. Linn. Soc.* 103: 696–706. [246]

Miller, A. H. 1956. Ecologic factors that accelerate formation of races and species of terrestrial vertebrates. *Evolution* 63: 884–897. [177]

Millington, S. J., and P. R. Grant. 1983. Feeding ecology and territoriality of the cactus finch *Geospiza scandens* on Isla Daphne Major, Galápagos. *Oecologia* 58: 76–83. [209]

Millington, S. J., and P. R. Grant. 1984. The breeding ecology of the Cactus Finch *Geospiza scandens* on Isla Daphne Major, Galápagos. *Ardea* 72: 177–188. [43]

Millington, S. J., and T. D. Price. 1985. Song inheritance and mating patterns in Darwin's finches. *Auk* 102: 342–346. [250]

Miner, B. E., L. de Meester, M. E. Pfrender, W. Lampert, and N. G. Hairston, Jr.

2012. Linking genes to communities and ecosystems: *Daphnia* as an ecogenomic model. *Proc. R. Soc. B* 279: 1873–1882. [99]

Morton, N. E., J. F. Crow, and H. J. Muller. 1956. An estimate of the mutational damage in man from data on consanguineous marriages. *Proc. Natl. Acad. Sci. USA* 42: 855–863. [107]

Moy, C. M., G. O. Seltzer, D. T. Rodbell, and D. M. Anderson. 2002. Variability in El Niño/Southern Oscillation activity at millennial timescales during the Holocene epoch. *Nature* 420: 162–165. [272]

Moyle, R. G., C. E. Filardi, C. E. Smith, and J. Diamond. 2009. Explosive Pleistocene diversification and hemispheric expansion of a "great speciator." *Proc. Natl. Acad. Sci. USA* 108: 1863–1868. [238]

Navarro, C., T. Pérez-Contreras, J. Avilés, K. G. McGraw, and J. J. Soler. 2010. Beak color reflects circulating carotenoid and vitamin A levels in spotless starlings (*Sturnus unicolor*). *Behav. Ecol. Sociobiol.* 64: 1057–1067. [314]

Nelson, D. A., and P. R. Marler. 1990. The perception of birdsong and an ecological concept of signal space. In W. C. Stebbins and M. A. Berkeley, eds., *Comparative Perceptions. Complex Signals*, vol. 2, 443–478. John Wiley and Sons, New York, NY. [241, 291]

Nemeth, E., N. Pieretti, S. A. Zollinger, N. Geberzahn, J. Partecke, A. C. Miranda, and H. Brumm. 2013. Bird song and anthropogenic noise: vocal constraints may explain why birds sing higher-frequency songs in cities. *Proc. R. Soc. B* 280: 20122798. [241]

Newton, I., ed. 1989. *Lifetime Reproduction in Birds*. Academic Press, London. [87]

Nieminen, M., M. C. Singer, W. Fortelius, K. Schöps, and I. Hanski. 2001. Experimental confirmation that inbreeding depression increases extinction risk in butterfly populations. *Amer. Nat.* 157: 237–244. [265]

Nilsson, J-Å., M. Åkesson, and J. F. Nilsson. 2009. Heritability of resting metabolic rate in a wild population of blue tits. *J. Evol. Biol.* 22: 1867–1874. [73]

Nolte, A. W., and D. Tautz. 2009. Understanding the onset of hybrid speciation. *Trends Genetics.* 26: 54–58. [246]

Nosil, P. 2012. *Ecological Speciation*. Oxford University Press, New York, NY. [298]

Nosil, P., and J. L. Feder. 2012. Genomic divergence during speciation: causes and consequences. *Philos. Trans. R. Soc. B* 367: 332–342. [285]

Nosil, P., and T. E. Reimchen. 2005. Ecological opportunity and levels of morphological variance within freshwater stickleback populations. *Biol. J. Linn. Soc.* 86: 297–308. [179]

Nunney, L. 1993. The influence of mating system and overlapping generations on effective population size. *Evolution* 47: 1329–1341. [83]

O'Connor, J. A., F. J. Sulloway, J. Robertson, and S. Kleindorfer. 2010. *Philornis downesi* parasitism is the primary cause of nestling mortality in the critically endangered Darwin's medium tree finch (*Camarhynchus pauper*). *Biodivers. Conserv.* 19: 853–866. [284]

O'Connor, R. J., M. P. Usher, A. Gibbs, and K. C. Brown. 1986. Biological characteristics of invaders among bird species in Britain. *Philos. Trans. R. Soc. B* 314: 583–598. [106]

Ochman, H., J. G. Lawrence, and E. A. Groisman. 2000. Lateral gene transfer and the nature of bacterial innovation. *Nature* 405: 299–304. [245]

Olson, J. A. 1984. Serum levels of vitamin A and carotenoids as reflectors of nutritional status. *J. Natl. Cancer Inst.* 73: 1439–1444. [315]

Orr, R. T. 1945. A study of captive Galápagos finches of the genus *Geospiza*. *Condor* 47: 177–201. [140]

Ozgul, A., S. Tuljapurkar, T. G. Benton, J. M. Pemberton, T. H. Clutton-Brock, and T. Coulson. 2009. The dynamics of phenotypic change and the shrinkage of sheep of St. Kilda. *Science* 325: 464–467. [310]

Packer, C., R. Hillborn, A. Mosser, B. Kissui, M. Borner, G. Hopcraft, J. Wilmshurst, S. Mduma, and A.R.E. Sinclair. 2005. Ecological change, group territoriality, and population dynamics in Serengeti lions. *Science* 307: 390–393. [310]

Päckert, M., J. Martens, L. Hering, L. Kvist, and J. C. Illera. 2013. Return flight to the Canary Islands—the key role of peripheral populations of Afrocanarian blue tits (Aves: *Cyanistes teneriffae*) in multi-gene reconstruction of colonization pathways. *Mol. Phylogenet. Evol.* 67: 458–467. [104]

Panov, E. N. 1989. *Natural Hybridisation and Ethological Isolation in Birds*. Nauka, Moscow, Russia. [166]

Parmesan, C. 2006. Ecological and evolutionary responses to recent climate change. *Annu. Rev. Ecol. Evol. Syst.* 37: 637–669. [276]

Pelletier, F., T. Clutton-Brock, J. Pemberton, S. Tuljapurkar, and T. Coulson. 2007. The evolutionary demography of ecological change: linking trait variation and population growth. *Science* 315: 1571–1574. [78]

Pemberton, J. M. 2010. Evolution of quantitative traits in the wild: mind the ecology. *Philos. Trans. R. Soc. B* 365: 2431–2438. [213]

Pereira, R. J., and D. B. Wake. 2009. Genetic leakage after adaptive and non-adaptive divergence in the *Ensatinia escholtzii* ring species. *Biol. J. Linn. Soc.* 98: 2288–2301. [246]

Peters, S., E. P. Derryberry, and S. Nowicki. 2012. Songbirds learn songs least degraded by environmental transmission. *Biol. Lett.* 8: 736–739. [241]

Petit, K. E., and K. A. Tarvin. 1990. First record of Prothonotary Warbler from Galapagos. *American Birds* 44 (5). [328]

Petren, K. 1998. Microsatellite primers from *Geospiza fortis* and cross-species amplification in Darwin's finches. *Mol. Ecol.* 7: 1782–1784. [42, 46, 111]

Petren, K., B. R. Grant, and P. R. Grant. 1999a. Low extrapair paternity in the cactus finch (*Geospiza scandens*). *Auk* 116: 252–256. [50]

Petren, K., B. R. Grant, and P. R. Grant. 1999b. A phylogeny of Darwin's finches based on microsatellite DNA length variation. *Proc. R. Soc. B* 266: 321–329. [233, 307]

Petren, K., P. R. Grant, B. R. Grant, and L. F. Keller. 2005. Comparative landscape genetics and the adaptive radiation of Darwin's finches: the role of peripheral isolation. *Mol. Ecol.* 14: 2943–2957. [231, 233, 238, 240, 304]

Pfennig, D. W., and K. S. Pfennig. 2010. Character displacement and the origin of diversity. *Amer. Nat.* 176: S26–S44. [122]

Pfennig, D. W., and K. S. Pfennig. 2012 *Evolution's Wedge. Competition and the Origins of Diversity.* University of California Press, Los Angeles, CA. [122]

Philander, S.G.H. 1990. *El Niño, La Niña and the Southern Oscillation.* Academic Press, New York, NY. [5, 274, 275]

Piertney, S. B., R. Summers, and M. Marquiss. 2001. Microsatellite and mitochondrial DNA homogeneity among phenotypically diverse crossbill taxa in the U.K. *Proc. R. Soc. B* 268: 1511–1517. [238]

Podos, J. 1997. A performance constraint on the evolution of trilled vocalizations in a songbird family (Passeriformes: Emberizidae). *Evolution* 51: 537–551. [241, 292]

Podos, J. 2001. Correlated evolution of morphological and vocal signal structure in Darwin's finches. *Nature* 409: 185–188. [241, 292]

Podos, J. 2010. Acoustic discrimination of sympatric morphs in Darwin's finches: a behavioral mechanisms for assortative mating? *Philos. Trans. R. Soc. B* 365: 1031–1039. [148, 292]

Porter, D. M. 1967. Another *Tribulus* adventive in the New World. *Rhodora* 69: 455–456. [298]

Porter, D. M. 1977. Typification of *Tiquilia darwinii* and *Tiquilia fusca* (Boraginaceae). *Rhodora* 79: 288–291. [322]

Postma, E., and A. J. van Noordwijk. 2005. Gene flow maintains a large genetic difference in clutch size at a small spatial scale. *Nature* 433: 65–68. [78]

Poulakakis, N., M. Russello, D. Geist, and A. Caccone. 2012. Unravelling the peculiarities of island life: vicariance, dispersal and the diversification of the extinct and extant Galápagos tortoises. *Mol. Ecol.* 21: 160–173. [303]

Prager, E. M., and A. C. Wilson. 1980. Phylogenetic relationships and rates of evolution in birds. In *Proc. XVII International Ornithological Congress, Berlin*, 1209–1214. Verlag der Deutsche Ornithologon-Gesellschaft. [291]

Pratt, H. D. 2005. *The Hawaiian honeycreepers.* Oxford University Press, Oxford, U.K. [307]

Price, T. D. 1984a. Sexual selection on body size, territory and plumage variables in a population of Darwin's finches. *Evolution* 38: 327–341. [43, 92]

Price, T. D. 1984b. The evolution of sexual size dimorphism in Darwin's finches. *Amer. Nat.* 123: 500–518. [74, 92, 263]

Price, T. D. 1985. Reproductive responses to varying food supply in a population of Darwin's finches: clutch size, growth rates and hatching asynchrony. *Oecologia* 66: 411–416. [44]

Price, T. D. 1987. Diet variation in a population of Darwin's finches. *Ecology* 68: 1015–1028. [30, 59, 66, 127, 172]

Price, T. 2008. *Speciation in Birds.* Ben Roberts, Greenwood, CO. [148, 289, 291, 300]

Price, T. D. 2010. The roles of time and ecology in the continental radiation of the Old World leaf warblers (*Phylloscopus* and *Seicercus*). *Philos. Trans. R. Soc. B* 365: 1749–1762. [239, 305]

Price, T. D., and P. R. Grant. 1984. Life history traits and natural selection for small body size in a population of Darwin's finches. *Evolution* 38: 483–494. [74, 92]

Price, T. D., and P. R. Grant. 1985. The evolution of ontogeny in Darwin's finches: a quantitative genetics approach. *Amer. Nat.* 125: 169–188. [234, 236]

Price, T. D., P. R. Grant, and P. T. Boag. 1984a. Genetic changes in the morphological differentiation of Darwin's Ground Finches. In K. Wöhrmann and V. Loeschke, eds., *Population Biology and Evolution*, 49–66, Springer, Berlin. [63, 230]

Price, T. D., P. R. Grant, H. L. Gibbs, and P. T. Boag. 1984b. Recurrent patterns of natural selection in a population of Darwin's finches. *Nature* 309: 787–789. [61, 63, 232]

Price, T., S. Millington, and P. Grant. 1983. Helping at the nest in Darwin's finches as misdirected parental care. *Auk* 100: 192–194. [327]

Price, T., A. B. Phillimore, M. Awodey, and R. Hudson. 2010. Ecological and geographical influences on the allopatric phase of island speciation. In P. R. Grant and B. R. Grant, eds., *In Search of the Causes of Evolution*, 251–273. Princeton University Press, Princeton, NJ. [301, 304, 305]

Price, T. D., P. J. Yeh, and B. Harr. 2008. Phenotypic plasticity and the evolution of a socially selected trait following colonization of a novel environment. *Amer. Nat.* 172: S49–S62. [78]

Pritchard, J. K., M. Stephens, and P. Donnelly. 2000. Inferences of population structure using multilocus genotype data. *Genetics* 155: 945–959. [114, 186, 247]

Pritchard, J. K., X. Wen, and D. Falush. 2007. Documentation for structure software: version 2.2. [186, 247]

Qvarnström, A., A. M. Rice, and H. Ellegren. 2010. Speciation in *Ficedula* flycatchers. *Philos. Trans. R. Soc. B* 365: 1841–1852. [310]

Rabosky, D. L., and R. E. Glor. 2011. Equilibrium speciation dynamics in a model adaptive radiation of island lizards. *Proc. Natl. Acad. Sci. USA* 107: 22178–22183. [306, 308]

Rabosky, D. L., and I. J. Lovette. 2008. Explosive evolutionary radiations: decreasing speciation or increasing extinction through time? *Evolution* 62: 1866–1875. [306, 308]

Raby, P. 2001. *Alfred Russel Wallace: a life*. Princeton University Press, Princeton, NJ. [383]

Randler, C. 2002. Avian hybridization, mixed pairing and female choice. *Anim. Behav.* 63: 103–119. [142]

Rands, C., A. Darling, M. Fujita, L. Kong, M. T. Webster, and 14 others. 2013. Insights into the evolution of Darwin's finches from comparative analysis of the *Geospiza magnirostris* genome sequence. *BMC Genomics* 14: 95. DOI:1186/1471-2164-14-95. [284]

Rannala, B. 1996. The sampling theory of neutral alleles in an island population of fluctuating size. *Theoret. Pop. Biol.* 50: 91–104. [115]

Ratcliffe, L. M., and P. R. Grant. 1983. Species recognition in Darwin's Finches (*Geospiza*, Gould), I: Discrimination by morphological cues. *Anim. Behav.* 31: 1139–1153. [143, 148]

Ratcliffe, L. M., and P. R. Grant. 1985. Species recognition in Darwin's Finches

(*Geospiza*, Gould), III: Male responses to playback of different song types, dialects and heterospecific songs. *Anim. Behav* 33: 290–307. [143, 148]

Reddy, S., A. Driskell, D. L. Rabosky, S. J. Hackett, and T. S. Schulenberg. 2012. Diversification and the adaptive radiation of the vangas of Madagascar. *Proc. R. Soc. B* 279: 2062–2071. [306]

Reich, D., R. E. Green, M. Kircher, J. Krause, N. Patterson, E. Y. Durand, B. Viola, A. W. Briggs, U. Stenzel, P.L.F. Johnson, and others. 2010. Genetic history of an archaic hominin group from Denisova Cave in Siberia. *Nature* 468: 1053–1060. [246]

Reid, J. M., E. M. Bignal, D. I. Cracken, and P. Monaghan. 2003. Environmental variability, life-history covariation and cohort effects in the red-billed chough *Pyrrhocorax pyrrhocorax*. *J. Anim. Ecol.* 72: 36–46. [77, 95]

Rein, B., A. Luckage, L. Reinhardt, F. Sirocko, A. Wolf, and W-C. Dullo. 2005. El Niño variability off Peru during the last 20,000 years. *Paleoceanography* 20: 2004PA001099. [272]

Renne, P. R., A. L. Deino, F. J. Hilgen, K. F. Kuiper, D. F. Mark, W. S. Mitchell III, L. E. Morgan, R. Mundil, and J. Smit. 2013. Time scales of critical events around the Cretaceous-Paleogene boundary. *Science* 339: 684–687. [315]

Restrepo, A., P. Colinvaux, M. Bush, A. Correa-Metrio, J. Conroy, M. R. Gardner, P. Jaramillo, M. Steinitz-Kannen, and J. Overpeek. 2012. Impacts of climatic variability and human colonization on the vegetation of the Galápagos Islands. *Ecology* 93: 1853–1866. [272, 275]

Reyer, H.-U., K. Bollmann, A. R. Schläpfer, A. Schymainda, and G. Klecak. 1997. Ecological determinants of extra-pair fertilizations and egg-dumping in alpine water pipits (*Anthus spinoletta*). *Behav. Ecol.* 8: 534–543. [326]

Rheindt, F. E., and S. V. Edwards. 2011. Genetic introgression: an integral but neglected component of speciation. *Auk* 128: 620–632. [246, 294]

Rhymer, J. M., and D. Simberloff. 1996. Extinction by hybridization and introgression. *Annu. Rev. Ecol. Syst.* 27: 83–109. [139]

Richter, I. A., ed. 1985. *The Notebooks of Leonardo da Vinci*. World's Classics paperback, Oxford University Press, Oxford, U.K. (Quotation is from the Codex Trivulzi, Castello Sforzesco, Milan, published by Luca Beltrami, Milan, 1891.) [245]

Ricklefs, R. E. 1980. On geographical variation in clutch size among passerine birds: Ashmole's hypothesis. *Auk* 97: 38–49. [79]

Ricklefs, R. E. 2006. Global variation in the diversification rate of passerine birds. *Ecology* 87: 2468–2478. [306, 308]

Ricklefs, R. E. 2007. Estimating diversification rates from phylogenetic information. *Trends Ecol. Evol.* 22: 601–610. [296]

Ricklefs, R. E., and E. Bermingham. 2007. The causes of evolutionary radiations in archipelagoes: passerine birds in the Lesser Antilles. *Amer. Nat.* 169: 285–297. [238]

Ricklefs, R. E., and G. W. Cox. 1972. Taxon cycle in the West Indian avifauna. *Amer. Nat.* 106: 195–219. [308]

Riedinger, M. A., M. Steinitz-Kannen, W. M. Last, and M. Brenner. 2002. A ~6100 ^{14}C yr record of El Niño activity from the Galápagos Islands. *J. Paleolimnology* 27: 1–7. [272]

Rieseberg, L. H., O. Raymond, D. M. Rosenthal, Z. Lai, K. Livingstone, T. Nakazato, A. E. Schwarzbach, L. Donovan, and C. Lexer. 2003. Major ecological transitions in wild sunflowers facilitated by hybridization. *Science* 301: 1211–1216. [246]

Rieseberg, L. H., A. Widmer, A. M. Arntz, and J. M. Burke. 2003. The genetic architecture necessary for transgressive segregation is common in both natural and domesticated populations. *Philos. Trans. R. Soc. B* 358: 1141–1147. [265]

Rijssel, J. C., and F. Witte. 2013. Adaptive responses in resurgent Lake Victoria cichlids over the past 30 years. *Evol. Ecol.* 27: 253–267. [310]

Ríos-Chelén, A. A., C. Salaberria, I. Barbosa, C. Macías Garcia, and D. Gil. 2012. The learning advantage: bird species that learn their song show a tighter adjustment of song to noisy environments than those that do not learn. *J. Evol. Biol.* 25: 2171–2180. [241]

Robinson, B. W., and D. Schluter. 2000. Natural selection and the evolution of adaptive genetic variation in northern freshwater fishes. In T. A. Mousseau, B. Sinervo, and J. Endler, eds., *Adaptive Genetic Variation in the Wild*, 65–94. Oxford University Press, Oxford, U.K. [179]

Rosindell, J., and A. B. Phillimore. 2011. A unified model of island biogeography sheds light on the zone of radiation. *Ecol. Lett.* 14: 552–560. [304]

Rothstein, S. I. 1973. The niche-variation model—is it valid? *Amer. Nat.* 107: 598–620. [56]

Roughgarden, J. 1972. Evolution of niche width. *Amer. Nat.* 106: 683–718. [139]

Roughgarden, J. 1995. *Anolis lizards of the Caribbean: Ecology, Evolution, and Plate Tectonics.* Princeton University Press, Princeton, NJ. [135]

Russel, R. M. 1999. The vitamin A spectrum: from deficiency to toxicity. *Am. J. Clin. Nutr.* 71: 878–884. [314]

Rutschmann, S., M. Matschiner, M. Damerau, M. Muschick, M. F. Lehmann, R. Hanel, and W. Salzburger. 2011. Parallel ecological diversification in Antarctic notothenoid fishes as evidence for adaptive radiation. *Mol. Ecol.* 20: 4707–4721. [296]

Ryan, P. G., P. Bloomer, C. L. Moloney, T. J. Grant, and W. Delport. 2007. Ecological speciation in South Atlantic island finches. *Science* 315: 1420–1423. [262, 303]

Ryan, P. G., C. L. Moloney, and J. Hudon. 1994. Color variation and hybridization among *Nesospiza* buntings on Inaccessible island, Tristan da Cunha. *Auk* 111: 314–327. [303]

Saccheri, I. J., and P. M. Brakefield. 2002. Rapid spread of immigrant genomes into inbred populations. *Proc. R. Soc. B* 269: 1073–1078. [282]

Sæther, B-E., S. Engen, and E. Matthysen. 2002. Demographic characteristics and population dynamical patterns of solitary birds. *Science* 295: 2070–2073. [280]

Sæther, B-E., S. Engen, A. P. Møller, M. E. Visser, E. Matthysen and 12 others. 2005. Time to extinction of bird populations. *Ecology* 86: 693–700. [280]

Salvin, O. 1876. On the avifauna of the Galápagos archipelago. *Trans. Zool. Soc. Lond.* 9: 447–510. [6]

Santisteban, L., C. W. Benkman, T. Fetz, and J. W. Smith. 2012. Survival and population size of a resident bird species are declining as temperature increases. *J. Anim. Ecol.* 81: 352–363. [276]

Santos, M. E., and W. Salzburger. 2012. How cichlids diversify. *Science* 338: 619–621. [316]

Sardell, R. J., P. Arcese, L. F. Keller, and J. T. Reid. 2012b. Are there indirect fitness benefits of female extra-pair reproduction? Lifetime reproductive success of within-pair and extra-pair offspring. *Amer. Nat.* 179: 779–793. [326]

Sardell, R. J., P. Arcese, and J. T. Reid. 2012a. Offspring fitness varies with parental extra-pair status in song sparrows, *Melospiza melodia. Proc. R. Soc. B* 279: 4078–4086. [326]

Sasakai, A., and S. Ellner. 1997. Quantitative genetic variance maintained by fluctuating selection with overlapping generations: variance components and covariances. *Evolution* 51: 682–696. [179]

Sato, A., C. O'hUigin, F. Figueroa, P. R. Grant, B. R. Grant, and J. Klein. 1999. Phylogeny of Darwin's finches as revealed by mtDNA sequences. 2001. *Proc. Natl. Acad. Sci. USA*. 96: 5101–5106. [233]

Sato, A., H. Tichy, C. O'hUigin, P. R. Grant, B. R. Grant, and J. Klein. 2001. On the origin of Darwin's finches. *Proc. Natl. Acad. Sci. USA* 18: 299–311. [305]

Schalansky, J. 2009. *Atlas of Remote Islands*. Penguin Books, London, U.K. [122]

Schemske, D. W., and M. T. Morgan. 1990. The evolutionary significance of hybridization in *Eucalyptus. Evolution* 44: 2150–2151. [166]

Schlesinger, M. E., and N. Ramankutty. 1994. An oscillation in the global climate system of period 65–70 years. *Nature* 367: 723–726. [5]

Schluter, D. 1984. Morphological and phylogenetic relations among the Darwin's finches. *Evolution* 38: 921–930. [230–232]

Schluter, D. 1994. Experimental evidence that competition promotes divergence in adaptive radiation. *Science* 266: 798–801. [126]

Schluter, D. 1996. Ecological causes of adaptive radiation. *Amer. Nat.* 148 (suppl.): S40–S64. [231, 298]

Schluter, D. 2000. *The Ecology of Adaptive Radiation*. Oxford University Press, New York, NY. [4, 230, 231, 298, 306]

Schluter, D. 2009. Evidence for ecological speciation and its alternative. *Science* 323: 737–741. [298]

Schluter, D., and P. R. Grant. 1984. Determinants of morphological patterns in communities of Darwin's Finches. *Amer. Nat.* 123: 175–196. [34, 38, 39, 290]

Schluter, D., T. D. Price, and P. R. Grant. 1985. Ecological character displacement in Darwin's Finches. *Science* 227: 1056–1059. [34, 127]

Schoener, T. W. 1984. Size differences among sympatric, bird-eating hawks: a worldwide survey. In D. R. Strong, Jr., D. Simberloff, L. G. Abele, and A. B. Thistle, eds., *Ecological Communities: Conceptual Issues and the Evidence*, 254–281. Princeton University Press, Princeton, NJ. [238]

Schoener, T. W. 2011. The newest synthesis: understanding the interplay of evolutionary and ecological dynamics. *Science* 331: 426–429. [99]

Schoener, T. W., D. A. Spiller, and J. B. Losos. 2004. Variable ecological effects of hurricanes: the importance for timing of survival of lizards on Bahamian islands. *Proc. Natl. Acad. Sci. USA* 101: 177–181. [78]

Schwarz, D., B. M. Matta, N. L. Shakir-Botteri, and B. A. McPheron. 2005. Host shift to an invasive plant triggers rapid animal hybrid speciation. *Nature* 436: 546–549. [246]

Schwarzer, J., E. R. Swartz, E. Vreven, J. Snoeks, F.P.D. Cotterill, B. Misof, and U. K. Schliewen. 2012. Repeated trans-watershed hybridization among haplochromine cichlids (Cichlidae) was triggered by Neogene landscape evolution. *Proc. R. Soc. B* 279: 4389–4398. [316]

Schwenk, K., N. Brede, and B. Streit. 2008. Introduction. Extent, processes and evolutionary impact of interspecific hybridization in animals. *Philos. Trans. R. Soc. B* 363: 2805–2811. [245]

Seddon, N. 2005. Ecological adaptation and species recognition drives vocal evolution in neotropical suboscine birds. *Evolution* 59: 200–215. [292]

Seehausen, O. 2004. Hybridization and adaptive radiation. *Trends Ecol. Evol.* 19: 198–207. [316]

Seehausen, O. 2006. Review. African cichlid fish: a model system in adaptive radiation research. *Proc. R. Soc. B* 273: 1987–1998. [316]

Seehausen, O., G. Takimoto, D. Roy, and J. Jokela. 2008. Speciation reversal and biodiversity dynamics with hybridization in changing environments. *Mol. Ecol.* 17: 30–44. [183, 184]

Shapiro, M. D., M. E. Marks, C. L. Peichel, B. K. Blackman, J. S. Nereng, B. Jónsson, D. Schluter, and D. M. Kingsley. 2004. Genetic and developmental basis of evolutionary pelvic reduction in threespine sticklebacks. *Nature* 428: 717–723. [234]

Siepielski, A. M., J. D. DiBattista, J. A. Evans, and S. M. Carlson. 2011. Differences in the temporal dynamics of phenotypic selection among fitness components in the wild. *Proc. R. Soc. B* 278: 1572–1580. [78]

Simpson, G. G. 1953. *The Major Features of Evolution*. Columbia University Press, New York, NY. [205, 306]

Slabbekoorn, H., and T. B. Smith. 2002. Bird song, ecology and speciation. *Philos. Trans. R. Soc. B* 357: 493–503. [148, 292]

Slagsvold, T., B. T. Hansen, L. E. Johannessen, and J. T. Lifjeld. 2002. Mate choice and imprinting in birds studied by cross-fostering in the wild. *Proc. R. Soc. B* 269: 1449–1456. [152]

Slatkin, M. 1996. In defense of founder-flush theories of speciation. *Amer. Nat.* 147: 493–505. [115]

Slud, P. 1960. The birds of Cocos Island (Costa Rica). *Bull. Amer. Mus. Nat. Hist.* 134: 263–295. [79]

Smith, B. T., A. Amei, and J. Klicka. 2012. Evaluating the role of contracting and expanding rainforest in initiating cycles of speciation across the Isthmus of Panama. *Proc. R. Soc. B* 279: 3520–3526. [304]

Smith, J.N.M., P. R. Grant, B. R. Grant, I. J. Abbott, and L. K. Abbott. 1978. Seasonal variation in feeding habits of Darwin's ground finches. *Ecology* 59: 1137–1150. [23, 24, 33, 123]

Smith, J.N.M., L. F. Keller, A. B. Marr, and P. Arcese. 2004. *Conservation and*

Biology of Small Populations: The Song Sparrows of Mandarte Island. Oxford University Press, Oxford, U.K. [78]

Snodgrass, R. E. 1902. The relation of the food to the size and shape of the bill in the Galapagos genus *Geospiza. Auk* 19: 367–381. [17]

Sobel, J. M., G. F. Chen, L. R. Watt, and D. W. Schemske. 2010. The biology of speciation. *Evolution* 64: 295–315. [262]

Sol, D., R. P. Duncan, T. M. Blackburn, P. Cassey, and L. Lefebvre. 2005. Big brains, enhanced cognition, and responses of birds to novel environments. *Proc. Natl. Acad. Sci. USA* 102: 5460–5465. [106]

Sol, D., J. Maspons, M. Vall-lloser, I. Bartomeus, G. E. García-Peña, J. Piñol, and R. P. Freckleton. 2012. Unraveling the life history of successful invaders. *Science* 337: 580–583. [117]

Solomon, S., D. Quin, M. Manning, Z. Chen, M. Marquis, and 3 others. 2007. *Climate Change 2007: The Physical Science Basis; Contributions of the Working Group I to the Fourth Assessment report of the Intergovernmental Panel on Climate Change*. Cambridge University Press, Cambridge, U.K. [276]

Soulé, M. 1971. The variation problem: the gene-flow-variation hypothesis. *Taxon* 20: 37–50. [139]

Soulé, M. 1972. Phenetics of natural populations, III: Variation in insular populations of a lizard. *Amer. Nat.* 106: 429–446. [139]

Sprunt, A. 1953. Newcomer from the old world. *Audubon Magazine* 55: 178–181. [104]

Stelkens, R., and O. Seehausen. 2009. Genetic distance between species predicts novel trait expression in their hybrids. *Evolution* 63: 884–897. [265]

Stern, D. L. 2000. Perspective: Evolutionary developmental biology and the problem of variation. *Evolution* 54: 1079–1091. [167]

Stern, D. L. 2011. *Evolution, Development, & the Predictable Genome*. Roberts and Co., Greenwood Village, CO. [300, 301]

Strayer, D., J. S. Glitzenstein, C. G. Jones, J. Kolasa, G. E. Likens, M. J. McDonnell, G. G. Parker, and S.T.A. Pickett. 1986. *Long-term ecological studies: An illustrated account of the design, operation, and importance to ecology*. Occasional Publication of the Institute of Ecosystem Studies, no. 2. New York Botanical Garden, Millbrook, NY. [310]

Surai, P. F. 2002. *Natural antioxidants in avian nutrition and reproduction*. Nottingham University Press, Nottingham, U.K. [314]

Svärdson, G. 1970. Significance of introgression in coregonid evolution. In C. C. Lindsey and C. S. Woods, ed., *Biology of coregonid fishes*, 33–59. University of Manitoba Press, Winnipeg, Canada. [177]

Svenson, H. K. 1946. Vegetation of the coast of Ecuador and Peru and its relation to that of the Galápagos Islands, II: catalogue of plants. *Amer. J. Botany* 33: 427–498. [298]

Svensson, E., and R. Calsbeek. 2012. *The Adaptive Landscape in Evolutionary Biology*. Oxford University Press, Oxford, U.K. [34]

Swarth, H. S. 1931. The avifauna of the Galápagos islands. *Occas. Pap. Calif. Acad. Sci.* 18: 1–299. [6]

Swarth, H. S. 1934. The bird fauna of the Galápagos Islands in relation to species formation. *Biol. Revs.* 9: 213–234. [138]

Tajima, S., T. Goda, and S. Takase. 2001. Co-ordinated induction of β-carotene cleavage enzyme and retinal reductase in the duodenum of the developing chicks. *Comp. Biochem. Phys. B* 128: 425–434. [315]

Taylor, E. B., J. W. Boughman, M. Groenenboom, M. Sniatynski, D. Schluter, and J. L. Gow. 2006. Speciation in reverse: morphological and genetic evidence of the collapse of a three-spined stickleback (*Gasterosteus aculeatus*) species pair. *Mol. Ecol.* 15: 343–355. [183]

ten Cate, C., and C. Rowe. 2007. Biases in signal evolution: learning makes a difference. *Trends Ecol. Evol.* 22: 380–387. [241]

ten Cate, C., M. N. Verzijden, and E. Etman. 2006. Sexual imprinting can induce sexual preferences for exaggerated parental traits. *Curr. Biol.* 16: 1128–1132. [241]

ten Cate, C., and D. R. Vos. 1999. Sexual imprinting and evolutionary processes in birds: a reassessment. *Adv. Stud. Behav.* 28: 1–31. [148, 299]

Thesiger, W. 1959. *Arabian Sands*. Book Club Assoc., London, U.K. [298]

Tingley, M. W., W. B. Monahan, S. R. Beissinger, and C. Moritz. 2009. Birds track their Grinnellian niche through a century of climate change. *Proc. Natl. Acad. Sci. USA* 106: 19637–19643. [276]

Tokinaga, H., S-P Xie, C. Deser, Y. Kosaka, and Y. Okumura. 2012. Slowdown of the Walker circulation driven by tropical Indo-Pacific warming. *Nature* 491: 439–443. [276]

Tonnis, B., P. R. Grant, B. R. Grant, and K. Petren. 2005. Habitat selection and ecological speciation in Galápagos warbler finches (*Certhidea olivacea* and *C. fusca*). *Proc. R. Soc. B.* 272: 819–826. [104]

Trenberth, K., and T. Hoar. 1996. The 1990–95 El Niño–South Oscillation event: longest on record. *Geophys. Res. Lett.* 23: 57–60. [276]

Twain, M. 1883. *Life on the Mississippi*. Dawson Bros., Montreal, Canada. [271, 331]

Uyeda, J. C., T. F. Hansen, S. J. Arnold, and J. Pienaar. 2011. The million-year wait for macroevolutionary bursts. *Proc. Natl. Acad. Sci. USA* 108: 15908–15913. [239, 240]

Vähä, J-P., and C. R. Primmer. 2006. Efficiency of model-based Bayesian methods for detecting hybrid individuals under different hybridization scenarios and with different numbers of loci. *Mol. Ecol.* 15: 63–72. [188]

van der Pol, M., L. Brower, B. J. Ens, K. Oosterbeek, and J. M. Tinbergen. 2010. Fluctuating selection and the maintenance of individual and sex-specific diet specialization in free-living oystercatchers. *Evolution* 64: 836–851. [310]

Van Doorn, G. S., P. Edelaar, and F. J. Weissing. 2009. On the origin of species by natural and sexual selection. *Science* 326: 1704–1707. [262]

Van Valen, L. 1965. Morphological variation and the width of the ecological niche. *Amer. Nat.* 99: 377–390. [139, 140]

Van Valen, L. 1976. Ecological species, multispecies, and oaks. *Taxon* 25: 233–239. [177, 298]

Van Valen, L., and P. R. Grant. 1970. Variation and niche width reexamined. *Amer. Nat.* 105: 589–590. [140]

Veron, J.E.N. 1995. *Corals in Space and Time: The Biogeography and Evolution of the Scleractinia.* Comstock, Ithaca, NY. [245]

Verzijden, M. N., C. ten Cate, M. R. Servedio, G. M. Kozak, J. W. Boughman, and E. Svensson. 2012. The impact of learning on sexual selection and speciation. *Trends Ecol. Evol.* 27: 511–519. [148, 292]

Vilà, C., A-K. Sundqvist, O. Flagstad, J. Seddon, S. Björnerfeldt, I. Kojola, A. Casulli, H. Sand, P. Wabkken, and H. Ellegren. 2003. Rescue of a severely bottlenecked wolf (*Canis lupus*) population by a single immigrant. *Proc. R. Soc. B.* 270: 91–97. [118]

Visscher, P. M. 2008. Sizing up human height variation. *Nat. Genet.* 40: 489–490. [48]

Vonlanthen, P., D. Bittner, A. G. Hudson, K. A. Young, R. Müller, B. Lundsgaard-Hansen, D. Roy, S. Di Piazza, C. R. Largiarder, and O. Seehausen. 2012. Eutrophication causes speciation reversal in whitefish adaptive radiations. *Nature* 482: 357–362. [184]

Wagner, W. H., Jr., and V. A. Funk, eds. 1996. *Hawaiian Biogeography: Evolution on a Hot Spot Archipelago.* Smithsonian Institution Press, Washington, DC. [5]

Wallace, A. R. 1847. Letter to H. W. Bates, 11 October 1847 (Quoted in Raby 2001, p. 1). [3]

Walsh, B. 2004. Population- and quantitative-genetic analysis of selection limits. *Plant Breed. Rev.* 24: 177–225. [231]

Walsh, B., and M. W. Blows. 2009. Abundant genetic variation + strong selection = multivariate genetic constraints: a geometric view of adaptation. *Annu. Rev. Ecol. Evol. Syst.* 40: 41–59. [75]

Walsh, N., J. Dale, K. J. McGraw, M. A. Pointer, and N. I. Mundy. 2011. Candidate genes for carotenoid coloration in vertebrates and their expression profiles in the carotenoid-containing plumage and bill of a wild bird. *Proc R. Soc. B* 279: 58–66. [311, 314]

Warner, R. E. 1968. The role of introduced diseases in the extinction of the endemic Hawaiian avifauna. *Condor* 70: 101–120. [5]

Warren, B. H., E. Bermingham, R. P. Prys-Jones, and C. Thébaud. 2006. Immigration, species radiation and extinction in a highly diverse songbird lineage: white-eyes on Indian Ocean islands. *Mol. Ecol.* 15: 3769–3786. [296, 303]

Webb, W. C., J. A. Marzluff, and K. E. Omland. 2011. Random interbreeding between cryptic lineages of the Common Raven: evidence for speciation in reverse. *Mol. Ecol.* 20: 2390–2402. [183]

Weber, K. E. 1996. Large genetic change at small fitness cost in large populations of *Drosophila melanogaster* selected for wind tunnel flight: rethinking fitness surfaces. *Genetics* 144: 205–213. [231]

Weider, L. J., W. Lampert, M. Wessels, J. K. Colbourne, and P. Limburg. 1997. Long-term genetic shifts in a micro-crustacean egg bank associated with anthropogenic changes in the Lake Constance ecosystem. *Proc. R. Soc. B* 264: 1613–1618. [310]

Weiner, J. 1994. *The Beak of the Finch*. A. Knopf, New York, NY. [vii, 287]

Weir, J. T., and T. D. Price. 2011. Limits to speciation inferred from times to secondary sympatry and ages of hybridizing species along a latitudinal gradient. *Amer. Nat.* 177: 462–469. [239, 305]

Weir, J. T., and D. Schluter. 2007. The latitudinal gradient in recent speciation and extinction rates of birds and mammals. *Science* 315: 574–576. [304]

Wesolowski, T. 2011. Blackcap *Sylvia atricapillus* numbers, phenology and reproduction in a primeval forest—a 33-year study. *J. Ornithol.* 152: 319–329. [78]

Weston, J. 2005. *Chasing the Hoopoe*. Peterloo Poets, Calstock, Cornwall, U.K. [vii]

Wheelwright, N. T., C. R. Freeman-Gallant, and R. A. Mauck. 2007. Asymmetrical incest avoidance in the choice of social and genetic mates. *Anim. Behav.* 71: 631–639. [78]

White, G. 1789 (1877). Letter XLIX, May 7, 1779, in *The Natural History of Selborne*. Walter Scott, London. [310]

Whiteman, N. K., S. J. Goodman, B. J. Sinclair, T. Walsh, A. A. Cunningham, L. D. Kramer, and P. G. Parker. 2005. Establishment of the avian disease vector *Culex quinquifasciatus* Say 1823 (Diptera: Culicidae) on the Galápagos Islands, Ecuador. *Ibis* 147: 844–847. [284]

Whittaker, R. H. 1960. Vegetation of the Siskiyou Mountains, Oregon and California. *Ecol. Monogr.* 30: 279–338. [24]

Wiggins, I. L., and D. M. Porter. 1971. *Flora of the Galápagos Islands*. Stanford University Press, Stanford, CA. [13, 322]

Wikelski, M., L. Spinney, W. Schelsky, A. Scheuerlein, and E. Gwinner. 2003. Slow pace of life in tropical sedentary birds: A common-garden experiment on four stonechats populations from different latitudes. *Proc. R. Soc. B* 270: 2383–2388. [79]

Wiley, C., N. Fogelberg, S. A. Sæther, T. Veen, N. Svedin, J. Vogel Kehlenbeck, and A. Qvarnström. 2007. Direct benefits and costs for hybridizing *Ficedula* flycatchers. *J. Evol. Biol.* 20: 854–864. [143]

Wiley, R. H. 1991. Associations of song properties with habitats for oscine birds in eastern North America. *Amer. Nat.* 138: 973–993. [241]

Wilkins, M. R., N. Seddon, and R. J. Safran. 2013. Evolutionary divergence in acoustic signals: causes and consequences. *Trends Ecol. Evol.* 28: 156–166. [292]

Williams, E. E. 1972. The origin of avifaunas: Evolution of lizard congeners in a complex island fauna; a trial analysis. *Evol. Biol.* 6: 47–89. [5]

Williams, J. B. 1996. A phylogenetic perspective of evaporative water loss in birds. *Auk* 113: 457–472. [73]

Williams, J. B., and B. I. Tieleman. 2005. Physiological adaptation in desert birds. *BioScience* 55: 416–425. [73]

Williams, S. T., and T. F. Duda, Jr. 2008. Did tectonic activity stimulate Oligo-Miocene speciation in the Indo-West Pacific? *Evolution* 62: 1618–1634. [205]

Willis, B. L., M.J.H. van Oppen, D. J. Mills, S. V. Vollmer, and D. J. Ayre. 2007.

The role of hybridization in the evolution of reef corals. *Ann. Rev. Ecol. Syst.* 37: 489–517. [245]

Willis, K. J., and G. M. MacDonald. 2011. Long-term ecological records and their relevance to climate change predictions for a warmer world. *Ann. Rev. Ecol. Syst.* 42: 267–287. [276]

Woram, J. 2013. Las Encantadas: Human and Cartographic History of the Galápagos Islands. Table of Galápagos Island names. http://www.galapagos.to/table.php. [13]

Wu, P., T-X. Jiang, S. Suksaweang, R. B. Widelitz, and C-M. Chuong. 2004. Molecular shaping of the beak: a paradigm for multiple primordial morphologenesis. *Science* 305: 1465–1467. [234]

Yeh, S. W., J-S. Kug, B. deWitte, M-H Kwon, B. P. Kirtman, and F-F Jin. 2009. El Niño in a changing climate. *Nature* 461: 511–514. [276]

Zhang, G., P. Parker, B. Li, H. Li, and J. Wang. 2012. The genome of Darwin's finch (*Geospiza fortis*). *GigaScience*. http://dx.doi.org/10.5524/100040. [284]

Zhang, R. H., L. M. Rothstein, and A. J. Busalacchi. 1998. Origin of upper-ocean warming and El Niño change in decadal scales in the tropical Pacific Ocean. *Nature* 391: 879–883. [276]

Zhang, X-S. 2012. Fisher's geometrical model of fitness landscape and variance in fitness within a changing environment. *Evolution* 66: 2350–2368. [167, 179]

Zhen, Y., M. L. Aardema, E. M. Medina, M. Schumer, and P. Andolfatto. 2012. Parallel molecular evolution in an herbivore community. *Science* 337: 1634–1637. [296]

Subject Index

Geospiza magnirostris, large ground finch, passim

Geospiza scandens, cactus finch, passim

glacial/interglacial cycles, 21, 304, 309

global climate, 276

granivore, 6, 37, 132, 172, 230, 235–36, 239, 243, 247

growth: of finches, 44, 46, 63, 73–74, 76, 99, 169, 219, 234–38; of plants, 57, 68–69, 80, 124, 321–22; of populations, 117, 264. *See also* embryo, embryonic

gulls. See *Larus* gulls

habitats: of finches, 11, 18, 40, 48–49, 53, 78–79, 104, 118, 140, 179, 238, 241, 246, 276, 284, 301–5, 324; outside Galapagos, 139, 304, 305; effect of on song transmission, 291; and hatching success, 42, 81, 86, 107, 160

Hawaiian archipelago, Hawaii, 5, 238, 307

hawk, Galápagos. See *Buteo galapagoensis*

heat dissipation, 59, 72, 118, 128

heritability, 43–53, 62, 65, 97, 99–100, 132–33, 211, 232, 247. *See also* variation

heterospecific song, 148, 150–51, 243

heterozygousness, 111–13, 117–18, 198, 259, 264, 311, 335

H.M.S. *Beagle*, xxvii

H.M.S. *Daphne*, 13

homozygous loci, 252–53, 264–65, 335

honeycreeper finches (family Drepanidae), 5, 307

house sparrow, *Passer domesticus*. See sparrow: house

Hubbs principle, 141–42, 347

human activities, 4–5, 11, 139, 184, 298

hybridization, 139–43, 166, 167, 237, 290; effect of on allometry, 176, 194–97; causes of, 141, 143–56, 293; convergence from, 184–86, 189–90, 196–99, 204, 244, 293; frequency of,

141, 191, 192, 203, 339; heritable variation in, 169; introgressive, 16, 139, 142, 158–68, 170–80, 184–86, 188–90, 197, 201, 219, 238–39, 280, 285, 289, 301, 303, 305, 315; versus mutation, 174; phenotypic implications of, 294–96, 297; and selection, 213, 216, 228, 234, 275, 289, 298; influenced by song, 143–56, 176; influence of on speciation, 245–68, 272, 275, 281–82, 286, 288–90, 293–98, 303, 305, 309, 315; effect of on variation, 169–80, 184, 192–93, 203–4, 285, 328–29; variation of in time, 193–96

hybrids and backcrosses, 58, 140–42, 154–55, 160, 164, 169, 341; in assignment tests, 187, 188; diets of, 158, 265, 267, 291; fitness of, 156–57, 159, 160–65, 170, 275; heterozygosity of, 201; mating patterns of, 163–64; morphology of, 169; plumage and behavior of, 201

Ice Age, 20, 21

immigrant, 8, 16, 24, 39, 95, 105–15, 118–20, 127, 140, 142, 174–76, 180, 186, 192, 193, 219, 223–24, 247–50, 254–58, 262, 268, 293, 323

immigration, 107–10, 114, 117–18, 121, 173, 176, 184, 198, 206, 216, 219, 223, 228, 247, 282, 308, 335, 336

imprinting, sexual, 141, 143, 148, 149, 151–53, 165, 291

inbreeding: avoidance of, 96–97; coefficient of, 96, 107, 162; occurrence of, 96–97, 106–7, 162, 198, 249, 250, 259–261, 263, 268, 285, 301, 303; reduced fitness (depression) with, 96, 104, 107, 109, 117–18, 121, 161–62, 197, 265, 281, 282, 301, 302, 309, 332, 337

inheritance, 39, 65, 143, 149, 169, 201, 252, 311; Mendelian, 186, 311, 335

inornata, Pinaroloxias. See *Pinaroloxias inornata*

interbreeding. *See* hybridization